Pandemics and Epidemics in Cultural Representation

Sathyaraj Venkatesan · Antara Chatterjee ·
A. David Lewis · Brian Callender
Editors

Pandemics and Epidemics in Cultural Representation

Springer

Editors
Sathyaraj Venkatesan
Department of Humanities and Social Sciences
National Institute of Technology
Tiruchirappalli, Tamil Nadu, India

A. David Lewis
School of Arts and Sciences
MCPHS University
Boston, MA, USA

Antara Chatterjee
Department of Humanities and Social Sciences
Indian Institute of Science Education and Research
Bhopal, Madhya Pradesh, India

Brian Callender
Biological Sciences Division
University of Chicago
Chicago, IL, USA

ISBN 978-981-19-1295-5 ISBN 978-981-19-1296-2 (eBook)
https://doi.org/10.1007/978-981-19-1296-2

This Springer imprint is published by the registered company Springer Nature Singapore Pte Ltd.
The registered company address is: 152 Beach Road, #21-01/04 Gateway East, Singapore 189721, Singapore

Preface

This book emerged out of thoughts and discussions generated by the catastrophic arrival of the COVID-19 pandemic in 2020. The global upheaval triggered by this once-in-a-century phenomenon (the last such major pandemic being the 1918 "Spanish" Flu) caused death and suffering, panic, isolation, and uncertainty on a scale which contemporary society has not experienced. The pandemic also ushered in paradigmatic shifts in the ways in which we live, work, interact, travel, socialize, and communicate, as people across the globe scrambled to adjust to the "new normal" of masks, sanitization, and "social distancing." The last two years of our negotiation with COVID-19 have also more powerfully highlighted the deep schisms and tensions existing within local communities and nations as well as between countries as the pandemic, through its socioeconomic impacts across nation, race, class, gender, ethnicity, and politics, has revealed unequal human vulnerabilities, exposing the fraught fault lines within these different formations. Into its third year now, with new variants appearing at regular intervals, like the current Omicron spreading globally since the end of 2021, the COVID-19 pandemic continues to powerfully (re)shape our lives, poignantly revealing both scientific inadequacy and human fragility and resilience.

While attempting to grapple with the global upheaval and uncertainty caused by the pandemic, we turned to history and to past epidemics/pandemics to seek comfort in and lessons from the past. In our attempts to make sense of and come to terms with the crisis through which we are living, we also turned to cultural and creative expressions engendered by past pan/epidemics. Through successive periods of "lockdowns" and isolation, many of us were drawn to literary and cultural works that grappled with ideas of contagion and community in moments of devastating disease outbreak, both historical and imagined, through their iconic representations of pandemics. Some of these milestones in literary and cultural representation serve as historical records of similar crises in times past, while exploring disease outbreaks as complex, multi-faceted sociopolitical, and cultural experiences, expanding their existences beyond medical and biological phenomena. They also offer crucial insights into human precarity as well as collective resilience and strength in the face of shared trauma.

Thus, our individual and collective explorations and conversations surrounding contagion, community, culture, and representation in the wake of the COVID-19 pandemic led to the genesis of this book in 2020, which reviews artistic and literary responses to notable epidemics in world history, across a wide range of geographical spaces and time periods and across a spectrum of media from poetry, drama, comics and fiction to films, memes, and visual art. From its inception in 2020, the collection's journey and consolidation over the past two years has been a learning and enriching experience for all of us.

We would like to express our thanks to all the contributors of this collection for their insightful and thought-provoking essays which delve into the complex and multifarious social and cultural meanings of epi/pandemics through the lens of cultural representation. Thanks also to the editorial team at Springer Nature for their support and commitment toward this effort. We would also like to thank the reviewers of our book proposal and manuscript for their fruitful comments on how to improve the collection and make it more holistic in its approach and coverage. We hope that this collection adds value to our knowledge and understanding of epidemics and pandemics in the realms of culture and representation.

Tiruchirappalli, India — Sathyaraj Venkatesan
Bhopal, India — Antara Chatterjee
Boston, USA — A. David Lewis
Chicago, USA — Brian Callender

Contents

About the Editors

Sathyaraj Venkatesan Ph.D. is an Associate Professor of English in the Department of Humanities and Social Sciences at the National Institute of Technology (NIT), Tiruchirappalli, India. His research concentrates on health humanities, comics studies, and graphic medicine. He is the author of eight books and over ninety-five research publications that span American literature, health humanities, graphic medicine, film studies, and other literary and culture studies disciplines. His recent co-authored books are Metaphors of Mental Illness in Graphic Medicine (Routledge, 2021) and India Retold (Bloomsbury, 2021).

Antara Chatterjee Ph.D. is an Assistant Professor of English in the Department of Humanities and Social Sciences at the Indian Institute of Science Education and Research, (IISER), Bhopal, India. Her research interests include Indian writing in English, South Asian diasporic literatures, trauma, violence and cultural memory, medical and environmental humanities. She has received grants and fellowships from the University Grants Commission, India, the Charles Wallace India Trust, the Indian Council of Historical Research, and the Indian Institute of Advanced Study, Shimla, India. She has publications in South Asian Review, Humanities, and in an edited collection of essays The Postcolonial Short Story published by Palgrave Macmillan. She has chapters forthcoming in edited collections to be published by Bloomsbury and Routledge. She is set to be the inaugural Strauss Fellow at the Cedars-Sinai Center for Medicine, Holocaust and Genocide Studies at the Cedars-Sinai Medical Center, LA, USA, in summer 2022.

A. David Lewis Ph.D. is the Eisner Award-nominated author of American Comics, Religion, and Literary Theory: The Superhero Afterlife as well as co-editor of both Graven Images: Religion in Comic Books and Graphic Novels and Muslim Superheroes: Comics, Islam, and Representation. Featured on numerous podcast and television programs, Dr. Lewis is currently program director for the MHS degree at the MCPHS University School of Arts and Sciences where his teaching and research focus on graphic medicine, specifically the depiction of cancer in comic books and graphic novels. Finally, he is the acclaimed author of such comics as The Lone and Level Sands and Kismet, Man of Fate.

Brian Callender M.D. is an Associate Professor of medicine at the University of Chicago. He is a faculty member of the MacLean Center for Clinical Medical Ethics and co-organized the 2020–2021 lecture series, ethics and the COVID-19 pandemic: Medical, Social, and Political Issues. As a core faculty member of the Institute on the Formation of Knowledge, he teaches health humanities courses, including: The Body in Medicine and the Performing Arts; Graphic Medicine: Concepts and Practice; The Art of Healing: Medical Aesthetics in Russia and the U.S.; Death Panels: Exploring Dying and Death through Comics; and The Narratives and Aesthetics of Contagion: Knowledge Formation and the COVID-19 Pandemic. He has also curated two exhibits at the University of Chicago's Special Collections Research Center that examine the visual culture of medicine: Imaging/Imagining: The Body in Anatomical Representation and The Fetus in Utero: From Mystery to Social Media.

Chapter 1
Introduction

Sathyaraj Venkatesan, Antara Chatterjee, A. David Lewis, and Brian Callender

Pandemics and epidemics have always shaped our history. Throughout recorded human history, periodic outbreaks of catastrophic pandemics and epidemics have threatened human existence on this planet and have been a regular reminder of our mortality and vulnerability in the face of nature's calamities, including deadly diseases. These episodes have often shown the limitations of human knowledge, particularly advancements in science to fortify against the violence of new and often deadly pathogens. At the same time, pandemics and epidemics have also been formative moments of human history, ushering in momentous transformations in the ways in which we live and relate to each other. They have provided opportunities to gain new insights into the human body and how it functions, as well as reflect on the ways in which societies and communities form and interact.

As Priscilla Wald has argued, "Contagion is more than an epidemiological fact. It is also a foundational concept in the study of religion and of society" (2). Pandemics and epidemics have, thus, been moments to understand afresh the human body as well as societies and collectives like communities and nations. They have offered opportunities to rethink, reconfigure, and renegotiate our ideas of the individual, the communal, the social, the private and public, and of health, illness, hygiene, safety,

S. Venkatesan (✉)
NIT, Tiruchirappalli, India
e-mail: sathya@nitt.edu

A. Chatterjee
IISER, Bhopal, India
e-mail: achatterjee@iiserb.ac.in

A. D. Lewis
MCPHS University, Boston, USA
e-mail: a.lewis@mcphs.edu

B. Callender
The University of Chicago, Chicago, USA
e-mail: bcallend@medicine.bsd.uchicago.edu

S. Venkatesan et al. (eds.), *Pandemics and Epidemics in Cultural Representation*,
https://doi.org/10.1007/978-981-19-1296-2_1

the body, distance, intimacy, borders, and boundaries (bodily, social, communal, political, national). They raise questions about the larger social meanings of health, disease, and death and provide opportunities to reflect on power, control, exclusion, inequality, and marginalization within these experiences. The current COVID-19 pandemic that the world is living through is one such momentous turn in history that bears testimony to the above.

Be it the Athenian plague, the catastrophic Black Death of the fourteenth century or the current COVID-19 outbreak, pandemics and epidemics continue to surprise humanity and cause immense harm and upheaval in human life, while simultaneously being moments that inscribe potential for great change. The emergence of an epidemic quickly challenges established medical knowledge and, thus, has "an unusual power to provoke panic, hysteria, and dread" (Honigsbaum, 2019, xvi).

However, as pointed out, pandemics and epidemics are not only medical or biological phenomena. Disease and the body's fight against it are embedded within contexts and couched in a vocabulary that is socio-political. As W. J. T. Mitchell points out, "the very notion of immunity [...] is originally based in a sociopolitical discourse, not a biological one"; "the immune system and the discipline of immunology is riddled with images drawn from the sociopolitical sphere—of invaders and defenders, hosts and parasites, natives and aliens, and of borders and identities that must be maintained" (282). Donna Haraway too asserts, "the immune system is both an iconic mythic object in high-technology culture and a subject of research and clinical practice of the first importance" (205). The immune system is, thus, a biological and a socio-cultural entity. Disease outbreak can be read simultaneously as resonating with social and political, not just biological connotations. Pandemics and epidemics are, therefore, complex, multifaceted social, political, economic, and cultural experiences, embedded within intersecting socio-political and cultural contexts. They raise questions about and compel us to rethink notions of the human body as well as the body of collectives like cities, communities, nations, and the world and the ways in which these bodies are understood and regulated.

While every pandemic is a metaphorical "living with death," what is intriguing and fascinating is how such crises engender an outbreak of creativity, expressed through a wide-ranging variety of forms of art, literature, performance, and oral tradition. Examples range from Thucydides' account of the Athenian plague, Daniel Defoe's *A Journal of the Plague Year* which narrates the impact of the bubonic plague on London in 1655, Igor Stravinsky's "The Firebird Suite" (1919), Sergei Prokofiev's "The Love for Three Oranges" (1918), Katherine Ann Porter's *Pale Horse, Pale Rider* (1939) and William Maxwell's *They Came Like Swallows* (1937) in response to the 1918 influenza pandemic, Tony Kushner's "Angels in America'" on HIV/AIDS, films like *Outbreak* (Wolfgang Petersen, 1995), and the fantasy film series *Twilight* (based on Stephenie Meyer's 2005 novel of the same name), to television series like the historical fiction series *Downton Abbey*. Put together, from the classical times to the contemporary moment, cultural texts have regularly engaged with ideas of contagion and community through representations of historical and imagined apocalyptic catastrophes in the form of species-threatening epidemics/pandemics.

Cultural works produced in the context of pandemics not only serve as historical records but also offer hope and teach resilience. They reflect on human precarity and throw into relief "existing cleavages in societies and shortcomings in their organization" (Vrdoljak & Bauer, 2020, 441). More importantly, they help us "make sense" of the crisis and, thus function as a diary of human history. Put differently, the cultural work of the arts and literature during or after pandemic times cannot be underestimated as they enable, reimagine, and exercise transformational impact on social life.

Taking these cues, "Pandemics and Epidemics in Cultural Representation" comprehensively reviews the artistic and literary responses to some of the most influential epidemics in world history, covering a wide range of geographical spaces and time periods in history. These range from the medieval Black Death and cholera in eighteenth-century Europe and nineteenth-century Cuba, to the Spanish flu and AIDS in the twentieth century and Ebola and COVID-19 in the twenty-first century. More specifically, this project seeks to examine how artists, authors, and cultural practitioners, starting from Chaucer in the fourteenth century to meme producers on Instagram in 2020, have responded to and represented these episodes which have ushered in paradigmatic shifts in the ways in which we live and interact and the ways in which we individually and collectively configure the contours of community, communicability, and contagion. A wide range of artistic and literary genres is examined in this collection of essays, from the novel, drama poetry, and short fiction to visual art, memes, and films. In doing so, the collection foregrounds the richly diverse, multifarious cultural responses to epidemics and pandemics through cultures across time and space, filtered through multiple cultural and artistic perspectives.

While there is a universal grammar in the way humanity has responded to pandemics in terms of the double bind of fear and hope, there are also indigenous and local responses to pandemics making pandemics simultaneously global phenomena which are also locally experienced, grappled with, and negotiated. Taking into account such "glocal" responses to pandemics, "Pandemics and Epidemics in Cultural Representation" offers a rich array of local, regional, national, and global cultural responses to pandemics. That is, the curated essays in the book also demonstrate how specific social, cultural, and environmental factors and one's situatedness within specific social, cultural, geographical, political contexts influence the understanding, meaning making, and responses to pandemics.

Do we learn from one pandemic to the next? Certainly, medicine and science make advances, but they can be just as easily taken unawares or unprepared for the onset of the next uncontrollable disease. If we gain any advantage over contagion from one iteration to the next, it may only be in the chronicles of past eras lending wisdom, comfort, or simply warning to the next set of hapless peoples in the face of a new disease. These cultural representations are more than sterile artifacts; they are testimony to the human ability to endure and the social empowerment to overcome. The dead speak, not just those whose lives were cut short by illness, but also those who came through a time of calamity and left insightful works for posterity. If, in fact, we do not advance each time, global populations are stricken with some malady that may only be due to the cold comfort that such a familiar circumstance provides. These works tell us there will be a future because, for them, we are that future.

How works about pandemics resonate with the future raises questions about how pandemic knowledge is formed, circulated, and used to inform attitudes and behaviors, both contemporaneously and historically. In an era of misinformation and disinformation that has negatively impacted the response to and mitigation of the pandemic, impactful knowledge that reduces the global morbidity and mortality of pandemics is vital to pandemic preparedness and response. In reading the works in this collection, one must ask not only what we may learn as individuals, but importantly, what are we collectively learning? And how do we use that knowledge? The process of knowledge formation during the COVID-19 pandemic remains rapid and dynamic, but in the longue duree of pandemics, this acute process both mirrors and becomes integral to the cycles of knowledge formation. The current pandemic allows us to rethink past pandemics while forming useful knowledge for the future, and we must read these works as important contributions to the long history of cultural responses to pandemics.

This collection, thus, brings together scholarly inquiry that examines cultural configurations and representations of devastating pandemic outbreaks to understand how cultural works have negotiated the fraught territories of human contact and contagion, fraught with both possibility and danger. As pointed out before, essays examining cultural representations of pandemics and epidemics in different contexts, periods, languages, media, and genres are brought together. Through its focus on representations and narratives of life-threatening and indeed civilization-threatening diseases, the collection examines both the potential and the limitations of narrative with reference to bodily illness and suffering and large-scale human catastrophe. Proponents of "narrative medicine" like Rita Charon, for instance, have pointed out the fruits of "narrative knowledge" in medicine: "Medicine can benefit from learning that which literary scholars and psychologists and anthropologists have known for some time—that is, what narratives are, how they are built, how they convey their knowledge about the world, what happens when stories are told and listened to, how narratives organize life" (9). Yet, critics like Charon, Wald, and Elizabeth Outka, for instance, have also commented on how difficult it is to represent such outbreaks and how part of that difficulty lies in reckoning with "non-human actors and non-human violence" (Outka, 30). As John Barry has observed, "People write about wars [...] They write about the horrors that people inflict on people. Apparently they forget the horrors that nature inflicts on people" (920). This collection brings together significant research on these complex intersections between pandemics/epidemics, representation, narrative, and cultural memory as a means of negotiating and looking both into our past and future from the perspective of our current pandemic.

Layout of the Book

This book comprises fifteen essays, arranged in thematic sections. In Dilip K. Das's essay "Epidemic Narrative: Two Paradigms," the author examines the broad trajectory of illness narratives across a spectrum book ended by two paradigms: the biomedical and the cosmological. Das builds a framework for this essay by focusing first on illness narratives and epidemic narratives, noting an important distinction between accounts of personal suffering and collective suffering, respectively. The epidemic narrative is further elaborated upon to note the diverse forms and ways in which we tell and use these stories before establishing a conceptual framework of narrative resolution incorporating biomedical and cosmological explanations. Drawing upon historical examples, Das subsequently explores the tension between the natural and the supernatural to address the question of how "epidemic narratives construct the events and the time–space in which they unfold." Working with Rosenberg's understanding of epidemics, Das stresses how the construction of epidemic narratives, tethered between the medical and the mythic, contributes to our collective understanding of epidemics.

Be it historical tragedies, wars, or pandemics, their impact has never been restricted to the generation that suffered them but are carried forward for decades as stories and images leaving definitive traces in every human activity. Lorenz A. Hindrichsen's "Chaucer's *Canterbury Tales* as a Postpandemic Text" is an examination of the lasting intergenerational trauma of the Black Death and how *The Canterbury Tales* processes it at multiple levels. Black Death swept westward through Europe and had enormous consequences for social, economic, and cultural life. The Black Death which continued in intermittent cycles in Europe spurred literary imagination or influenced literary works produced during that period—for instance, Giovanni Boccaccio's *Decameron*, Geoffrey Chaucer's *The Canterbury Tales*, William Langland's *Piers Plowman* among others. Inspired by Decameron, *The Canterbury Tales* by Geoffrey Chaucer is a collection of stories within a frame story written in Middle English in c.1387–1400. A group of characters narrate these stories as they undertake a pilgrimage to the shrine of Thomas Beckett in Canterbury, Kent. Deftly navigating history and literature and borrowing from economic history, narratology, discourse analysis, iconography, and emerging theories on care, Hindrichsen reads Chaucer's *The Canterbury Tales* as a post-pandemic text. As a post-pandemic text, *The Canterbury Tales* reenacts structures of collective trauma, cultural anxieties of the itinerant first and second generation survivors, and pandemic's deathly fecundity itself. Given the cyclical nature of the Black Death, Hindrichsen also examines how such cyclicality structures and constitutes the narratological framework of *The Canterbury Tales* and, hence, into the nature of storytelling itself.

Rebecca Welshman's essay "'The Invisible Operator': Plague, Corruption and Conspiracy in Renaissance Drama" addresses the "invisibility" of plague in English Renaissance drama, arguing that plague's presence in the drama of this period was latent, rather than explicit, and expressed through metaphor, wordplay, and ambiguous symbolism. Through a close and detailed reading of plays by Shakespeare, Jonson, Dekker, and Middleton, she situates the multilayered understandings of plague in early modern English drama within the intersecting contexts of corruption, conspiracy, and the fear of the unknown, arguing that contemporary anxieties surrounding contagion were revealed in an increasing pre-occupation with the corrosive effects of "unseen agencies." The notion of an "invisible" threat was repeatedly invoked in these plays through linguistic experimentation and choice of layered metaphors, reflecting the contested notions of contagion within contemporary medical and social discourses. Welshman, thus, uncovers, underneath the apparent invisibility, a "deeper endemic presence of plague" with a proliferation of meanings in early modern drama.

Aureo Lustosa Guerios's essay "Three facets of the Literary Imagination of Cholera: Hysteria, Ridicule, and the Rise of Bacteriology" traces three cultural and literary impulses to the cholera pandemics of the nineteenth century. The first one expressed the cultural shock of the cholera outbreaks in the 1800s through a proliferation of apocalyptic texts, which represented cholera as a pestilence far more deadly than plague and which would lead to the annihilation of the world and the human race. These texts included Poe's *The Mask of the Red Death*, Dostoyevsky's *Crime and Punishment*, Mary Shelly's *The Last Man*, Flaubert's *Le Femme du Monde*, Pushkin's *A Feast in Time of Plague*. The second impulse was one of denial and dismissal of the threat of cholera through ridicule and humor, leading to strategies of presumed inoculation against the fear of the disease and healing. These texts included Colerige's "Cholera Cured Before Hand," G. G. Belli's sonnets on cholera and Heine's *Cholera in Paris*. As familiarity with the disease increased, both the tendencies abated; the fear of cholera subsided with sanitary reforms and advances in bacteriology, while ridicule of cholera was considered insensitive and distasteful. Eventually, these contradictory responses converged in a cultural response which celebrated scientific progress to vanquish cholera and the potential of healing and renewal.

Bethany M. Wade's "The Blue Death: Cholera and Reimagined Community in Nineteenth-century Havana" is a wide-ranging account of how emerging and evolving fields of scientific inquiry and social orders "embodied new, untested ways of representing the world" in 1800's Havana to shape collective consciousness. After outlining the two competing models of cholera during that era (germ theory and miasmatic) from both biological and sociological lenses, Wade details how these theories and perspectives of cholera were "integrated into growing concerns about race, class, and power in the existing colonial system." A notable response to these concerns was the emergence of an intellectual movement and formation of a Cuban literary community that addressed these issues through prose fiction. Wade adeptly elaborates on the impact of this community's contribution to collective understanding by comparing and contrasting the works *El Cólera en la Habana* by Ramón de Palma and Villaverde's *Cecilia Valdés*.

Cheryl Thompson and Lucy Wowk engage with the contemporaneous media publishing during a pandemic in "Reading Toronto's Response to Spanish Influenza: *The Globe* and *Daily Star* Report on the 1918–20 Pandemic's Second Wave." While the major Canadian newspapers both reported factually and responsibly about the illness, they featured different angles in their coverage. The *Daily Star* offered more sensationalistic and personal accounts of the pandemic's cost, while *The Globe* focused on more data-driven and official accounts. Thompson and Wowk attribute this to the different audiences served by each newspaper and their corresponding socio-economic status. Lengthier obituaries and celebrity deaths could be found in the *Daily Star*, as could be a more pronounced focus on women's roles and rights. A degree of partisanship (and xenophobia) could be found in both publications over this dire period.

Pei-chen Liao's essay "Outbreak Narratives of the 1918 Influenza Pandemic: Contagion, Community, and Politics in Myla Goldberg's *Wickett's Remedy* and Thomas Mullen's *The Last Town on Earth*" examines the resurgence of literary narratives of the 1918 influenza pandemic in the twenty-first century, focusing particularly on two novels Myla Goldberg's *Wickett's Remedy* (2005) and Thomas Mullen's *The Last Town on Earth* (2006), reading them within contemporary discourses of contagion, community, and communicability. Liao first analyzes the "amnesia" about the 1918 pandemic in literary narratives and popular and official discourses in the 1920s and 1930s, attributing this amnesia to several factors: public attention at the time and memory in its immediate aftermath being dominated by WWI (despite having less casualties than the pandemic), war-time military censorship of flu casualties, and post-war optimism about the successes of modern science and efforts to rebuild a "healthy" nation. The few literary narratives, which appeared in the wake of the pandemic, were all focused on the trauma of illness and the horror of the deaths, experienced by ones who had immediate contact with it. On the other hand, literary narratives of the twenty-first century, which appeared around the centenary of the pandemic, like Goldberg's and Mullen's novels, revisit the 1918 influenza with a contemporary lens, which look at configurations and intersections of contagion and community, transcending the personal trauma of illness in the earlier narratives. These novels examine the constructions of imagined community and immunity, the potential and dangers of contact, and how ideas of disease and immunity overlap with discourses of the nation, community, borders, insider/outsider, and so on. Reading the novels with the Foucauldian lens of "biopower," Liao analyzes how these "outbreak narratives" (Wald) configure notions of disease, contact, community, and immunity.

Gaana Jayagopalan's essay "Spatial Pathologies: The Biopolitics of Disease, Death, and the Caste-body in Ananthamurthy's *Samskara: A Rite for a Dead Man*" reads plague in U. R Ananthamurthy's Kannada novel *Samskara* as a "narrative landscape" foregrounding the body politics of the Brahmin community in which the narrative is situated. Plague breaks out in a Brahmin colony in Ananthamurthy's novel through its first victim, the self-proclaimed anti-Brahmin Narannapa. Jayagopalan situates the "diseased/deceased body" of Narannapa within the triangulating discourses of caste, disease, and death; the spatialization of this body in the Brahmin settlement reveals the diseased body politic of the insular and spiritually and morally decadent community. Offering a close reading of the text through the lens of "autoimmunity," conceptualized by Baudrillard and Derrida, Jayagopalan reflects on how the Brahmin space becomes a site of the "spatial pathologies," both biomedical and moral that frame Naranappa's death by plague. While Baudrillard's concept of sterile prophylaxis brings the annihilation of the community, Derrida's model of the liberatory potential of that moment of destruction is inscribed in the crisis and conflict in the devout, "pure" Brahmin Praneshacharya.

Jessica Hume's essay titled "'No Country for the Infirm': Reading *Angels in America* During COVID-19" revisits Tony Kushner's landmark two-part play "Angels in America: A Gay Fantasia on National Themes" in the light of COVID-19 to demonstrate the fault lines and inconsistencies of the universal components of Americanism such as future orientation, equity, and equality among others. Hume after reviewing what constitutes the American value system shows how America operates on the mechanisms of bifurcation (image versus reality) and religious nationalism. As such, public discourse around illness and pandemics makes this opposition evident. Drawing insights from literary analysis, health policy, health law, news media, and psychology and close reading Kushner's play in the context of the COVID-19, Hume shows the influence of Christian morality on health policy decision-making and how illness and marginalization are often constitutive of one another. The article concludes with how American's toxic future orientation and thanatophobia succeed in "erasing the experiences" of "the ill and dying fellow citizens" and, thus, diminishing "Americanness" itself.

Can imagined pandemics affect the treatment and diagnosis of real-life ones? That is the question central to Rebecca Henderson's "'Bleeding' into Reality: Popular Representations of Ebola and the 2013–2016 Ebola Epidemic." Her chapter highlights the popularity of Robin Cook's novel *Outbreak*, its film adaptation, and Richard Preston's *The Hot Zone* as influencing African triage of the potentially contaminated, pregnant women in particular. Specifically, the pronounced bleeding that is associated with Ebola only occurs in a small minority of cases, yet it has become the hallmark of media depictions of the illness and, thus, a key symptom for practitioners. This fetishization of blood and bleeding could, arguably, be influenced by the HIV/AIDS epidemic as much as the perceived otherness of the pregnant body.

Akinori Hamada explores an artistic response to governmental policy in the essay "Wearing Masks: Living and Coping "WITH CORONA" in Japan under the Pandemic." The historical and evolving work of the Japanese artist Tadanori Yokoo is the beginning focal point of this essay, tracing the artist's new series of works "WITH CORONA" that were created as part of dynamic exhibition in response to the Japanese government's pandemic mitigation policy, "With Corona." Contrasting the visual works of Yokoo with another important visual representation of the pandemic—epidemiologic graphs depicting case counts—Hamada outlines individual and communal response to these works within the broader context of the "With Corona" policy.

Echoing Thompson and Wowk's work, Julie M. Powell assesses published responses to a pandemic, albeit a century later and in the native digital sphere of Instagram. In "<3 I AM TRACY: Meme Culture, Coping, and Community during the COVID-19 Pandemic," Powell wades into the communal dark humor of Millennials during the 2020 lockdown and the strange resilience it suggests. A popular and prolific platform, Instagram, allows for expressions of grim camaraderie, of wry coping mechanisms, and of candid precariousness that are trackable and measurable. The platform has become "an important lifeline" during the isolation and loneliness of the pandemic response, with tens of thousands of users responding and adding to psychologically dark memes. For a generation that has experienced repeated disaster within a very short period, Instagram has become an important tool for Millennials in their negotiation of identity against a backdrop of calamity.

People dealt with the COVID-19 lockdown however they could, and contributor Mónica Lalanda responded by collecting the comics and editorial cartoons that were being produced in Spain during that period. Slowly, she amassed a sizeable collection and began to find patterns and categories that she details in "Comics, Cartoons, and Vignettes: The Graphic Narrative of the COVID-19 Pandemic." Notably, the virus itself took on an imagined shape, and visual metaphors (e.g., healthcare workers as superheroes, death as the Grim Reaper, pigeons confused by humans' absence) emerged in common, without any coordinated editorial effort. Lalanda credits social media as a major driver of the shared images; even as so much of the illness and death remained largely unseen, these expressions of graphic medicine worked to depict the widespread social response to intangible experiences of anxiety, isolation, and fear.

Though COVID-19 is touted as a great leveler, it has disproportionately impacted the Black, Indigenous, and People of Color (BIPOC) communities putting them more at risk of getting sick and dying from COVID-19. Taking this cue, Sathyaraj Venkatesan's "'Their Lives Just Don't Matter?': 'Racing' COVID-19 and Graphic Medicine" through close reading selects graphic medical narratives (intersection of comics and healthcare), foreground health inequities faced by the BIPOCs in the time of COVID-19. In order to do so, the article utilizes the concepts of precarity (Judith Butler), syndemics (Merrill Singer), and intersectionality. The chapter concludes in showing how graphic medical narratives through using various affordances of comics expose the complex architectures of politics, racism, and colonialism and their impact on health.

While Venkatesan's essay dwells on health inequities faced by the BIPOC communities in the time of COVID-19, Antara Chatterjee's essay "Caste in Epidemics in India: A Historical and Literary Reading" offers some iterations on health disparities in the context of caste in epidemics in India. The essay examines how caste and its constructions and practices, like untouchability, spatial and social segregation, and caste-based allocation of work, form a crucial subtext in perceptions and understandings of infectious disease outbreaks in India, in history as well as in cultural representation. Reading multiple epidemic outbreaks, like the 1896 bubonic plague outbreak and the 2020–21 COVID-19 pandemic, intersectionally with literary texts featuring epidemics, the essay uncovers how caste has configured meanings and trajectories of infectious disease in India and, in turn, been perceived within epidemic outbreaks and their representations. Chatterjee delves into how notions of the body and its boundaries, touch, defilement, contamination, and (im)purity figure centrally in both caste and infectious disease discourses and intersect within epidemic disease constructions in India. In foregrounding caste as a crucial index in experiencing and understanding epidemic disease differentially, the essay highlights health inequities based on socio-economic markers like caste, which in turn are intimately connected with parameters like class, economic status, and labor.

The wide cross-section of scholarly insights into the multifarious and complex cultural meanings and connotations of epidemics and pandemics, both historical and imagined, brought together in this collection, as detailed above, reveals these episodes as complex and textured biocultural and medico-social phenomena, which are not only medical and epidemiological but also social, economic, political, and cultural, in their ramifications. As the COVID-19 pandemic enters its third year with an uncertain future ahead of us, we are called upon by our shared humanity to vigilantly learn from past and current outbreaks of infectious pathogens and contribute cultural artifacts to the collective grand narrative of contagion. We hope that this work, while a very small contribution to this narrative, contributes to our understanding of epidemics and pandemics, both individually and collectively.

Notes

1. While epidemic calls attention to an outbreak of a disease in a certain geographical area, pandemic refers to an outbreak of a disease that has spread globally.

Works Cited

Barry, J. M. (2004). *The great Influenza: The epic story of the deadliest plague in history*. Viking Press.

Charon, R. (2006). *Narrative medicine: Honoring the stories of illness*. Oxford University Press.

Haraway, D. (1991). The biopolitics of postmodern bodies: constitution of self in immune system discourse. In *Simians, cyborgs, and women: The reinvention of nature*. Routledge.

Honigsbaum, M. (2019). *The pandemic century: One hundred years of panic, hysteria and hubris*. Hurst & Company.

Mitchell, W. J. T. (2007). Picturing terror: Derrida's autoimmunity. *Critical Inquiry, 33*(2), 277–290.

Outka, E. (2019). *Viral modernism: The Influenza pandemic and interwar literature*. Columbia University Press.

Vrdoljak, A., & Bauer, A. (2020). Pandemics and the role of culture. *International Journal of Cultural Property, 27*(4), 441–448.

Wald, P. (2008). *Contagious: Cultures, carriers, and the outbreak narrative*. Duke University Press.

Part I
Framing Pandemics/Epidemics

Chapter 2
Epidemic Narrative: Two Paradigms

Dilip Kumar Das

Abstract The suffering caused by disease and what it portends have been a major theme of narratives, from stories of illness in fiction and autobiography to those of epidemic outbreak. Disease, whether epidemic or non-epidemic, invariably calls for a narrative resolution, because of the way it is experienced as an event and not a condition of being. Within the genre of disease narratives, there has been since the second half of the twentieth century a greater critical focus on stories of personal illness rather than those of epidemic, two of the most well-known works being Arthur Kleinman's *The Illness Narratives* (1988) and Arthur W. Frank's *The Wounded Storyteller* (1995). This essay attempts to shift the focus from stories of personal health to public health, which seem pertinent in the context of the emerging epidemics that have broken out with alarming frequency since the beginning of the present century. It proposes that stories of epidemic are structured by two major paradigms, the *biomedical-realist* and the *cosmological-mythic*. The former views epidemic disease as naturalistic, a biological event caused by pathogen or environment, and recounts its unfolding in real time and space. The latter views it as the outcome of cosmic disorder, attributing it to supernatural agency, and recounting its unfolding in mythic time–space. With examples from narratives pertaining to India, the essay draws on arguments from medical anthropology and from the narrative theories of Mikhail Bakhtin and Paul Ricoeur to examine generic distinctions in terms of their emplotment and chronotopic forms.

D. K. Das (✉)
Department of Cultural Studies, School of Interdisciplinary Studies, EFL University, Hyderabad, India
e-mail: dilipdas@efluniversity.ac.in

S. Venkatesan et al. (eds.), *Pandemics and Epidemics in Cultural Representation*,
https://doi.org/10.1007/978-981-19-1296-2_2

Introduction

The outbreak of novel coronavirus disease in December 2019, which has since spread to more than 200 countries worldwide and has claimed a total of 3.5 million lives,[1] is the latest in a series of new viral disease outbreaks at the turn of the twenty-first century. The first to emerge was HIV disease, the first cases of which were reported in the USA in 1981. This was followed in quick succession by bird flu, severe acute respiratory syndrome (SARS), H1N1 and swine flu, dengue hemorrhagic fever, Middle Eastern Respiratory Syndrome (MERS), Ebola, Zika, Nipah, and currently COVID. One outcome of this succession of both older and new viral infections, which international travel has carried widely beyond their points of emergence, has been to bring public health issues into the forefront once again. This comes after a long gap when antibiotics and advances in therapeutic technologies seemed to have made epidemics a thing of the past. The pandemic of AIDS, as Mark Honigsbaum notes, "forced scientists to confront the hubristic assumption that medicine was on the verge of conquering infectious disease" (Honigsbaum, 2019, 165). This was especially so after the confident declaration in 1980 that smallpox had been eradicated globally.[2] The AIDS pandemic, Honigsbaum continues, also pointed to the fact that pathogens could mutate in unforeseen ways and that human intervention into natural environments and changing socio-cultural practices could be an important factor in such mutation. The turn of the century provided dramatic proof of this as new zoonotic viruses that had been confined to animal reservoirs in the wild evolved into human pandemics.

These outbreaks have had another consequence as well, a resurgence of the horrors of epidemic in the popular imagination, especially in narratives where they acquire the form of myth. To mention a few recent examples, Robin Cook's novel *Pandemic* (2018) is about a disease that leads to respiratory failure and death, like SARS, but is caused by an HIV-like retrovirus present in pigs. Cook's protagonist, a forensic pathologist discovers it as he investigates a mysterious death on a New York subway. Though it is not about epidemic but the ethics of medical research, zoonotic disease is the central event. Steven Soderburg's film *Contagion* (2011) is about an epidemic of viral respiratory disease that breaks out in Los Angeles, and it foregrounds the heroic efforts of the Centers for Disease Control and the FBI in containing it. *Contagion* emphasizes a theme that has become, as Priscilla Wald points out, a staple in epidemiological narratives: the work of disease detection (Wald, 2008, 36–39). Aashique Abu's film *Virus* (2019) is a fictionalized account of the Nipah virus epidemic that broke out in Kerala, India, in 2018. Like *Contagion*, it highlights the efforts of the public health authorities in tracing its origin and eventually stopping its spread. As

[1] COVID-19 Dashboard, World Health Organization, https://covid19.who.int/?gclid=EAIaIQobChMIpPKQp9C77wIVCVVgCh1ZlAjCEAAYASACEgIt_PD_BwE, accessed on 25 May 2021.

[2] This declaration was made in the 33rd. World Health Assembly on 8 May 1980, which in its resolution called it an "unprecedented achievement in the history of public health." https://apps.who.int/iris/bitstream/handle/10665/155528/WHA33_R3_eng.pdf?sequence=1&isAllowed=y, accessed on 23 March 2021.

Wald argues, popular accounts of pandemic fuse biomedical theories of emerging viruses with mythical plots of disaster caused by human interference in nature, genetic engineering, bioterrorism and international travel, highlighting as well the heroism of communities and individuals in containing the crisis. "Microbes, spaces, and interactions blend together as they animate the landscape and motivate the plot of the outbreak narrative: a contradictory but compelling story of the perils of human interdependence and the triumph of human connection and cooperation, scientific authority and the evolutionary advantages of the microbe, ecological balance, and impending disaster" (Wald, 2008, 2).

An epidemic outbreak is understood as a crisis, and people need to make collective sense of it. Culture as collective memory provides pre-existing frames for this, with narrative being the preferred mode of representation. Perhaps the most notable modern example is Albert Camus's *The Plague* (1947), about a fictional outbreak of the disease in the Algerian city of Oran. There had actually been a number of minor outbreaks of plague in Oran in 1921, 1931, and 1944, though the last major event that had devastated the city was way back in 1678 (Bertherat et al., 2007, 1459). The story of the "plague-stricken" people of Oran, as Camus writes at the end of the novel, is a "record of what had had to be done, and what assuredly would have to be done again in the never-ending fight against terror and its relentless onslaughts, despite their personal afflictions, by those who, while unable to be saints but refusing to bow down to pestilences, strive their utmost to be healers" (Camus, 1960, 297).

In the following section, I will look at the emergence of a genre of disease narratives against which I define the form and significance of epidemic narratives, the stories of personal illness. After a brief section on the AIDS narrative, which may be about epidemic but adopts the format of the illness story, I propose two paradigms for the analysis of epidemic narratives based on the way they explain disease causation (biomedical/cosmological) and the mode of narration (realist/mythic) that they adopt. In the fourth section, I look at the narrative construction of epidemic and its unfolding in time and space, drawing on the theories of Charles Rosenberg, Paul Ricoeur, and Mikhail Bakhtin, with examples from stories of epidemics in India. I conclude with the importance of stories as a source of a cultural knowledge that both represents and shapes show we respond to epidemics.

Illness Narratives

Mid-twentieth-century developments in antibiotic therapy made "infectious disease a relatively minor inconvenience in the global North" (Wald, 2008, 30)—as well as in the South, though to a lesser extent—and attention came to be directed to chronic and lifestyle diseases like cancer or heart disease. There appear to be two reasons for this. One is an actual increase in the incidence of such diseases, to the extent that they have become the new epidemics. Thus, a recent article in *BMJ* states: "The data suggest a growing epidemic of chronic disease." But the authors also argue that this could be because of "more vigilant recording of diagnoses, lowering diagnostic thresholds, and increasing intensity of services" (McGrail et al., 2016, 2). New diagnostic techniques and therapies, argues Ulrich Beck, have resulted in a greater visibility of "illnesses that can be diagnosed thanks to the more acute medical and technical sensory system"(Beck, 1992, 204).Whatever the reason, chronic disease has effectively come to supplant epidemic as the cultural symbol of human vulnerability today. This is evident in a large corpus of narratives about personal illness, which have emerged in the historical conjuncture of two events. The first is the development of new therapeutic procedures that can prolong life in even terminal cases, so that one lives long enough to tell the story. The second is a techno-scientific turn in clinical practice that privileges objective evidence of disease over the patient's subjective experience: It is as if the machine and the doctor speak to each other and the patient is factored out of the transaction. As Arthur Kleinman, a doctor as well as anthropologist, states in *Writing at the Margin*: "The patient's and family's complaints are regarded as *subjective* self-reports, biased accounts of a too-personal somewhere. The physician's task, wherever possible, is to replace these biased observations with *objective* data: the only valid sign of pathological processes, because they are based on verified and verifiable measurements." The result, he continues, "is a split between the constructed object of biomedical cure, which is the dehumanized disease process, and the constructed object of most other healing systems, which is the all-too-humanly narrated pathos and pain and meaning-directed perplexity of the experience of suffering" (Kleinman, 1995, 32, italics in original). Kleinman's point is that the discourse and practice of modern biomedicine has instituted in general a split between disease as an objective condition and illness as subjective experience, which an ethically responsive healthcare must overcome. Cecil B. Helman, another doctor-anthropologist, writes: "The doctor's view of the disease process must be reconciled with the patient's subjective view of his own illness and contradictions between the two models must be resolved by the process of negotiation" (1981, 551).[3] When clinical practice fails to do this it becomes imperative for patients to tell their own stories. The illness narrative, writes Joseph Frank, "opens a space for the particular experience of individuals" (Frank, 1993, 40). New forms of patient socialization contributed to the collective circulation of personal stories, which soon became popular enough for a broader readership (Frank, 1993, 39–40). As a genre, it has become an important subject of academic research, in studies such as Kleinman's *The Illness Narratives* (1988), Arthur Frank's *The Wounded Storyteller* (1995, 2nd. edn. 2013), and Einat Avrahami's *The Invading Body* (2007).

[3] Helman defines disease as "the named pathological entities that make up the medical model of ill-health," and illness as "the subjective response of the patient to being unwell" (Helman, 1981, 548). This distinction has now become standard in medical anthropology: see Strathern and Stewart (2010, 6–7).

The AIDS Narrative

The illness narrative serves as a background for my discussion of epidemic narratives. With disease as their common theme, illness narratives recount personal suffering while epidemic narratives recount collective suffering. This difference underlies differences in their form and cultural significance. The new viral pandemics, as I have stated at the outset, have resulted in a significant number of epidemic narratives in literature, cinema and the media. While most of the narratives written at first were on HIV/AIDS, other epidemics have become prominent in the last two decades, replacing AIDS as collective horror in the public perception.[4] Let me first discuss the AIDS narrative, which I argue is distinct from other epidemic narratives. It constitutes a large corpus of fictional and non-fictional stories in different media, told both in the first-person and the third. The number of such narratives has been greatest in the USA, where a project of cultural self-restitution was started by an active gay community in dealing with institutional homophobia and government inaction; and the community had the cultural capital to undertake such a project. Novels, short stories and memoirs were a part of this undertaking, as were poems, plays, films, paintings, photographs, and the AIDS Memorial Quilt.[5] The number of AIDS stories is far less in India for two possible reasons: (i) AIDS has disproportionately affected the poor, who lack access to cultural capital and publishing, and (ii) those that do have access hesitate to come out in the open because of the intense stigma. Stigmatization was intense in the USA too, compounding the stigma of disease with homophobia, but its targets were already organized into a community that had been vocal since the 1970s. Community-formation among persons with AIDS in India has, in contrast, been slight, perhaps because of the problematic role of the syndrome as identity-marker. AIDS narratives in India comprise a handful of fictional and non-fictional accounts in print and cinema, with few of the non-fictional accounts being in the first-person. In form, the AIDS narrative draws primarily on the format of the illness narrative with its focus on personal rather than collective suffering. Even a narrative like Kalpana Jain's *Positive Lives*, which is a journalistic report on the epidemic as she travels from one "hot-spot" to another, includes in large part the numerous personal stories that she gathers: "the high point of the book [is] the stories of courageous men and women with HIV/AIDS and their caregivers". Because of the way it

[4] With the development of anti-viral therapy or ART, HIV disease is now understood as manageable rather than fatal, though its narrative capacity for eliciting personal tragedy has not thereby diminished.

[5] The AIDS Memorial Quilt was a way of publicly remembering and honouring those who died of AIDS, its aim being to counter the stigma and secrecy associated with such a death. It started as a candle march on November 27, 1985, to protest against the killing of George Moscone and Harvey Milk, two San Francisco public officials who supported gay rights, and protestors displayed placards with names of people who had died of AIDS. Dubbed the Names Project, it led to the creation of a memorial quilt in 1987. Each panel tells the story of the life it commemorates: "These," as *The Quilt* puts it, "are not stories of an illness. Rather, they are stories of courage, fear, and anger, and mostly they are stories of love. They tell of people who worked and played, who laughed and fought, and who are finally remembered" (Ruskin et al., 1988, 13).

personalizes the epidemic, the AIDS narrative also serves as the preferred format for projects of health pedagogy, especially in cinema. For instance *Nidaan*, a commercially released film about a teenage girl who is infected through blood transfusion "teaches" the viewer about HIV, its modes of transmission and prevention, in shots where a doctor directly addresses the camera. *Phir Milenge* (2004) and *My Brother Nikhil* (2005) are about the rights of HIV-positive people and the need to counter stigma and discrimination (Das, 2019, 96–106). While the core object of these films is the "human interest" story rather than awareness creation, others like *Saavdhaan* (2007) are explicitly pedagogic. Made by UNAIDS and the Border Security Force, it uses the personal story format to teach BSF soldiers about risk prevention and the need to test (Das, 2019, 179–180). In focusing on personal suffering, therefore, the AIDS narrative departs significantly from the typical structure of epidemic stories that I discuss in the next section.

Biomedical and Cosmological Paradigms

Epidemic stories come in diverse forms: fictional and non-fictional accounts, epidemiological and journalistic reports, memoirs, and as episodes embedded in larger narratives. As narrative, however, three elements are essential: a set of events, their plotting in a causal sequence, and their unfolding over time and space. This is also how epidemics are understood in social experience, as events happening in real time–space and signifying a meaning. We tell stories of epidemic in order to make sense of it, to understand why it happened and what it portends. Epidemic signifies a crisis that may be environmental, social, or cosmological, a sudden and inexplicable disruption of the course of everyday existence. Broadly speaking, we look for two kinds of explanation leading to two kinds of narrative resolution, biomedical and cosmological. By biomedical, I mean an explanation-form that attributes epidemic to a natural cause, such as a pathogen that acts according to natural laws and can be contained by applying the same laws. By cosmological, I mean an explanation-form that attributes it to magical causes, such as the work of a deity or a spirit, and relief is sought through the performance of appropriate rites. The first type of explanation generally adopts a realist mode of narration while the second adopts a mythic mode. Following from this, we may distinguish between two paradigmatic forms of epidemic narrative, the biomedical-realist and the cosmological-mythic. Actual narratives, however, do not fit readily into this binary scheme. Rather, the scheme may be understood as constituting two poles of a spectrum, at one end of which is the biomedical report and at the other liturgical story. How each particular narrative constructs its epidemic event would depend on the accentuation it gives to either of the two paradigms.

Having defined my conceptual frame, let me now discuss two examples of the limit types. The first, which is biomedical, is a report on the cholera pandemics of the nineteenth century, "Oriental Pilgrimages and Cholera" by William G. Eggleston, published in 1892 in *The North American Review*. Eggleston reviews scientific literature on the spread of the disease in the Middle East, Africa, Europe, and America in successive waves from the 1820s to 1880s. He tracks the course of the disease from its origins in India through pilgrimage routes to Mecca and Egypt and thence through trade routes to Europe and America, recording the number of deaths each wave caused in the lands through which it passed. Eggleston's objective is to highlight the role of pilgrims in transmission of the disease, while underplaying that of colonial trade.[6] The mapping of spaces and recording of morbidity-and-mortality data were two of the major techniques by which the discourse of medical modernity constructed the time–space of epidemic as "real" time and space (Das, 2019, 76–77). Disease cartography was probably introduced in the 1690s, when the Neapolitan administrator Filippo Arieta included maps in his report on a plague epidemic at Bari (Koch, 2005, 19–24). But it came into prominence when John Snow undertook the mapping of a cholera outbreak in Soho in 1854, linking the spread of bacteria to the water supply system (Koch, 2005, 75–103). In the nineteenth century, medical mapping became an important source of knowledge in colonial public health, an early example being James Ranald Martin's *Notes on the Medical Topography of Calcutta* (1837). Curiously, however, underlying the realistic construction of the time–space of disease in Eggleston's narrative is a geography *imagined* through the trope of orientalism. Highlighting the role of pilgrim routes serves to construct India and Egypt as twin sources of contagion. "It is beyond human possibility to put India in good sanitary condition in any reasonable time ... The most riotous imagination could scarcely exaggerate the filth of India and Egypt and of the Hindoo and Mohammedan pilgrims" (Eggleston, 1892, 128). Contagion, as Mary Douglas points out, is both a medical and moral idea—or, more precisely, it is the medicalization of an originally moral idea. Eggleston's condemnation of Hindu and Muslim pilgrims expresses both sanitary concern and moral outrage, in an epidemic topography that is both real and imagined. To cite another example that brings out the point more explicitly, Alfred Reed's "Medical Sketches of the Orient," a disease travelog published three decades later, describes the scene in Meenakshi Temple of Madurai thus: "Inside is a howling, shouting pandemonium, a jostling, crowding melange of pilgrims, priests, ascetics, holy men, fakirs, beggars, sooth-sayers, thieves, panderers, children, and aged—all mixed up together with sacred animals and birds. From elephants and camels down to the leprous dogs and sleek rats that scuttle about in the semi-darkness, all is filth, vileness, and obscenity" (Reed, 1930, 201–202). The imputation of promiscuous intermixing and moral contamination is unmistakable in Reed's description. Though written in a realist mode premised on a naturalistic etiology, Eggleston's and Reed's texts reveal the pressure of mythicization that epidemic stories call for, in their attempt to comprehend what appears to be otherwise incomprehensible.

---

[6] Underplaying the role of colonialism in the spread of disease was characteristic of much medical literature on epidemics in the nineteenth century, which emphasized instead the colonies as sources of contagion and sites of intervention. This patho-geography of blame served two functions in colonialist discourse: It contributed to the presumed superiority of Europe and justified the exercise of medical power in the colonies (Arnold, 1993, 290–294).

In contrast to Eggleston's account of epidemic cholera, is a cosmological narrative, the liturgical story of Sitala, the smallpox goddess of South Asia. The worship of Sitala as a healing deity is an age-old tradition in most parts of India, except in the South where another goddess called Mariamma takes her place. Both are believed to protect their devotees from smallpox, which appeared in epidemic form almost annually during the hot months of March–May (Bhattacharya, 1999; Ferrari, 2015; Hanumanthan, 1980; Jash, 1982; Misra, 1969; Nicholas, 1981; Wadley, 1980).[7] Traditionally, stories or *kathas* of Sitala were publicly recited during her ritual worship, both in times of epidemic and on days attributed to her in the ritual calendar. In one of the stories, transcribed by Susan Wadley, a king's son and the son of a poor farmer were infected with smallpox. The farmer invoked the protection of Sitala, following the ritual practices, and offered only cold food without salt or spices which the goddess relished. The king's priest, too, offered prayers to the goddess but served her rich foods cooked hot and spicy. Soon, the farmer's son got well but the king's son continued to suffer. The king tried all ways of appeasing Sitala, but to no avail. Eventually, the goddess appeared to him in a dream and instructed him to do what the farmer had done, whereupon his son soon recovered. Recounted as myth rather than reality, without the time–space markers typical of the biomedical report, and set in a world where gods partake of human food and appear to them in dreams, the story nevertheless indexes the *real* world of the devotee in the practical instructions it integrates: "to keep houses very clean, to spread cowdung everyday (on the floor), to do the *puja* of Bhagavati [Sitala], to not eat salt, to not season foods, to not put anything on the pot, to not yourself eat hot things nor feed them to a pox victim, to serve only very cold things, etc." (Wadley, 1980, 38–39). The juxtaposition of the real and the mythic that we see here constitutes a cosmological sphere that includes natural and supernatural beings in their ordered relations with each other. Epidemic disease signifies a collapse of this order, which must then be restored through appropriate rituals. Embedded in performance, the stories themselves performatively produce the authority that they invoke, the goddess whose protection ensures the continuance of life. The difference between Eggleston's report and the Sitala story lies not just in the way they use realism or myth, but equally in the relative weight they give to each.

---

[7] The idea that goddesses can protect people during epidemics continues in contemporary times. Elsewhere I have discussed one such goddess, AIDS Amma to whom a shrine is dedicated in a village in South Karnataka (Das, 2019, 189–222). In April 2020, at the height of the COVID epidemic in India, media reported worship of a new goddess called Corona Devi or Corona Mai in widely scattered parts of the country (Das, 2021). It was a spontaneous phenomenon that attests to the continued belief in disease goddesses and ritual practices associated with them. In May 2021, a temple in Coimbatore (Tamil Nadu) consecrated and installed an idol of Corona Devi, to protect people in the second wave of the epidemic. Belief in protective deities, which the rituals articulate, authorize and reproduce, has been a widespread cultural response to epidemic in South Asia. It may be dismissed as "superstition" from a scientific or rational point of view, but it certainly allays fear of the outbreak and the moral uncertainty it causes. Its efficacy, as Marcel Mauss says of magic, is of an order different from the "mechanical" (Mauss, 1972, 25).

Emplotment, Chronotope, and the Significance of Epidemic

Within these broad distinctions between the biomedical-realist and cosmological-mythic modes, how do epidemic narratives construct the events and the time–space in which they unfold? Charles E. Rosenberg, in an essay on the social understanding of epidemic, outlines a three-stage schema in which the first refers to the event of outbreak and its public acknowledgment, the second to the various frames within which it is comprehended, and the third to the measures taken to contain it and its eventual decline.[8] The outbreak is socially understood in terms of "spiritual assumptions" of divine intervention or "mundane biological processes" that rule health and disease. "This eclectic mixture of moral assumption and mechanistic pathology provided the style of explanation that has been fundamental to the social management of epidemics in the West for the past three centuries". Public response to the crisis, following either of these explanations, takes the form of a "ritual," as collective acts that are coded and reiterated[9]: "the imposition of a quarantine, … the burning of tar to clear an infected atmosphere, the gathering of men and women in churches for days of fasting and prayer, all play a similar role—the visible acting out of community solidarity." The rituals serve to affirm belief in the explanation, and the belief in turn affords "a measure of control over an intractable reality". With the end of the epidemic comes a retrospective assessment of the event, the strategic and moral lessons that it has taught, and their implications for the future. The "implicit moral structure" of traditional knowledge, Rosenberg claims, is not incompatible with the public health responses of modernity but accompanies and even supplements them. Epidemic outbreak, in other words, is comprehended in ways both medical and moral, both realist and mythic.

Rosenberg's essay is instructive in the way it presents epidemic knowledge as the sedimented history of frames of understanding, with culture being the repository of this collective memory. By culture, I mean a system of meaning-making through which we understand events and actions. Culture, states Clifford Geertz, comprises the "webs of significance" that give meaning to our everyday acts (1973, 5). We make sense of events in the world through "ordered clusters of significant symbols," and culture is "the accumulated totality" of such clusters. Rosenberg's schema, which is about clusters of meaning attributed to epidemic outbreak and corresponding social responses, is relevant to the way epidemic narratives are plotted to produce and communicate meaning. The memory of past events and the way they are framed in culture provide epidemic narratives their source of meaning and their authenticity for readers. This is precisely the point that Paul Ricoeur makes about narratives in general, through his reformulation of Aristotle's concept of

[8] Rosenberg uses the metaphor of a three-act play to show how the stages are logically connected: "Just as a playwright chooses a theme and manages plot development, so a society constructs its characteristic response to an epidemic." But what he really emphasizes is the "predictable narrative sequence," or the ordering of events into a plot, which is common to both drama and story.

[9] Roy A. Rappaport defines ritual as "the performance of more or less invariant sequences of formal acts and utterances not encoded by the performers" (Rappaport, 1979, 175).

mimesis. Defining mimesis as the activity by which authors creatively order events and actions, he divides it into three stages of pre-composition, composition, and reception. "Mimesis$_1$" is the author's act of drawing on the meanings of human action in culture, meanings that both author and readers share, in order to compose convincing events. "Mimesis$_2$" is the putting together of the events in an order that is more causal than chronological, in which temporality is subordinated to the rule of the probable. "Mimesis$_3$" refers to the reader's activity of interpreting the plot in terms of both the plausibility of events narrated and their consonance with the meanings that culture attributes to human acts (that is, Mimesis$_1$). Narrative emplotment thus links the meanings *configured* in culture to their *refiguration* in the story and *transfiguration* in the process of reading, marking "the intersection of the world of the text and the world of the reader or hearer; the intersection, therefore, of the world configured by the poem and the world wherein real action occurs and unfolds in its specific temporality". Fictional stories, thereby, allude to reality even as they imagine it, thus investing the lived world with significances that, in the reader's interpretation, may confirm dominant assumptions or question them. Corollarily, true stories reimagine the reality they refer to, attributing to it meanings taken from culture to make it comprehensible.

Epidemic stories, whether fictional or factual, invest with significance the event of outbreak, incorporating it into the cultural memory of suffering. As Catherine Belling writes: "The medium that carries and communicates the burdens and lessons of past suffering is narrative. … Without narration, our accounting of the past transmits suffering only abstractly, as inexpressible inference or as deceptively simple, abstract numerical data." Stories, in other words, make suffering comprehensible and relatable to our lived experience, in a way that morbidity-and-mortality figures do not, providing ways of affectively dealing with it in the present and future. Let me take as example an extract from Suryakant Tripathy's memoir, *A Life Misspent*. This is an episode in which Tripathy (or "Nirala" as he was popularly known) recounts the outbreak of Spanish flu in his ancestral village in 1918. People die in such numbers that they cannot be properly cremated, and half-burnt corpses are thrown into the Ganga: "The Ganga was swollen with dead bodies." The river's bank is full of corpses waiting to be cremated. The epidemic claims his wife, uncle, cousin, sister-in-law, and infant niece: "My family disappeared in the blink of an eye." All their farmhands, too, die, and "[in] whichever direction I turned, I saw only darkness." Numbers are not mentioned, what is emphasized instead is their subjective impact. Recalling the event beggars description, even as he describes it: "Words cannot describe how pitiful the scene was, how helpless, how tender." To convey the scale of the tragedy, Tripathy alternates between personal loss and collective calamity, and the combination of objective fact and subjective perception constructs the epidemic as a *human* event. In recurrent shifts between past and narrative present, memory is reconfigured through reflection and invests the events with mythic significance. He recalls sitting on a mound by the Ganga to calm his disturbed mind, gazing at an abandoned ascetic's hut atop it and "the file of corpses" on the river bank below: "Sometimes I would think of the ascetic sadhus, sometimes of the ephemerality of the world." Real space merges with cosmic space, the *topos* of myth, as the river "swollen" with corpses is linked in his recalling with the "mound" of ascetics, two perspectives on the epidemic that signify both the transience of the body and the undying spirit of humanity. "'Those who are dead have found peace already,'" his friend Kullu Bhat tells him. "'Those who have beheld death seek peace as you do.'" And the mound is their symbolic "abode of peace" (Tripathy, 2016, 53–55).

Epidemics unfold in time as well as space: How do narratives represent this unfolding in the way they organize these two dimensions of human experience? They do not simply recount events, they locate them in a specific time–space that is both real and symbolic, or what Mikhail Bakhtin calls the "chronotope":

> In the literary artistic chronotope, spatial and temporal indicators are fused into one carefully thought-out, concrete whole. Time, as it were, thickens, takes on flesh, becomes artistically visible; likewise, space becomes charged and responsive to the movements of time, plot and history. (1981, 84)

The chronotope of epidemic is the narrative correlate of medical cartography and timelines of spread. In the Spanish flu episode, Tripathy juxtaposes memory and attention, past and present, as well as the real spaces he traverses and their symbolic meaning, to construct the mythic chronotope of epidemic. The point is explicitly brought out in a YouTube video on Spanish flu that opens with a quotation from this passage, followed by news footage of the epidemic in India as it spreads across the country ("Spanish Flu"). The video was uploaded in May 2020, at the height of the COVID-19 pandemic, which the Spanish flu story frames. The past serves to signify the present, and the scenes from 1918—the Ganga with its dead bodies, hospital morgues and mass cremation sites—function as the retrospective landscape of an epidemic that has both unfolded and is unfolding now. Another YouTube video documents the revival of nature following the global lockdown in 2020, with shots of wild animals exploring city streets, birds flocking to water bodies that are no longer polluted, and an inter-title that announces: "One of the positives within all this negativity of pandemic is … its giving our mother nature a break" ("Corona Pandemic Lockdown Effect on Wild Animals"). Natural and urban spaces become indistinct in a return to an imagined past of cosmic harmony, thus investing the epidemic (or its consequence) with a new significance as the vindication of "mother nature." The artistic chronotope, as Bakhtin explains, is what links the constructed world of the text to that which we inhabit: "The work and the world represented in it enter the real world and enrich it, and the real world enters the work and its world as part of the process of its creation, as well as part of its subsequent life, in a continual renewing of the work through the creative reception of listeners and readers" (1981, 254). Like Ricoeur's cycle of three-stage mimesis, therefore, the chronotope is for Bakhtin the site of ongoing exchanges between literature and life. In investing real time–space with the significance of myth, epidemic narratives like the ones I have discussed above merge biomedical and cosmological perspectives in a figure that illumines the meaning and value of human suffering.

My final example is U.R. Ananthamurthy's *Samskara*, a novel about community and caste in a Brahmin *agrahara* that centers on a plague epidemic. When plague breaks out in the fictional village of Durvasapura, one of its first victims is

Naranappa, a Brahmin who has willfully turned against the community and its traditions, alienated his kinsmen, consumed meat and alcohol, befriended Muslims, and lived openly with a low-caste concubine. Is he any longer a Brahmin, and would anyone who cremates him lose caste purity? Yet till he is cremated no one in the *agrahara* can eat. "Alive, Naranappa was an enemy; dead, a preventer of meals; as a corpse, a problem, a nuisance" (Ananthamurthy, 1976, 5). The community turns to Praneschacharya, its most revered member and an expert on the shastras, to provide a direction. But the holy texts are silent on the matter, and he goes to a temple of Maruti in the forest for a sign that never comes. Desperate and unable to see in the dark, he stumbles on Naranappa's concubine, loses his self-control in a moment of confusion and has sex with her. He can no longer claim the moral right to lead the community, as he discovers in himself a bodily desire that he had thought to have renounced. Ethically, he finds himself no different from Naranappa and others who had abandoned themselves to carnal passion. "*Samskara* is about the *embodiment* of abstract cultural norms, the *samskara* of the Brahmin community, and the disease that results when the body fails in fulfilling it" (Das, 2020, 42). Ananthamurthy uses the figure of the plague-stricken village, overrun by rats and vultures "as if the protecting gods had left them to their own devices" (1976, 65), as the symbol of a cosmic disorder that can only be remedied by a rite of expiation, for both the community and its fallen leader. As Praneshacharya muses: "'The *agrahara* stands there, explicit form for what I am facing within, an entire chapter on the verse that's me'" (Ananthamurthy, 1976, 87).

In *Samskara*, to put it briefly, the plague epidemic is constructed in cosmological terms, unfolding in the realm of myth and linking the dis-ease within to the disease outside, the body's disorder and that of the soul. Durvasapura with its succession of deaths serves as the chronotope of both. But Ananthamurthy does not concede the resolution that such emplotment demands: Naranappa's cremation, which would release both Praneshacharaya and the *agrahara* from their respective crises, is secretly performed by the fish merchant Ahmad Bari, one of Naranappa's Muslim friends, without the appropriate rituals (Ananthamurthy, 1976, 61–62). It is through Manjayya that Ananthamurthy counterposes the biomedical view of epidemic to the cosmological, in an attempt to mute its significance. Manjayya mocks the Brahmins who see it as a sign of collective pollution and ritual expiation and thinks instead of more "practical" steps like informing the municipality, getting people vaccinated and arranging treatment for those infected, exterminating the rats, and evacuating the village (Ananthamurthy, 1976, 90). Elsewhere, I have argued that Ananthamurthy expresses a postcolonial view of Indian society, opposing modernity to tradition and science to superstition (Das, 2020, 44). Another instance of such counterpointing is an1898 short story by Fakir Mohan Senapati, "Rebati" that narrates an outbreak of cholera in the fictional village of Patapur in Odisha. The story offers multiple cosmological views of the epidemic—as the visitation of a "demonic deity," the "mysterious ways" of Providence, and cosmic punishment for the transgression of norms (Senapati, 2019, 208–210).[10] Senapati's own view, though nowhere explicitly stated except in scattered references to cholera and its symptoms, is biomedical, which undercuts through irony the cosmological explanations. Yet in invoking the cosmos both texts reveal a cultural predisposition that cannot be canceled out by undermining it. Rather, it shows how both paradigms coexist in tension as historically contingent narrative structures through which we make collective sense of epidemics. They shape the stories we tell in intricate if ambiguous ways.

Conclusion

Stories constitute an important way in which we respond culturally to the crisis of epidemic. As events, epidemic outbreaks call to be narrated, to recount their unfolding in time and space, the suffering they cause and the struggle, the need to comprehend their utter incomprehensibility. The two broad ways in which we construct epidemic in our stories are as biomedical or cosmological crises, quite often merging the two perspectives. Biomedical narratives view epidemic as natural phenomena, occurring in real time–space and requiring interventions based on science. Cosmological narratives view epidemic as phenomena that signify a disruption of the cosmic order, unfolding in mythic time–space and requiring ritual interventions to restore it. But, as we have seen, narration often merges the two forms, investing realist emplotment and its chronotopes with symbolic significance, making epidemic what it is to phenomenal experience—at once an outbreak of disease and a collective calamity, a trial both of body and spirit, underlining both our vulnerability and our will to survive, with a pattern of recurrence that links the past to its significance for the present and the future. Stories are a cultural resource for these meanings of epidemic outbreak, directing the way we socially experience and respond to it. They are a form of knowledge equal in value to the scientific, not devoid of it: They may not explain the pathogenesis of epidemic disease, but they illumine its human significance.

Bibliography

Abu, A. (2019). dir. *Virus*. OPM Cinemas.

Anananthamurthy, U. R. (1976). *Samskara: A rite for a dead man* (A. K. Ramanujan, Trans.). Oxford University Press.

Arnold, D. (1993). *Colonizing the body: State medicine and epidemic disease in nineteenth-century India*. University of California Press.

[10] In her grandmother's view, it is Rebati's transgressive desire for education that is the cause. Senapati wrote "Rebati" at a time when a patriarchal norm denied women access to education, and the story was his way of debunking it.

Avrahami, E. (2007). *The invading body: Reading illness autobiographies*. University of Virginia Press.
Bakhtin, M. M. (1981). *The dialogic imagination: Four essays*. In Holquist, M. (Ed.), (C. Emerson, M. Holquist, Trans.). University of Texas Press.
Beck, U. (1992). *Risk Society: Towards a new modernity*. Sage.
Bertherat, E., et al. (2007). Plague reappearance in Algeria after 50 years, 2003. *Emerging Infectious Diseases, 10*, 1459–1462. https://doi.org/10.3201/eid1310.070284
Bhattacharya, N. N. (1999). *The Indian mother goddesses* (Third enlarged). Manohar.
Camus, A. (1960). *The Plague* (S. Gilbert, Trans.). Penguin.
Cook, R. (2018). *Pandemic*. Penguin Random House.
Corona Pandemic Lockdown Effect on Wild Animals. (2020). You Tube Video, 2:49, Verily Vagabonding, April 22, https://www.youtube.com/watch?v=WYB8febAorM
Das, D. K. (2019). *Teaching AIDS: The cultural politics of HIV disease in India*. Springer Nature.
Das, D. K. (2020). Cultural narratives of epidemic: Ananthamurthy's *Samskara*." *Anekaant: A Journal of Polysemic Thought*, (11), 41–44.
Das, D. K. (2021). Corona Devi: Epidemic and the disease goddess tradition. *Sambalpur Studies in Literatures & Cultures, Series, 6*, 11–24.
Eggleston, W. G. (1892). Oriental Pilgrimages and Cholera. *The North American Review, 155*(428), 126–128.
Ferrari, F. M. (2015). *Religion, devotion and medicine in North India: The healing power of Sitala*. Bloomsbury.
Frank, A. W. (1993). The rhetoric of self-change: Illness experience as narrative. *The Sociological Quarterly, 34*(1), 39–52.
Frank, A. W. (2013). *The wounded storyteller: Body, illness, and ethics* (2nd ed.). University of Chicago Press.
Hanumanthan, K. R. (1980). The Mariamman cult of Tamil Nadu—A case study in cultural synthesis. In *Proceedings of the Indian History Congress* (pp. 97–103).
Honigsbaum, M. (2019). *The pandemic century: A history of global contagion from the Spanish Flu to Covid-19*. Penguin.
Jash, P. (1982). A note on the cult of Sitala. In *Proceedings of the Indian History Congress* (pp. 189–197).
Kleinman, A. (1988). *The illness narratives: Suffering, healing, and the human condition*. Basic Books.
Kleinman, A. (1995). *Writing at the margin: Discourse between anthropology and medicine*. University of California Press.
Koch, T. (2005). *Cartographies of disease: Maps, mapping, and medicine*. ESRI Press.
McGrail, K., Lavergne, R., & Lewis, S. (2016). The chronic disease explosion. *BMJ: British Medical Journal, 352*. https://doi.org/10.1136/bmj.i1312
Misra, B. (1969). Sitala: The smallpox goddess of India. *Asian Folklore Studies, 28*(2), 133–142.
Nicholas, R. W. (1981). The Goddess Sitala and epidemic smallpox in Bengal. *The Journal of Asian Studies, 41*(1), 21–44.
Manjrekar, M. (2000). dir. *Nidaan*. Pan Pictures.
Mauss, Marcel. 1972. *A general theory of magic* (R. Brain, Trans.). Routledge.
Rappaport, R. A. (1979). *Ecology, meaning, and religion*. North Atlantic Books.
Reed, A. C. (1930). Medical sketches of the orient. *The Scientific Monthly, 31*(3), 193–216.
Revathy, dir. (2004). *Phir Milenge*. Percept Picture
Ruskin, C., Herron, M., & Zemke, D. (1988). *The quilt: Stories from the NAMES project*. Pocket Books.
Senapati, F. M. (2019). "Rebati." In *The greatest Odia stories ever told*. (L. Mohapatra, P., St-Pierre, K. K. Mohapatra, Trans.). Aleph.
Shamberg, M., Sher, S., Jacobs, G. (Producers), & Soderbergh, S. (Director). (2011). Contagion [Motion Picture] Participant Media.

"Spanish Flu: Corona Virus *se bhi khatarnak who bimari …*," You Tube video, 7:41, BBC Hindi, May 3, 2020, https://www.youtube.com/watch?v=WiDGE_Zut18
Strathern, A., & Stewart, P. J. (2010). *Curing and healing: Medical anthropology in global perspective* (2nd ed.). Carolina Academic Press.
Onir, dir. (2005). *My Brother … Nikhil*. Four Front Films.
Tripathy, S. ("Nirala"). (2016). *A life misspent* (S. Khanna, Trans.). Harper Perennial.
UNAIDS and BSF. (2007). *Saavdhaan—It's a battle against AIDS* [Motion Picture]. India.
Wadley, S. S. (1980). Sitala: The cool one. *Asian Folklore Studies, 39*(1), 33–62.
Wald, P. (2008). *Contagious: Cultures, carriers, and the outbreak narrative*. Duke University Press.

Part II
Charting Medieval and Early Modern Pestilences

Chapter 3
Chaucer's *Canterbury Tales* as a Postpandemic Text

Lorenz A. Hindrichsen

Abstract Despite few overt references to the Second Plague Pandemic (1348–1840), *The Canterbury Tales* (c.1387–1400) offer a compelling psychogram of a diverse community processing massive demographic shifts in the wake of recurrent epidemic waves. Building on Boccaccio's *The Decameron* (1349–52), which tells of a group of noblemen escaping plague-stricken Florence, Geoffrey Chaucer imagines a motley crew of survivors journeying toward the shrine of a local plague saint, Thomas Becket at Canterbury. Echoes of the Black Death reverberate throughout the text, notably in the *Pardoner's Tale*, which allegorizes the plague's lethality, but also in the shape of satirical pandemic archetypes such as a greedy plague doctor (the Physician), contagious commoners (the Cook and the Summoner), fallen socialites (the Knight and the Prioress), wealthy inheritors (the Wife of Bath), rivaling opportunists (the Miller and the Reeve), and corrupt clergy (the Pardoner). Postpandemic social change is also reflected in volatile social and linguistic codes (such as skewed forms of address), while socioeconomic grievances are delivered through maladaptive child modes reminiscent of trauma literature. Significantly, egocentric coping mechanisms dominating the pilgrims' discourse are regularly countered by visions of collaborative inclusivity (the Host's), mutual altruism (the Franklin's), and genuine piety (the Parson's).

Postpandemic Chaucer

Few studies have considered reading Chaucer against the Second Plague Pandemic (1348–1840), which in its first wave (1348–50), known as "The Black Death", killed approximately 40% of the population in Europe and the Middle East (Horrox, 1994: 3), and irrevocably altered the region's demographics, power structures, forms of worship, and cultural expression. This omission is quite surprising, given how strongly Chaucer's life and career were shaped by the pandemic's sociocultural repercussions. Much of Chaucer's writing seems not only *post-pandemic*, written

L. A. Hindrichsen (✉)
Copenhagen, Denmark
e-mail: lorenz.hindrichsen@cis.dk

S. Venkatesan et al. (eds.), *Pandemics and Epidemics in Cultural Representation*,
https://doi.org/10.1007/978-981-19-1296-2_3

in the wake of a major outbreak, but also *postpandemic*, evaluating its fallout, and envisioning a beyond while acknowledging the possibility of future outbreaks.[1] Born around 1340, Chaucer experienced the "Black Death" in his childhood, followed by later waves in his early twenties (Second Pestilence, 136061), his late twenties (Third Pestilence, 1369), his mid-thirties (Fourth Pestilence, 1374–79), and early fifties (Fifth Pestilence, 1390–93) (Horrox, 1994: 85–92). Despite lower mortality rates (10–15% on average), these later outbreaks were economically and psychologically significant,[2] making the prospect of a new wave a constant concern.

Fourteenth-century sources reveal an intense preoccupation with the pandemic in various contexts. Plague sermons urged parishioners to atone for their sins and join prophylactic processions.[3] Universities elaborated on outbreaks arising from cosmic humoral imbalances—a topic of great relevance for an astrologer like Chaucer.[4] Mystery Plays compared the impact of the Plagues of Egypt and the Flood to the 1348–50 wave,[5] while urban legislators responded to outbreaks by tightening regulations on cleanliness and hygiene, measures which Chaucer—son of a shop-owning family, and responsible for repairing sewers in the early 1390s—would have been aware of.[6] Socioeconomically, pandemic repercussions were impossible to overlook. Losing forty percent of the population upended feudal power structures, and handed laborers unheard-of bargaining power. What seigneurial records describe as a disruption of social order offered new levels of social mobility (Horrox, 1994: 238–244).

Chaucer's own nuclear family were among the biggest beneficiaries of the 1348–50 outbreak, with Geoffrey's parents inheriting more than 20 shops, along with property, from wealthy relatives (Selby, 1900: 2.149–150, 2.157). The Chaucers—nominally shoemakers—intensified their contacts with the court (Selby, 1900: vii, xi), resulting in lifelong public service appointments for Geoffrey. Although undoubtedly a traumatic event, the Second Plague Pandemic thus offered the family new

---

[1] I am using the hyphenated and non-hyphenated forms analogous to *post-colonial/postcolonial*, whereby *post*-indicates a temporal rupture (life *after* an era) while *post* (non-hyphenated) represents the envisioning of a life *beyond* (West-Pavlov, 2013: 161).

[2] Exact figures for these later outbreaks are hard to come by. Senn (2003: 572) assumes a 20% mortality rate for the Second Pestilence, with gradually diminishing lethality for later waves. Similarly falling curves are reported from France, where a papal doctor at Avignon estimates a 66% infection rate (with almost no survivors) for 1348, a 50% rate for 1361 (with some survivors), a 10% rate for 1371 (with many survivors), and a 5% rate for 1382, where almost everyone survived (Cohn & Samuel, 2002: 727). The economic toll of these later waves is asserted by Bridbury, who notes that in terms of grain prices, the plague "left a far deeper mark in 1361–62 than it did when it first devastated the country in mid-century" (1973: 584).

[3] See the sampling of sermons, prayers and masses from 1348 to 1382 excerpted by Horrox (1994: 111–125).

[4] The Paris medical faculty's finding in October 1348 that the plague was triggered by humoral imbalances (Horrox, 1994: 158–163) was widely echoed in the following centuries.

[5] As a Lincolnshire Cistercian wrote in 1413: "[S]o great a multitude of people [as died in the pandemic] were not killed in Noah's Flood" (Horrox, 1994: 66–67).

[6] See the London butchery legislation of 1371 reproduced in Horrox (1994: 203–205), and Chaucer's assignment as Clerk of the Works in the *Life-records of Chaucer* (Selby, 1900: 2.283–85).

opportunities, while presumably eliciting a strong sense of guilt (Turner, 2019: 36). In the light of these various contexts, what are the chances that Chaucer would have read Boccaccio's poignant dramatization of the Florentine plague in the *Decameron* without responding to it in some shape or form?

A closer look at *The Canterbury Tales* does indeed reveal many "spectral traces" (Outka, 2020: 2) of the pandemic throughout the text. Religious and astrological allusions (especially in the *General Prologue*), epidemic threats (embodied in the Cook), shameless profiteering (perfected by the Physician), and inclusive visions (notably the Host's) make pandemic discourse a pervasive subtext. The frame narrative itself, imagining a motley crew of pilgrims embarking on a common journey, seems unmistakably postpandemic by design. To close-read these and other traces, the following discussion shall consider a range of critical lenses: economic history, narratology, discourse analysis, iconography, and emerging theory on care (Care Collective, 2020). As this blend of approaches suggests, pandemic discourse can manifest itself in many ways, opening up multiple avenues for postpandemic readings.

Framing the Postpandemic: *The General Prologue*

How can one describe a pandemic not limited to a single event but occurring in serial waves? Chaucer likely wrestled with this challenge, given the temporal overlap of the *Canterbury Tales* (c.1387–1400) with the Fifth Pestilence (1390–93), following hard upon the Fourth (1374–79). Apart from raising tricky etiological questions regarding origin and dissemination, these repeated outbreaks posed a major narratological challenge, as they undercut familiar archetypes of short-lived apocalyptic epidemics (like the Plagues of Egypt) or protracted, individualized cases (like Job's). While Boccaccio was able to build on familiar dystopian accounts in earlier sources—some of his most poignant lines are directly lifted from Paulus Diaconus's description of the Justinian plague in the *History of the Lombards* (8th c.) (Boccaccio, 2016: 6, footnote 1)—Chaucer needed to devise a new narratological framework that could capture the cyclical nature of gradually weakening waves.

What we find in the *General Prologue* is a careful echoing of the Florentine plague in a forward-looking *reverdie* celebrating the overcoming of winter and disease through spring. Chaucer specifically alludes to a key passage from the *Decameron*, the dramatic finale summing up the corrosive effects of the Florentine plague as follows:

> [T]he cruelty of the heavens - and perhaps, in some measure, that of men, too - was so great and so malevolent that from March to the following July, [...] more than one hundred thousand human beings were deprived of their lives within the city of Florence [...]. Oh, how many great palaces, beautiful houses, and noble dwellings, once filled with lords and ladies and their retainers, were emptied of all their inhabitants, down to the last serving boy! How many [...] judged perfectly healthy, dined in the morning with their families, companions, and friends, only to have supper that evening with their ancestors in the next world! (Boccaccio, 2016: 11)

The basic temporal, spatial, and social dimensions Boccaccio evokes here—the outbreak began in March, was triggered by heavenly forces, and uprooted community life—are uncannily echoed in the *General Prologue*, Chaucer's equivalent to Boccaccio's *cornice*. The borrowings start in the very first lines

> Whan that Aprill with hire shoures soote
> The drought of March hath perced to the roote. (Chaucer, 1987: 1.1–2)

where the "dryness" of March invokes dry soil for sowing (Daley, 1970: 171–179), but also references dryness in a humoral, Galenic sense (Lewis et al. 1952–2001:"drŏught(e n., 4"). The "piercing" of "dry" March by "wet" April, personified by the Zodiacal Ram (Chaucer, 1987: 1.8), dramatizes seasonal change as a cosmic spectacle where various actants—including Zephyr (Chaucer, 1987: 1.5), the fertile west wind—overthrow the month of Mars, an unwholesome planet triggering disease in a scenario echoing the attempted killing of pandemic Death in the *Pardoner's Tale* (Beidler, 1982).

Cosmic humoral imbalances were throughout the fourteenth century viewed as the main *natural* cause of the second plague pandemic, alongside *spiritual* factors such as sinful behavior. In October 1348, the medical faculty at Paris—referencing Albertus Magnus—declared that the plague had arisen from an unfortunate "conjunction of Mars [dry and cold] and Jupiter [wet and hot], caus[ing] a great pestilence in the air" (Horrox, 1994: 159), an explanation echoing throughout contemporary sources.[7] Reading the "dryness" of March as code for a pandemic outbreak aligns well with Chaucer's investment in astrology, and the harmful characteristics attributed to Mars in his writing.[8]

The scenario of the pilgrims escaping a humorally dry, infectious March—analogous to Boccaccio's "cruelty of the heavens"—dovetails nicely with the rationale as to why they flock to Canterbury:

> The hooly blisful martir for to seke,
> That hem hath holpen whan that they were seeke. (Chaucer, 1987: 1.16–18)

The narrator's pairing of quest (*seke*) with disease (*seeke*) articulates a common motivation of pilgrims to pray for a cure, or give thanks for successful recovery. The popularity of pilgrimages skyrocketed during the Second Plague Pandemic to the point where some local administrators feared human crushes or infections spreading across massive crowds.[9]

---

[7] See John of Reading, who worries in 1365 that the triple conjunction of Mars, Jupiter and Saturn might herald a new outbreak, or the Northern English chronicler attributing a 1379 outbreak to "a hostile configuration of the planets" (Horrox, 1994: 133, 89).

[8] See for example *Troilus and Cressida* (Chaucer, 1987: 3.716), where Troilus hypothesizes on being influenced by "[a]spectes badde of Mars or of Saturne".

[9] The Bishop of Coventry and Lichfield expresses his concerns that many pilgrims heading for Merevale "are brought to the point of death by the crush of people or by falling sick of the various illnesses prevalent in these days", referencing the summer of 1361 when the Second Pestilence rages. (Horrox, 1994: 148).

Chaucer's pilgrims, despite their irreverence, constantly endorse saint worship through their dress or speech,[10] thus sustaining the notion of a devotional journey restoring spiritual and physical health. The same idea is expressed through the pilgrims first stopping at the "Waterying of Seint Thomas" (Chaucer, 1987: 1.830), a landmark referencing the "Water of St. Thomas", a panacea issued at Canterbury that supposedly contained traces of the saint's blood (Sumption, 2003: 113). That the pilgrims' response falls into a larger pattern of epidemics triggering worship is signaled through a hypotaxis that links seasonal change ("Whan that", "Whan" (Chaucer, 1987: 1.1, 1.5)) to pilgrimages ("Thanne" (Chaucer, 1987: 12)), thus emphasizing the cyclical element underpinning epidemiological responses.

The *General Prologue* also engages with the sociocultural repercussions of the pandemic fallout, effectively replacing Boccaccio's scenes of social fragmentation with a vision of postpandemic unity. The narrator stresses the effortless bonding of a bunch of pilgrims "by aventure yfalle/In felaweship" (1.25–26). The "hostelrye" (1.23) at Southwark, a cauldron uniting different social strata, takes the place of Boccaccio's deserted "palaces" and "noble dwellings once filled with lords and ladies and their retainers." That Chaucer may have thought of Boccaccio when imagining Southwark is suggested by a prominent pair of analogous food metaphors linking the texts. Boccaccio concludes his apocalyptic vision of Florence by stating that "many [...] [...] judged perfectly healthy, dined in the morning with their families, companions, and friends" only to have "supper that evening with their ancestors in the next world!" (Boccaccio, 2016: 11). The grotesque metaphor of a community split into joining worldly and heavenly suppers is directly countered by Chaucer's Host, who unites the pilgrims by promising the best storyteller "a soper at oure aller cost/Heer in this place, sitting by this post" (Chaucer, 1987: 1.799–800).

That the Host's "soper" references the Last Supper emerges from the narrator customarily calling Harry Bailey "[oure] Hoost", a homonym with the Eucharist in Middle English (Lewis et al., 1952–2001:"hōst(e n.(2)", "hōst(e n.(4)"). The pun is visualized in later woodcuts of that imaginary supper with the Host presiding over the table in a Christ-like fashion (Archer et al., 2015: 3–4). Harry Bailey's sanctified worldly supper thus effectively rewrites Boccaccio's split metaphor, suggesting an overcoming of a destructive pandemic. While Boccaccio arguably uses his suppers to euphemize the plague's impact, Chaucer's metaphor is decidedly *post*pandemic as it envisions a shared journey and common narrative culminating in a return to "this place", "this post" at Southwark.

By evoking various scientific, religious, and literary subtexts, *The Canterbury Tales* thus articulate the desire to move beyond pandemic experience on multiple levels: temporally (overthrowing "sickly" March), spatially (abandoning crowded Southwark),[11] socially (bridging demographic divides), and intertextually (replacing

---

[10] See the Yeoman wearing a brooch of Saint Christopher (Chaucer, 1987:1.115), or Alison from the *Miller's Tale* swearing allegiance to her lover by "Seint Thomas of Kent" (Chaucer, 1987: 1.3291).

[11] The notion of leaving Southwark as an escape aligns well with "three golden rules" given by a 15th-century plague doctor: "[G]et out quickly, go a long way away and don't be in a hurry to come back" (Horrox, 1994: 108).

plague-stricken Florence with the road to Canterbury). Casting the pilgrims as actants heading toward Canterbury before returning to Southwark's "post" lends the *General Prologue*'s forward-looking reverdie a hopeful touch.

Carnivalesque Discourse: Conversation Among the Pilgrims

The feisty unruliness characterizing much of the pilgrims' interaction marks the *Canterbury Tales* as carnivalesque, a discursive mode Bakthin sees epitomized in pandemic life as dramatized in Boccaccio's *Decameron*:

> [The plague] grants the right to use other words, to have another approach to life and to the world. Not only have all conventions been dropped, but all laws 'both human and divine' are silenced. Life has been lifted out of its routine, the web of conventions has been torn; all the official hierarchic limits have been swept away. The plague has created its own unique atmosphere that grants both outward and inward rights. Even the most respected man may now wear his 'breeches for headgear.' And in accord with the general turnover, the problems of life may be discussed not in churches and schools but in whorehouses. (Bakthin, 1984: 273–74)

The sociopolitical shifts Bakhtin sees reflected in the *Decameron*—the subversion of hierarchies, the secularizing and democratizing of discourse, the triumph of silliness over the sacred—also govern interactions among Chaucer's pilgrims, whose questionable piety accentuates their carnivalesque status.

Not only brash commoners like the Miller subvert social order ("I kan a noble tale for the nones,/With which I wol now quite the Knyghtes tale." (Chaucer, 1987: 1.3126–27)); also the Host, "juge" (Chaucer, 1987: 1.814) of the storytelling contest, embraces Bakthin's new "approach to life", perhaps most memorably so when censuring the Pardoner for pressuring the other pilgrims to kiss a counterfeit relic:

> "Nay, nay!" quod he [the Host], "thanne have I Cristes curs!
> Lat be," quod he, "it shal nat be, so theech!
> Thou woldest make me kisse thyn olde breech,
> And swere it were a relyk of a seint,
> Though it were with thy fundement depeint!
> But, by the croys which that Seint Eleyne fond,
> I wolde I hadde thy coillons in myn hond
> In stide of relikes or of seintuarie.
> Lat kutte hem of, I wol thee helpe hem carie;
> They shul be shryned in an hogges toord!" (Chaucer, 1987: 6.946–55)

Blending the sacred (relics, Christ, saints) with the profane (cursing, private parts, excrement), the Host subverts social conventions, and normalizes a degree of profanity alien to courtly and religious discourse.

His expletive-laden rant starkly contrasts with the obsequious formality with which he approaches some nobles earlier:

> Sire Knyght, [...] my mayster and my lord,
> Now draweth cut, for that is myn accord.
> Cometh neer [...] my lady Prioresse.
> And ye, sire Clerk, lat be youre shamefastnesse (Chaucer, 1987:1.837–40)

The Host's substituting of the reverential for the profane, also when reproaching the Miller ("Tel on, a devel wey!" (Chaucer, 1987: 1.3135) or the Reeve ("The devel made a reve for to preche" (Chaucer, 1987: 1.3902–04)), celebrates a new linguistic code where deference and piety are suspended, and repartee and blasphemy become the norm. (That this change is spearheaded by the Host, whose name echoes the Eucharist, underscores the point.)

Many pilgrims transform accordingly and lay claim to new "outward and inward rights" in the absence of "laws 'both human and divine'" (Bakthin, 1984: 273). Many commoners self-select their conversational turn rather than waiting to be invited by the Host: the Cook ("a povre man" (Chaucer, 1987: 1.4341)), the Canon's Yeoman, the Shipman and most churls (Miller, Manciple, Reeve, and Summoner) boldly seize the floor, sometimes rudely interrupting another speaker. Only two "gentils" follow their example.[12] One aristocrat, the Monk "daun Piers" (Chaucer, 1987: 7.2792), visibly struggles to be heard (Chaucer, 1987: 1.3118–19, 7.1924–27), and declines to tell a second tale: "I have no lust to pleye./Now lat another telle, as I have toold" (Chaucer, 1987: 7.2806–07).

This iconoclastic, carnivalesque dynamic creates a social environment where privilege must be earned, and self-fashioning becomes the norm.[13] As a linguistic marker of this new social contract, the "requiting" of tales becomes the dominant speech act offering previously marginalized groups a voice: the Miller "quites" the Knight for being his a social superior (Chaucer, 1987: 1.3127), the Reeve "quites" the Miller for insulting Reeves (Chaucer, 1987: 1.3864), while the Cook "quites" "hostileer[s]" like the Host (Chaucer, 1987: 1.4360–62). Ironically, these commoners appropriate a phrase the Host initially uses to celebrate piety ("Ye goon to Caunterbury—God yow speede,/The blissful martir *quite* yow youre meede!" (Chaucer, 1987:1.770, emphasis added) and to preserve privilege ("Now telleth ye, sir Monk, if that ye konne,/Somewhat to quite with the Knyghtes tale" (Chaucer, 1987: 1.3118–19), thus openly usurping and mocking dominant discourse in a distinctly carnivalesque twist (Patterson, 1991: 244–45).

One character who masters "requiting" to perfection is the Wife of Bath, whose miraculous surviving of five husbands marks her as a physiological, and perhaps epidemiological, anomaly. Since women had a lower mortality rate in the Second

[12] The two self-selecting gentles are the Friar (Chaucer, 1987: 3.1278–79) and the Merchant (Chaucer, 1987: 4.1233–38). I am relying on Morgan's definition of gentils and commoners, which considers the first half of pilgrims introduced in the *General Prologue* (from the Knight to the Franklin) as gentils, while viewing the second half (starting with the Five Guildsmen) as commoners (Morgan, 1978: 486–98).

[13] Greenblatt's concept can be meaningfully applied to figures like the Wife of Bath, the Physician, the Pardoner or the Miller, who successfully reinvent themselves in this socially mobile community (Greenblatt, 2012).

Pestilence and successive waves,[14] are Alisoun's vanishing husbands casualties of the plague? She certainly invites such a reading when praying to outlive future partners, and cursing the stingy ones with the plague:

> [...] and Jhesu Crist us sende
> Housbondes meeke, yonge, and fressh abedde,
> And *grace t' overbyde hem* that we wedde;
> And eek I praye Jhesu *shorte hir lyves*
> That noght wol be governed by hir wyves;
> And olde and angry nygardes of dispence,
> God sende hem soone *verray pestilence!* (3.1261–64, emphasis added)

Tellingly, she also invokes Saint Judoc, whose body was supposedly "incorruptible" (Watson, 2021), when "quiting" her fourth, philandering husband:

> Now wol I tellen of my fourthe housbonde.
> I seye, I hadde in herte greet despit
> That he of any oother had delit.
> *But he was quit*, by God and by Seint Joce! (Chaucer, 1987: 3.480–83, emphasis added)

The levity with which she mocks her deceased fourth husband sets the stage for a grotesque funeral where—in an episode mimicking Boccaccio—she falls in love with the shapely legs of a coffin bearer, Jankyn, who will become her fifth husband (Chaucer, 1987: 3.593–99), thereby tearing up the "web of conventions" Bakthin defines as the carnivalesque.

The illustrator of the *Ellesmere Manuscript* expands on the Wife of Bath's irreverence by handing her a whip (Fig. 3.1) resembling the whips of pandemic flagellants in chronicles and contemporary illustrations (Fig. 3.2).[15] By turning Alisoun's metaphorical "whipp[ing]" of husbands (Chaucer, 1987: 3.175) into a barbed commentary on self-flagellation, the illustrator casts Alisoun as a provocateur extraordinaire who sets a new standard for postpandemic intersectional subversion. By undermining patriarchal, spiritual, and feudal hierarchies (she is a commoner), her discursive affronts signpost late fourteenth-century demographic shifts directly arising from the Second Plague Pandemic.

[14] Horrox lists various chroniclers describing the 1361 outbreak as a "great pestilence of men" [sic]; according to John of Reading, this triggered promiscuous behaviour in women, who "no longer worried about sexual lapses, for they now regarded fornication, incest or adultery as a game rather than a sin" (1994: 87–88). The 1390 outbreak also disproportionately affected "adolescents and boys" (1994: 91).

[15] According to Robert of Avesbury, each flagellant "carried in his right hand a whip with three thongs. Each thong had a knot in it, with something sharp, like a needle, stuck through the middle of the knot so that it stuck out on each side [...]." (Horrox, 1994: 153). Similar whips feature in the *Belles Heures of Jean de France, duc de Berry* (Anon, 2021).

Fig. 3.1 The Wife of Bath from the Ellesmere Manuscript, mssEL 26 C 9, f.72r (detail), The Huntington Library, San Marino, California

Suspect Bodies: The Summoner and The Cook

Another prominent postpandemic trope is the foregrounding of contagious bodies. Several pilgrims bear visible signs of past or present infections. A telling example is the Summoner, whose ruby cheeks are scarred by "saucefleem", a skin disease associated with leprosy:[16]

> No oynement that wolde cleanse and byte,
>
> [...] his whelkes white,
>
> Nor [...] the knobbes sittyinge on his chekes." (Chaucer, 1987: 1.631–33)

While the Summoner's symptoms clearly indict his "lecherous" character (Chaucer, 1987: 1.626), one may read his condition not just metaphorically as sexual deviance (Orkin & Joubin, 2019: 20–21), but also pathologically, as an infected body. Spots and

---

[16] "A skin ailment considered symptomatic of a type of leprosy originating in a humoral disorder, caused by sexual and dietary excess and characterized by red and black discoloration, pustules, swelling, loss of hair" (Lewis et al. 1952–2001:"saucefleume n.").

Fig. 3.2 The Flagellants at Doornik in 1349, miniature from the Chronicle of Aegidius Li Muisis (detail), reproduced in Helmholt (1902: 178–179)

other outward symptoms were greatly feared as possible indicators of a new epidemic outbreak.[17] Emphasizing the Summoner's condition and his failed treatments cast him as a potential epidemic threat, a quality the Ellesmere illustrator highlights by showing him delivering a courtly summons with a massive seal shaped like a large black spot, representing his contagious state (Fig. 3.3).

A pilgrim suffering from even more severe symptoms is the Cook, whose appearance greatly troubles the narrator:

> But greet harm was it, as it thoughte me,
> That on his shyne a mormal hadde he[.]. (Chaucer, 1987: 1.385–86)

While most pilgrims are introduced through several character traits, the Cook's ulcer is his only marker. The unusual term *mormal*, a transliteration of Old French *mort mal*, accentuates the lethality of this "mark of death" (Lewis et al., 1952–2001:"mormal(e n."). Open sores were greatly feared at the time for supposedly emitting

[17] An anonymous Canterbury chronicler states how during the Second Pestilence of 1361, many ""[m]embers of the religious orders and parish clergy and others died suddenly [...] when the first spots and the other signs of death appeared on their bodies" (Horrox, 1994: 86).

Fig. 3.3 The Summoner from the Ellesmere Manuscript, mssEL 26 C 9, f.81r (detail), The Huntington Library, San Marino, California

poisonous vapors.[18] As a source of contagion, the Cook's *mormal* thus poses a threat to the entire community of pilgrims, an association amplified through the dish, "mortreux" (Chaucer, 1987: 1.384), which echoes his open sore. Tellingly, the Host later reproaches him for preparing unwholesome food.[19]

The Ellesmere illustrator visualizes the Cook's contagious potential by depicting a poorly bandanged shin, whose red blotches are amplified through his carmine coat, red bridle, and ruby cheeks (Fig. 3.4). Replacing the "mormal" with smaller blotches only intensifies pandemic connotations, since small spots were considered deadlier than large ones (Cohn & Samuel, 2002: 716). The Cook's pronounced baring of his thigh references Saint Roche, an increasingly popular plague saint at the time, who is usually shown in a pilgrim's outfit lifting his gown to reveal a plague sore (Fig. 3.5).

Unlike the Summoner, whose incurable condition smacks of divine retribution, the Cook resembles a hapless Job deserving pity rather than reproach. The Ellesmere illustrator suggests as much through the iconographic referencing of Saint Roche, as well as the Cook's deferential greeting, which resembles a supplication. His plaintive

[18] As a mid-15th century Swedish bishop explains: "[P]estilence sores are contagious because of infectious humours, and the reek or smoke of such sores is venomous and corrupts the air. And therefore one should flee such persons as are infected" (Horrox, 1994: 175).

[19] "Of many a pilgrym hastow Cristes curs,/For of thy percely yet they fare the wors" (Chaucer, 1987: 1.4349–50).

Fig. 3.4 The Cook from the Ellesmere Manuscript, mssEL 26 C 9, f.47r (detail), The Huntington Library, San Marino, California

facial expression, and his eyeing of the hat's red inner lining (mirroring his symptoms), portrays him as acknowledging the condition is beyond his control. Such a sense of powerlessness is also verbalized in the line "[t]hat on his shyne a *mormal* hadde he" (emphasis added), where the stress on "mormal", contrasting with the anti-climactic "hadde he", accentuates the condition's hold over him. Verbal and visual cues, then, portray the Cook as a blameless casualty, unlike the Summoner, whose lechery seemingly provokes the disease. Taken together, Summoner and Cook embody contrasting ways in which the Second Plague Pandemic was conceptualized: as a self-inflicted purge or hapless circumstance.

Fig. 3.5 Saint Roche in a historiated initial from the Missal of Cardinal Albrecht von Brandenburg, Ms10, 420r. (detail), Hofbibliothek Aschaffenburg

Monetizing Suffering: The Physician and The Pardoner

Two pilgrims who exploit pandemic anxieties ruthlessly through elaborate scams are the Physician and the Pardoner. The "Doctour of Phisik", the narrator asserts, has profited greatly from the plague:

> His studie was but litel on the Bible,
> In sangwyn and in pers he clad was al,
> Lyned with taffata and with sendal.
> And yet he was but esy of dispence;
> He kepte that he wan in pestilence.
> For gold in phisik is a cordial,
> Therefore he lovede gold in special. (Chaucer, 1987: 1.438–44)

In typical estate satire fashion, initial praise is gradually eroded through sarcastic side remarks. His studies lack piety, his dress appropriates royal colors, his silks (taffata and sendal) are excessive. (Ironically, silks range among the very clothing some fourteenth-century writers blame for plague outbreaks.)[20] His unscrupulousness is effectively conveyed through the rapid transition from stinginess ('esy of dispense') to unfettered opportunism and greed ('He kepte that he wan in pestilence'), culminating in the metaphor of gold being his "cordial", which shows him living up to the Middle English moniker of being a "leech" (Lewis et al. 1952–2001:"lẹ̄che n.(3)").

The narrator continually highlights the Physician's salesmanship, suggesting that

> In al this world ne was ther noon hym lik,
> To *speke* of phisik and of surgerye
> For he was grounded in astronomye. (Chaucer, 1987: 1.411–13, emphasis added)

As the emphasis on "speaking" indicates, it is medical *discourse* rather than medical treatments he excels at:

> He knew the cause of everich maladye,
> Were it of hoot, or coold, or moyste, or drye,
> And where they engendred, and of what humour. (Chaucer, 1987: 1.417–21)

The Physician's mastery of Galenic humors gives him leverage to opine not just on his patients' diseases, but also on their origin—"where they engend[er]"—a phrase referencing the "engendr[ing]" of spring flowers in the opening of the *General Prologue* after "dry", infectious March has been defeated (Chaucer, 1987: 1.4).

That expertise, though, does not translate into effective treatments:

> He was a verray, parfit praktisour:
> The cause yknowe, and of his harm the roote,
> Anon he yaf the sike man his boote. (Chaucer, 1987: 1.422–24)

The speedy transition from diagnosis ("[t]he cause yknowe") to giving patients "his boote" exposes the materialism informing the Physician's practice. The narrator's critique here echoes widespread reservations about late fourteenth-century plague doctors, whose cures were often considered ineffective or overpriced (Orlemanski, 2014). The second half of the fourteenth century saw significant progress in diagnosing and treating plague victims (1348–50), resulting in a proliferation of plague tracts and a sharp rise in doctors' salaries (Cohn & Samuel, 2002: 708–10). During the plague year of 1369, the royal physician at Edward III's court ranked among the most highly remunerated officials, receiving far higher financial compensations than Geoffrey Chaucer and other squires.[21] Despite—or because of—this rise in status, much of the critique of physicians still survived. The ambivalence with which Chaucer's narrator and the Host treat the Physician matches this double-edged perception of doctors as well-remunerated potential frauds who tried to sustain "the illusion that the plague could be controlled" (Horrox, 1994: 109).

---

[20] Horrox, The Black Death, 132.

[21] On June 27, 1369 the royal surgeon received more than twice as much as Geoffrey Chaucer for his summer robe. He also received one third more than Chaucer as an advance prior to the war in France (Selby, 1900: 2.171, 2.175–76.).

An equally shrewd operator fond of subtle "praktike"[22] is the Pardoner, who peddles "hooly pardoun[s]" and solicits donations for "hooly bulle[s]" and kissing counterfeit "relikes" (Chaucer, 1987: 6.906, 6.909, 6.944). His prominence in *The Canterbury Tales*—he is introduced as the last pilgrim (Chaucer, 1987: 1.669714), delivers both a tale and a prologue, and interrupts another pilgrim twice (Chaucer, 1987: 3.164–68, 3.184–87)—underscores the significance of his trade. The Second Plague Pandemic intensified the issuing of pardons and the venerations of relics across Europe. Many English bishops granted special indulgences to parishioners participating in processions, masses, and prayers in response to advancing pandemic waves.[23] Chaucer's Pardoner visibly profits from that boom, carrying a pouch "bretful of pardon comen from Roome al hoot" (Chaucer, 1987: 1.687), along with him.

Pitching to a postpandemic community, the Pardoner praises "Jhesu Christ, that is oure soules leche" (Chaucer, 1987: 6.916), thus cleverly instrumentalizing the familiar metaphor of Christ the Physician (Jeffrey, 1992: 614), which permeates late medieval pandemic discourse.[24] The Pardoner leverages the topos by emphasizing that it is Christ the Physician who "graunt[s] [the pilgrims] his pardoun" (Chaucer, 1987: 6.916–17), thus anchoring the power of indulgences in Christ's healing capacities. Significantly, the Pardoner's services offer merely temporary respite, ensuring him a steady source of income. He urges the pilgrims to secure themselves a pardon immediately ("Com forth anon, and kneleth heere adoun" (Chaucer, 1987: 1.925)) or purchase one "[a]l newe and fressh" (Chaucer, 1987: 6.928) on the road. Indulgences must be administered regularly ("alwey newe and newe" (Chaucer, 1987: 6.929)), he claims, to ensure one's spiritual health, as a safeguard against unforeseen disaster:

> Paraventure ther may falle oon or two
> Doun of his hors and breke his nekke atwo
> Look whiche a seuretee is it to yow alle
> That I am in your felaweship yfalle[.] (Chaucer, 1987: 6.935–38)

His gleeful embracing of freak accidents, reflected in the disturbing rhyming of "oon or two" (note the non-chalant vagueness) with "breke his nekke *atwo*" (emphasis added), reveals a transactionality reminiscent of the Physician's attitude toward his patients. The Pardoner's claim that his presence offers "seuretee" unmasks him a sneaky hypocrite and self-absorbed profiteer matching the Physician in approach, sentiment, and skill. The Pardoner and the Physician, then, emerge as shameless profiteers who seize pandemic outbreaks as business opportunities, and perfect the monetizing of fear and suffering by twisting religious and scientific paradigms to their advantage.

---

22 The phrase appears when the Pardoner asks the Wife of Bath: "[T]eche us yonge men of your praktike" (Chaucer, 1987: 3.187).

23 See for example the special indulgences issued in the Archdiocese of Canterbury in 1348 and in 1375 during the Fourth Pestilence (Horrox, 1994: 89, 114, 117).

24 See for example the praising of "the most delightful and precious medicine: the body of our lord and saviour Jesus Christ" in a fifteenth-century Italian plague tract that promotes piety over human medicine (Horrox, 1994: 149).

Postpandemic Care: The Franklin, The Parson, and The Host

Contrasting these exploitative voices, several characters articulate visions of inclusivity, mutual respect, and care, notably the Franklin, whose Breton lai—an adaptation of Boccaccio's rendering of the 'rash promise'—compels the audience to reflect on which character is most generous ("the mooste fre" (Chaucer, 1987: 5.1622)): Aurelius dispensing his wife Dorigen from her marital vow, the suitor Arveragus freeing Dorigen from her reckless promise, or the Magician canceling Arveragus's financial debt. Reneging on one's pledges was a key concern in postpandemic England as countless court cases attest: Fear from pathogens and new economic opportunities made priests abandon their parishes, laborers deny their services, and plot holders leave their land (Horrox, 1994: 310–38). The Franklin's celebration of generosity arguably addresses those socioeconomic concerns, and suggests resolving contractual impasses collectively and altruistically, through dialogue and mutual understanding.

A second vision is articulated by the Parson, whose humble origins elevate him above a materialist fray. Introduced as a "povre Persoun of a Toun" (Chaucer, 1987: 1.478), his everyman status is evoked through the ambiguous term "persoun" identifying him both as clergy (a parson) and as a simple individual: a person (Lewis et al. 1952–2001:"persŏun(e n.(1)").[25] Unlike the genteel hare-hunting Monk (Chaucer, 1987: 1.165–207), the courtly-mannered Prioress (Chaucer, 1987: 1.140), or the materialist Friar who forsakes "sike lazars" (Chaucer, 1987: 1.245) for wealthier sinners, the Parson is "riche [...] of hooly thoght and werk" (Chaucer, 1987: 1.478), a phrase the narrator deploys without any hint of irony. Brother to the Plowman, whose humility and compliance makes him a postpandemic utopian ideal among rebellious laborers (Horrox, 1994: 312–338),[26] the Parson is no respecter of persons, and attends to all, "[w]hat so [they] were, of heigh or lough estat" (Chaucer, 1987: 1.522). Committed to visit "[t]he ferreste in his parrishe", "in siknesse" or in health (Chaucer, 1987: 1.493–94), he fills the lacunae left by careerist or abscondent priests, who flee from their flock as the plague strikes.[27] By offering the Parson the last word in a tale eclipsing all others in length, Chaucer elevates him to a postpandemic ideal embodying pastoral care beyond corruptibility or reproach.

A third embodiment of care is the Host Harry Bailey, whose booming voice resonates throughout the entire text. Unlike the egalitarian Parson, the Host embraces equity, answering each pilgrim—rambunctious churls and courtly-mannered gentils—in kind. A closer look at the Host's idiolect reveals nuanced code-switching from courtly ("Sire gentil maister" (Chaucer, 1987: 7.437)) to intimate ("my levee brother" (Chaucer, 1987: 1.3129)), derogatory ("thou Cook" (Chaucer, 1987: 9.15)) and mock-heroic language ("thou sir John" (7.2810). Many

[25] Chaucer's spellings for parson ('perso(u)n') and person ('persone') frequently overlap. See the Chaucer concordance on "perso(u)n" (Litwhiler, 2017).

[26] The Plowman is based on the Christian ideal of God's faithful steward (Jeffrey, 1992: 619).

[27] See, for example, the case from York during the Second Pestilence in Horrox (1994: 310).

pilgrims follow in turn, modulating their voice situationally (Manciple the Cook: "[S]ire Cook" (Chaucer, 1987: 9.26), "man" (Chaucer, 1987: 9.37), "stynkyng swyn" (Chaucer, 1987: 9.40)) and demographically (Franklin: "thou Somonour" (Chaucer, 1987: 3.1761), "Dame" (Chaucer, 1987: 3.1270)), thereby establishing a sustainable discourse cutting across all social strata.

The Host's leadership spearheads an inclusive language that takes the pilgrims beyond "carewashed markets" (The Care Collective, 2020: 9) and rigid egalitarianism. Nominally associated with healing (the host), he promotes a shared narrative that eclipses the silence of grief and takes the pilgrims beyond the "*banality* [...] of carelessness" (The Care Collective, 2020: 5) toward a world where everyone has a voice. Whereas in Boccaccio such a community is reserved for the nobles, the Host—and by implication Chaucer—envisages a heterogeneous society whose diversity articulates a powerful inclusive vision for postpandemic communities worldwide.

References

Anon. (2021). Belles Heures of Jean de France, duc de Berry, 1405–1409. The Metropolitan Museum of Art, The Cloisters Collection, 1954 (54.1.1), 74v.. Accessed May 15, 2021. https://blog.metmuseum.org/artofillumination/manuscript-pages/folio-74v/

Archer, J. E., Turley, R. M., & Thomas, H. (2015). 'Soper at Oure Aller Cost': The politics of food supply in the *Canterbury Tales*. *The Chaucer Review, 50*, 1–29.

Bakhtin, M. M. (1984). *Rabelais and his world*. Indiana University Press.

Beidler, P. G. (1982). The Plague and Chaucer's Pardoner. *The Chaucer Review, 16*, 257–269.

Boccaccio, G. (2016). *The Decameron* (W. Rebhorn, Trans.). W. W. Norton.

Bridbury, A. R. (1973). The black death. *The Economic History Review, 26*, 577–592.

Chaucer, G. (1987). *The riverside Chaucer*. In Benson, L. D. (Ed.), 3rd ed. Oxford University Press.

Cohn, Jr., & Samuel, K. (2002). The black death: End of a paradigm. *The American Historical Review, 107*, 703–738. https://doi.org/10.1086/532493

Daley, A. S. (1970). Chaucer's Droghte of March in medieval farm lore. *The Chaucer Review, 3*, 171–179.

Greenblatt, S. (2012). *Renaissance self-fashioning: From more to Shakespeare*. University of Chicago Press.

Helmholt, H. F. (1902). *The history of the world: A survey of man's record*. Dodd, Mead and Company.

Horrox, R. (1994). *The black death*. Manchester University Press.

Jeffrey, D. L. (1992). *A dictionary of biblical tradition in English literature*. Wm. B. Eerdmans Publishing.

Lewis, R. E., et al. (Eds). (1952–2001). *Middle English dictionary*. Ann Arbor: University of Michigan Press. Online edition in Middle English Compendium. McSparran, F., et al.(Eds.), 2000–2018. University of Michigan Library, Ann Arbor. Accessed September 14, 2021. http://quod.lib.umich.edu/m/middle-english-dictionary

Litwhiler, H. (2017). Chaucer concordance. Columbia University in the City of New York. Accessed May 14, 2021. http://www.columbia.edu/~hfl2110/cconcord.html

Morgan, G. (1978). The design of the general prologue to the *Canterbury tales*. *English Studies, 59*, 481–498.

Orkin, M., & Joubin, A. A. (2019). *Race. Routledge New Critical Idiom Series* (1st ed.). Routledge.

Orlemanski, J. (2014). Thornton's remedies and the practices of medical reading. In Fein, S. & Johnston, M.) (Eds.), *Robert Thornton and his books: Essays on the Lincoln and London Thornton manuscripts* (pp. 235–255). Boydell & Brewer.

Outka, E. (2020). *Viral modernism: The Influenza pandemic and interwar literature*. Columbia University Press.

Patterson, L. (1991). *Chaucer and the subject of history*. University of Wisconsin Press.

Selby, W. D. (1900). *Life-records of Chaucer*. Kegan Paul, Trench, Trübner & Co.

Senn, M. A. (2003). English life and law in the time of the black death. *Real Property, Probate and Trust Journal, 38*, 507–588.

Sumption, J. (2003). *The age of pilgrimage: The medieval journey to god*. Paulist Press.

The Care Collective. (2020). *The care Manifesto: The politics of interdependence*. Verso.

Turner, M. (2019). *Chaucer: A European life*. Princeton University Press.

Watson, T. (2021). Winchester's Lesser known saints. Accessed September 14, 2021. http://www.winchester-cathedral.org.uk/wp-content/uploads/Winchesters-Lesser-Saints-by-Professor-Tom-Watson.pdf

West-Pavlov, R. (2013). *Temporalities*. Routledge.

Chapter 4
"The Invisible Operator": Plague, Corruption, and Conspiracy in Renaissance Drama

Rebecca Welshman

Abstract The problem of unseen contagion was summed up by the preacher Benjamin Spencer in 1625 who described the plague as having "a hidden beginning" with a "visible" "effect," yet governed by an "invisible operator." Within the new cultural arena afforded by the advancement of medical understanding, material and non-material explanations of plague existed side by side. Girolamo Fracastoro's theory of contagion (1546) promoted the idea that illness could be spread through tiny particles or "seeds" in the air, which established a tangible connection between plague and the natural world, beyond spiritual or celestial explanations. At the same time, the more speculative elements of disease theory, including occult explanations, gave rise to increased fear of the unseen and unknown and to the suspicion of others. Through examining late sixteenth and early seventeenth century plays, this essay finds social anxieties associated with contagion to be reflected in an increasing preoccupation with the corrosive effects of unseen agencies, as in *Hamlet* where "rank corruption, mining all within,/Infects unseen." Playwrights invoked the notion of an invisible threat through linguistic experimentation and in their choice of metaphors drawn from the natural world. Through identifying a richer context for their language, including Biblical allusions, I argue for a deeper endemic presence of plague in early modern drama that textured lexicons and facilitated a form of dialogue between writers in their critique of the establishment. These layered understandings of plague in the intersecting contexts of corruption, conspiracy, and fear of the unknown serve to highlight the continued relevance of Renaissance drama to contemporary cultural debates.

R. Welshman (✉)
Independent Scholar, Edinburgh, England
e-mail: welshmanrebecca@gmail.com

S. Venkatesan et al. (eds.), *Pandemics and Epidemics in Cultural Representation*,
https://doi.org/10.1007/978-981-19-1296-2_4

Introduction

The problem of unseen contagion was summed up by the preacher Benjamin Spencer in 1625 who described plague as having "a hidden beginning" with a "visible" "effect," yet governed by an "invisible operator."[1] Known as The Black Death or simply "Death," plague had been intermittently present in England since its arrival in 1348. Its stealthy presence around the globe was a familiar enemy, often perceived as a malign force or "natural evil."[2] The apparent absence of plague in Renaissance drama has only fairly recently become the object of scrutiny. Jonson's comedy *The Alchemist,* for example, has been interpreted as an attempt to represent a London without plague as a purging form of remedy for the "potentially fatal fear" of the disease.[3] However, I argue that it is precisely the *invisibility* of plague in early modern drama that invites further consideration. In his analysis of Sontag on the "murky" origins of disease in *Illness as Metaphor* (1978), Ian Munro draws attention to the "transgressive and historical dimension" of plague that "makes metaphor an inadequate word to explain its productions of meaning."[4] Histories of the interrelation of contagion and culture reveal a "border between metaphorical and literal meanings of contagion" that is "contested, indistinct, and movable"[5]—creating a fertile imaginative field for the innovative dramatists of Renaissance England. As suggested by Erin Sullivan, the frequency with which early modern writers invoked physical illness to highlight the condition of sin indicates "a more complex and richer relationship between the two."[6] These new contexts show plague to be latent in Renaissance drama, rather than absent. Moreover, the presence of plague was so deeply embedded in culture that its articulation in dramatic works—perhaps in order to avoid a subject that had grown oppressive and tiresome—necessarily assumed more subtle forms, in particular those that focused on the corrosive effect of unseen agencies. Within the intersecting contexts of moral and societal degeneration, I thus argue for an endemic presence of plague in early modern drama that textured lexicons and facilitated a form of dialogue between writers in their critique of the establishment.

The essay first outlines how awareness of the invisibility of contagion was partly brought about by medical advancements of the mid-sixteenth century, whereby developments in disease theory exacerbated fear of the unseen and unknown and increased suspicion of others. Secondly, I provide a cultural context for the moral and societal degeneration during the Elizabethan period which precipitated civil unrest and challenges to the state. I then proceed to discuss examples from the works of Shakespeare and his contemporaries Thomas Dekker and Ben Jonson, written during a long

[1] Benjamin Spencer, *Vox civitatis*, 9.

[2] Henry Hexham, *A copious English and Netherduytch dictionarie* (no page number). John Withals, in *A dictionarie in English and Latine*, gives: "contagion or infection which hurt our neighbors [sic] bodies", 352.

[3] Patrick Phillips, in Nichole Dewall, "Sweet recreation barred": the case for playgoing, 133.

[4] Ian Munro, "The city and its double: Plague time in early modern London", 241.

[5] As noted by Martin S. Pernick, *Contagion and culture*, 858.

[6] Erin Sullivan, Physical and spiritual illness: Narrative appropriations of the bills of mortality, 88.

period of plague between 1603 and 1610 when theaters were mostly closed. Of this period, the historian J. Leeds Barroll has concluded that theaters opened for a few weeks each year and fewer than nine months in total.[7] After providing an overview of what was meant by "corrupted air," I discuss how this idea infused understandings of contagion in publications such as Thomas Dekker and Thomas Middleton's plague pamphlets. In my discussions of "Othello" and "The Alchemist", I explore how Shakespeare and Jonson invoke the notion of an invisible threat through mechanisms of linguistic playfulness and metaphors drawn from the natural world. Lastly, I turn to "The Witch of Edmonton", co-written by Dekker, John Ford, and William Rowley, where I discuss the scapegoating of Mother Sawyer in the context of state hypocrisy and corruption.

****

As a late fourteenth-century neologism "contagion" meant "a communicable disease; a harmful or corrupting influence," with a Latin origin meaning "contact." William Caxton's translation of "Reynard the Fox," first printed in 1481,[8] offers one of the earliest uses of "plaghe" as a verb, and by the sixteenth century, it had become a well-known verb in theatrical dialogue.[9] Consonant with the increasing cultural visibility of the plague through dramatic discourse and the printing press came a host of new associations that surrounded the entry of the word into everyday consciousness. English Renaissance theater did not tend to dramatize plague or its victims, and though plague is "notable… for its absence" in plot, it is "amply represented in dramatic metaphor."[10] Interpretations have tended to favor theatrical spatial dialectics over linguistic associations. Paula S. Berggren argues that Shakespeare's linguistic interest in plague indicates a "surprising nonchalance", and references to plague in his works as "an inescapable condition of his world have greater impact in the manipulation of theatrical and conceptual spaces than in any verbal echoes."[11] Berggren does note, however, that the phrase "a plague on't" is uttered in numerous instances by characters who "challenge the established order," thus suggesting an inherent connection between plague and state resistance.[12]

In the early sixteenth century, the work of Veronese poet and Doctor Girolamo Fracastoro suggested that illness could be spread through tiny particles or seeds in the air. His work pre-figured modern germ theory and first featured in his poem about syphilis (1530) and his prose treatise *De contagion et contagiosis morbis et*

[7] J. Leeds Barroll, Politics, plague, and Shakespeare's theater: the Stuart years, 99.

[8] Walter W. Skeat, *Principles of English etymology: the foreign element*, 156. Caxton was a merchant, diplomat, and writer who brought the printing press to England.

[9] See, for example, *Henry VI* (part III, v5, 1591) "I'll plague ye for that word"; *The merchant of Venice* (written 1596–1599): 'I'll plague him'; "The ambition in my love thus plagues itself" (*All's well that ends well*). See also Volumnia in *Coriolanus* (1605–08) "the gods will plague thee", and *Cymbeline* "The very devils cannot plague them better."

[10] Barbara H. Traister, ""A plague on both your houses" Sites of comfort and terror in early modern drama", 179.

[11] Paula S. Berggren, "Shakespeare's dual lexicons of plague", 150.

[12] Ibid.

eorum curatione (1546).[13] The idea that disease could be spread through human contact was not just a scientific breakthrough, but promised a temporal shift in the dynamic of how we encounter and perceive one another, one that played into the growing dramatic interest in trickery and the suspicion of others. Miasma theory had long pinpointed corrupted air as the cause of plague deaths and was the driving idea behind the purification of air and the cleansing or destruction of certain objects that had been exposed to that air.[14] Yet, Fracastoro's work was not simply disregarded in favor of the established miasma theories. As Vivian Nutton argues, "far from being neglected, his ideas on the causes of epidemic disease were widely known [and] part of the common stock of medical knowledge."[15] Within the new cultural arena afforded by the advancement of medical understanding, material and non-material explanations of plague existed side by side. Chalk and Floyd-Wilson note that although disease was often attributed to humoral imbalance within the body, transmission between people was a fertile area of enquiry:

> Early modern writers devoted constant attention to the possibility of contagious transmission, the notion that someone might be infected or transformed by the presence of others, through various kinds of exchange, or if exposed to certain ideas, practices, or environmental conditions, and they often did so in ways not limited to medical inquiry or the narrow study of a particular disease.[16]

This form of "pre-disciplinary fluidity" saw the entwined discourses of natural philosophy and drama "imbricated" with the "practices and vocabularies" of natural magic, alchemy, and demonology.[17] Importantly, Shakespeare and his contemporaries were working at a time when the first dictionaries were being written, and the meanings of words were not necessarily static. The Renaissance humanist, John Florio, for example, who was working in England, produced one of the earliest mixed language dictionaries of the time, which pre-dated Robert Cawdrey's *Table Alphabeticall*, the first English-only dictionary published in 1604. For dramatic writers, these developments forged new linguistic possibilities; the resonances of which are still yet to be fully recognized.

The relationship between seminal theories of contagion and how theater was considered an activity prone to spreading unwholesome ideas has been explored elsewhere.[18] Due to social and political tensions, the cultural environment of the time was particularly responsive to subversive tropes and satire. Queen Elizabeth

---

[13] Vivian Nutton, "The reception of Fracastoro's theory of contagion: the seed that fell among the thorns?", 196.

[14] Joseph P. Byrne, Encyclopedia of the black death, 120.

[15] Nutton adds that Fracastoro did not present the seed theory in direct opposition to miasma theory and such simplistic distinctions between miasma and contagion, which historians and humanities scholars continue to make, are "anachronistic or overrefined", 198.

[16] Darryl Chalk and Mary Floyd-Wilson, "Introduction: Beyond the Plague," in *Contagion and the shakespearian stage*, ed. D. Chalk and M. Floyd-Wilson, 1.

[17] John Shanahan, "Ben Jonson's "Alchemist" and early modern laboratory space", 37.

[18] Anti-theatrical writing posited the spread of seditious ideas, likening theatre to a plague that attacks its victims through infected, poisonous air. (See Darryl Chalk and Mary Floyd-Wilson, Introduction: Beyond the plague, 2).

had established the Church of England in 1559 and was responsible for developing a strict hierarchical society that inspired rebellion and civil unrest. Beneath its shining exterior and expansion of wealth and empire, a deep vein of corruption and hypocrisy was perceived to run through the Elizabethan regime that compromised the lives of ordinary people. This belief was symbolized in writings immediately following the Queen's succession in 1603 by the idea of Britain as a "half-sinking" island.[19] Playwrights promoted the idea of a rot-infested state that went deeper than what could be seen on the surface; a "city sin" and condition of moral decay that was believed by many to have been the cause of plague.[20] The endemic nature of state corruption is depicted in Hamlet's observation, for example, that "something is rotten in the state of Denmark" and that "rank corruption, mining all within,/Infects unseen"[21]. Due to problems of crowd control and fears of spreading corruptive ideas, censorship of public performances was deemed a necessity in order to maintain control over the populace. The offices of the Master of Revels oversaw and abridged all play scripts before they were performed. To circumnavigate these protocols, playwrights concealed ideas that critiqued the establishment through metaphors and word play.[22]

Shakespeare in particular wrote with an awareness of the fluidity of word meanings and employed this knowledge in his use of words that were deliberately ambiguous and had multiple meanings. As a linguistic innovator, Shakespeare has been credited with helping to shape the development of the English language, and it is often noted that there are more new words in Shakespeare's works than any other writer of the period.[23] His awareness of contagion has been given thorough treatment, yet studies that connect the two remain curiously few. In the reading of "Othello" as a "plague narrative," Jaecheol Kim notes that the play's "semantic features obsessively revolve around words such as "plague," "infection," "pestilence," and "contamination."[24] Beyond these literal terms, however, the corrupting presence of contagion—at both a personal and wider social level—is an undercurrent of the narrative, which surfaces into visibility at key moments. Iago's desire for revenge is a creeping pestilence that slowly poisons Othello's mind against his wife and ultimately results in their deaths. This idea is epitomized when Iago tells the audience about his plan: "I'll pour this

---

[19] See Hannah Ridge, "Medicine and medical practice in the works of Thomas Middleton and his contemporaries", 86.

[20] Dekker and Middleton, "News from nowhere". In *Thomas Middleton: the collected works*, 657.

[21] *Hamlet*, Act 1, scene 4, line 100, and Act 3, Scene 4, lines 169–70 respectively. The Folger Shakespeare. All references to Shakespeare's work in this essay are from texts on the Folger Shakespeare Database. No changes were made during attribution, and all quotations have been reproduced under the following Creative Commons License: https://creativecommons.org/licenses/by-nc/3.0/deed.en_US.

[22] Shakespeare often likens the spread of corruption to plague: in *Macbeth* (1.2.) "the multiplying villanies of nature" (13) are deemed to "swarm" (14), and in *Coriolanus* 3.1.395, Brutus says: "Pursue him to his house, and pluck him thence,/Lest his infection, being of catching nature,/Spread further." We might also remember Lady Macbeth's "out, damn spot" as a symbol of the irreversible course of corruption (*Macbeth*, 5.1.37).

[23] See Joseph Black et al., British literature: A historical overview, 153.

[24] Jaecheol Kim, "The plague and immunity in "Othello"", 25.

pestilence into his ear."[25] The act of seeding an idea in the mind of another in order to corrupt their outlook and bring about their downfall parallels the seed theory of contagion, wherein it was believed that "the seeds of diseases ... become sown and grown up together with the life [to] break the former harmony of health and life and shoot forth branches or symptoms answerable to their seeds or roots...subvert the whole frame ...put to flight the vital forces and strangle nature with her own cords."[26] In *De contagion et contagiosis morbis et eorum curatione* (1546), Fracastoro proposed that the seeds of disease could rapidly multiply. Similarly, in the discussion of "Hamlet," J. F. Bernard identifies "something invisible deep inside ... which could potentially break out, an image not altogether different from Fracastoro's seed-carrying model."[27]

In their plague pamphlet "News from Gravesend," written during the severe outbreak of 1603, Thomas Dekker and Thomas Middleton devote a section to the causes of plague, in which they ask:

Can we believe that one mans breath
Infected, and being blown from him,
His poison should to others swim:
For then who breathed upon the first?[28]

They go on to reject the explanations of medical science: "Her aphorisms proved a mockery"[29] because "whilst she's turning o'er her books/And on her drugs and simples looks,/She's run through her own armèd heart,/Th' infection flying above art."[30] As Hannah Ridge has noted previously, "aphorisms" refers to medical texts circulated under this title, including Hippocrates' *Aphorismes.*[31] In their collaboration "The Honest Whore" (1604), the "strumpet's breath" is "poisoned," and in "The Humorous Poet," a "poison" is deemed to be avoided by breathing "sweet spring" air: "This is the air I breathe. Corrupt it not." However, the notion of corrupted air had long been perceived to have celestial origins that were essentially beyond human control. In the early thirteenth century, Albertus Magnus, patron saint of natural scientists, revived the writings of Aristotle and brought Greek and Arab science to readers and scholars of his time in the Universities. In *On the Causes of the Properties of the Elements,* Albert writes that Jupiter in the sign of Gemini could raise "pestilent winds and corrupted airs," and that Mars could ignite the vapors to "infect the air with a sharp poison" that were fatal to "a multitude of men and animals."[32] For the

---

[25] *Othello*, 2.3, 376.

[26] William Simpson, Zenexton ante-pestilentiale, or, a short discourse of the plague: Its antidotes and cure, 20–21.

[27] J. F. Bernard, "Hamlet" story/stories of Hamlet: Shakespeare's theater, the plague, and contagious storytelling, 221.

[28] Thomas Middleton, *The collected works*, lines 578–581.

[29] Ibid., 487.

[30] Ibid., 488–91.

[31] Ridge, 86.

[32] Albertus Magnus wrote: "Air is corrupted in two ways: namely, sometimes from something below coming upon and corrupting it, and sometimes again from some superior celestial body

merchant Kitely in Ben Jonson's "Every Man in his Humour", the "pestiferous air" of a "new disease" symbolizes the progressive spread of suspicion. Akin to a "subtle vapor" or "contagion" that "may well be called poor mortals' plague," it soon infects every part of the brain, which according to the anatomical understanding of the time was represented by the "houses" of memory, reason, and imagination.[33]

Yet, while some professed the belief that plague struck due to a curse or unearthly event, this perception existed alongside an awareness that the seeds of disease could potentially be anywhere. As Shakespeare was writing "Othello" during a plague outbreak, it is not unreasonable to posit that the continual background presence of contagion not only proved disruptive to the theater, but also made its stamp upon the linguistic decisions made by the playwright at this time. As readers and interpreters, we might, then, consider the implications of aligning a deliberately ambiguous meaning with the unpredictable nature of contagion. With this in mind, we might turn to Roderigo's description of Othello as "an extravagant and wheeling stranger, Of here, and every where"; an allusion that certainly suggests a symbolic connection between the portrayal of Othello as an outsider and the affliction of disease. "Extravagant" had the Middle English meaning of divergent or unusual, from the Latin "extra" meaning "outside" and "vagan" meaning "wander." Kim suggests that a "wheeling stranger" was a symptom of the "downward social mobility and moral degeneration," associated with plague, not least because vagrants were often accused of spreading it. Kim also points out that representations of English politics were often depicted in an Italian setting in order to avoid casting suspicion, and that Othello's depiction as an "erring barbarian" associates him with plague carriers; not least because Italy was seen as a "cradle" of the plague at that time.[34]

However, as the appearance of "wheeling" in Othello is the single instance of the word across all Shakespeare's works, we may be invited to reflect upon its significance, over and above more commonly used terms. As discussed elsewhere, Othello's "hybrid language"—that pairs neologisms with known words—indicates his double identity.[35] Yet, I would argue for a compound identity, suggested by the multiple meanings of the terms used to define him. While it may be true that Othello's description as a "wheeling stranger" was designed to "trigger the anxiety of contagion lingering in the text,"[36] it would appear that Shakespeare went further in this analogy by directly associating Othello with the ephemeral nature of the airborne seed. In a culture that was far more attuned to natural phenomena than the present day, seed dispersal was a well-known stage of the life cycle of plants and trees. Seeds that traveled significant distances, such as dandelion, thistledown, and the winged

corrupting it", (Darryl H. Rutkin, *Sapientia astrologica: astrology, magic, and natural knowledge, Ca. 1250–1800*, 106).

[33] Ben Jonson, *Five Plays*. Edited by G. A. Wilkes, Act ii, scene iii. Lines 55–67.

[34] Kim, 28.

[35] See Novy's discussion of the work of Robert Watson, in Marianne Novy, *Shakespeare and outsiders*, 89.

[36] Kim, 28.

seeds of the sycamore, were regularly employed as metaphors.[37] A 1609 edition of *The Theosophist*, for example, warns of the spreading of malicious gossip: "Take great heed that thou spreadest not a slander, for it is like unto the seeds of a thistle-down." The imperviousness of seeds to conventional boundaries and the invisibility of what they might bring was, thus, an apt metaphor for the spread of anything that might be considered contagious.

Yet although, "here and everywhere" points to the unpredictable nature of seed dispersal and airborne disease, the multiple alternative meanings of "wheeling" invite further consideration. In Simpson's treatise on the plague, the word was used to depict the effect of fever upon the mind.[38] It may also be significant that "wheeling" appears in John Florio's seminal translation of Montaigne and three times in his earlier Dictionary of Italian-English.[39] Conversely, as Othello was a Moorish general in the Venetian army, we may also consider the military meaning:

> Wheeling is a Motion in Military Exercise which, by turning, brings the Head of a Battalion to the Flank ground, which is extremely necessary, if the Wing of a Battalion be attacked, or when you see it advantageous to attack the Enemy in the Flank.[40]

In a military context, the word assumes an opportunistic meaning; a specific manoeuver that would enable attack. We might also remember here Thomas Dekker's description of plague symptoms in "The Wonderful Year" (1603) as "the leaders, lieutenants, sergeants, and corporals" advancing for "The Plague" who is "muster-master and marshall of the field."[41] Significantly, all three meanings of "wheeling"—the unchecked progression of an airborne seed of disease, the restless meanderings of a traveler who might harbor disease, or the sudden turn of a military formation—converge in their connotations of an unpredictable threat. In his linguistic ambivalence, Shakespeare, thus encourages us to consider the unstable and changeable meanings of words that—like an army advancing in an opportune attack—can turn course depending on who is speaking and the context in which the words are articulated. This notion of unfixed or "corruptible" meaning was intrinsically related to the concepts of change and the alteration of state inherent in celestial philosophy of the time that challenged orthodox Christian narratives of eternal life.[42]

---

[37] Poplar seeds have been shown to disperse to a distance of 30 km, and sycamore up to 4 km (Peter Thomas, *Trees their Natural History*, 195).

[38] Simpson, Zenexton ante-pestilentiale: or, a short discourse of the plague, 35.

[39] "Wheeling" is found in Florio's *A Worlde of wordes*, 31, 582, and in his translation of Montaigne, where it refers to celestial motions (300): https://ufdc.ufl.edu/AA00004304/00001/321x?search=wheel

[40] By the late seventeenth century the sense to turn ("wheel") had been incorporated into some of the earliest known military dictionaries. See, for example, *The perfection of military discipline after the newest method*, 82. It is thus reasonable to assume that "wheeling" was already in use as a military term long before this.

[41] These military allusions are noted in Rebecca Totaro, "Introduction", in *Representing the plague in early modern England*, 6.

[42] See, for example, Father Joseph de Acosta, *The Natural and moral history of the Indies* (6–7), where he observes that the "space and region" where the stars "march and rowle is elementrie and

The technique of infusing ordinary phenomena with ambiguous symbolism to create textured meaning is also found in the work of Thomas Dekker, a prominent Jacobean plague writer whose work bears obvious hallmarks of the significant plague outbreak in London (1592-3).[43] In "*The Shoemaker's Holiday*" (1599), a shoe belonging to a lost love becomes an imagined route to rediscovering her, in response to which Firk mockingly replies: "Old shoe, that wert new—how a murrain came/this ague-fit of foolishness upon thee?"[44] The significance of the shoe is couched in terms of contagion—the medieval "murrain" meaning plague and the "ague" a word for fever. While physicians encouraged the burning of pungent ingredients, such as hops, frankincense, or herbs as a preventative means of keeping away corrupted air during plague time, for those who could not afford it, the burning of horn and old shoes was recommended.[45] Robert Smallwood and Stanley Wells note the insult "mustard tokens" to have a "compound" meaning as "token" also meant plague sore.[46] However, Dekker goes further in his analogy by corrupting the conventional idea of mustard as a condiment widely used at a meal table to accompany meat. The idea of mustard symbolizing the pus cankered sores on a plague-infested body is not only reviling to an audience, but also conjures the extreme reversal of order and control that severe disease brings with it. Mustard had its place at the table and was employed by the eater at their taste and convenience, in order to improve the experience of consumption, whereas plague sores erupted without warning and were a sign that the disease might rapidly be consuming the person.

In Jonson's "The Alchemist", staged in 1610, mustard again features as a symbol. When a plague outbreak provokes the wealthy owner of a London house to flee to the country, his servant—along with a fraudulent alchemist and his aide—seizes an opportunity to capitalize upon a devious idea to swindle ordinary people out of money in return for cures and portions of life-saving elixirs. In its setting in the middle of an actual plague outbreak, Jonson's play went against the usual custom by staging the threat of plague itself. Foley notes that seats upon the stage were sold for 25 times more than the one penny charge at the open-air Globe theater and suggests that the commonly held belief of the time that plague mortality was considered to be a poorer class problem was a reason for the deliberately provocative staging and seating arrangement. A challenging idea indeed when theories of the time had identified that plague transmission could and did take place in theaters.[47]

---

corruptible … the stars and planets of their owne motion should be by reason likewise corruptible, and so by consequence they must alter, change, and be finally extinct." Here Acosta also notes that "to say that the Celestiall bodies be corruptible, it agreeth not with the psalme, "That God made them for euer"".

[43] Munro, 241.

[44] Thomas Dekker, *The shoemaker's holiday*, lines 55–6.

[45] George C., Kohn, (ed.) Encyclopedia of plague and pestilence: From ancient times to the present, 227.

[46] Dekker, *The Shoemaker's holiday*, lines 21, 95.

[47] Christopher Foley, "Breathe less, and farther off": The hazardous proximity of other bodies in Jonson's "The Alchemist"", 507.

In Act Two, the elixir is deviously promoted as a means of renewing youth, protecting against "all infections" and curing "all diseases." Moreover, it is promised to drive the plague "out of the kingdom" in only three months:

Restore his years, renew him, like an eagle,
To the fifth age; make him get sons and daughters,
Young giants; as our philosophers have done,
The ancient patriarchs, afore the flood,
But taking, once a week, on a knife's point,
The quantity of a grain of mustard of it[48];

The presence of the mustard seed sits ambiguously within a web of material and religious connotations, designed to create a powerful and challenging analogy. As attested in books of the period, mustard was known to be remedial for a wide variety of health conditions and an antidote to poison.[49] That Jonson associates the mustard seed with plague clearly and deliberately subverts these conventional meanings. By the time "The Alchemist" was first performed in 1610, theaters had survived the period of extreme censorship and prohibition set by the city and religious authorities; plays were regularly performed at court and for large numbers of Londoners. "The Alchemist" was performed to a wealthy and pious audience.[50] Importantly, the mustard seed was a symbol of the Christian faith, known through the parable of the mustard seed in the Gospel of Matthew, in which a kingdom with small beginnings reaches a great size. A mustard seed, no bigger than a large pinhead, can grow a shrub of up to thirty feet. That scandalous conmen were exploiting Christian symbolism in this way passes comment upon the decayed nature of the society of the time. Jonson himself becomes an "invisible operator," subverting the holistic and holy associations of the mustard seed, which would likely have not gone unnoticed by an audience that had an acute ear and eye for satire. Moreover, Jonson's use of the symbol draws an important parallel between the mustard seed (as a symbol of the kingdom of God) and the airborne seed of the plague, both of which had humble beginnings that grew to global proportions. That the metaphor of the mustard seed was popular with readers and theater-goers of the time is suggested by its inclusion in *A Treasury, or storehouse of smiles* (1609), published a year before "The Alchemist" was performed.[51]

In its implicit suggestion of a remedy needed for a corrupt society, empty of moral worth, the promise of cure in "The Alchemist" extends far beyond a single disease. In his reference to the "ancient patriarchs, afore the flood" in the Book of Genesis, Jonson reminds the audience of the long catalog of sin that has ensued. It was in the consecutive Book of Exodus that God sent the Ten Plagues of Egypt—a further

[48] Ben Jonson, The Alchemist, in Ben Jonson's Plays and Masques, act 2, scene 1, 55–60.

[49] Mustard is listed as a treatment for a range of diseases, including epilepsy, and as a cleanser for the brain, heart and lungs. It "resisteth the force of poison, the malignity of Mushromes, and the venome of the Scorpion". John Parkinson, *Theatrum botanicum: The theater of plants*, 832–833.

[50] Foley, 507.

[51] Robert Cawdry, A treasury, or storehouse of similes, 389.

reminder of the idea of plague as a form of divine punishment. As noted by Berggren, The Ten Plagues are evoked in "Othello" in the reference to "summer flies … that quicken even with blowing."[52] The command "make him get sons and daughters" is a rendition of God's instruction to procreate, thus also invoking the idea of the original seeding of the human population.[53]

In "News from Gravesend," Dekker and Middleton cite overpopulation as a problem that plague has been sent to remedy or "purge." The line—"For every man within him seeds/A worm, which this contagion breeds"—posits an overseeded population, rife with corruption, that is continually fueled by its ability to keep breeding.[54] Furthermore, they draw attention to the hypocrisy of the patriarchal system that dominates government: "The courtier's pride, lust, and excess,/The churchman's painted holiness,/The lawyer's grinding of the poor."[55] Their theme of plague as a form of atonement is clearly articulated in the following lines that refer to almost half a century of corruption and misdeeds: "it may be, Jehovah looks/But now upon those audit books/Of forty-five years' hushed accounts/For hours misspent … Finding out grievous debts, doth clear/And cross them under his own hand,/Being paid with lives through all the land."[56]

When we consider that the plague pamphlet would not have been their publication of choice, but was rather an enforced necessity due to theater closures, it may be unsurprising that Dekker and Middleton's wording concerning corruption and hypocrisy is revisited in the play "The Witch of Edmonton" (1621). Co-written by Dekker, John Ford, and William Rowley and based on a true trial, the theme of judgment is returned to in more detail. After a conspiracy against Mother Sawyer within the community causes her to turn to witchcraft,

Sawyer challenges the establishment that persecutes her. When interrogated by magistrates, she draws attention to the hypocrisy and corruption of those in power:

> Men in gay clothes,
> Whose backs are laden with titles and honours,
> Are within far more crooked than I am.
> And if I be a witch, more witch-like.
> A witch! Who is not?
> What are your painted things in princes' courts,
> Upon whose eyelids lust sits, blowing fires
> To burn men's souls in sensual hot desires?[57]

---

[52] Berggren, 157.

[53] The life-promoting elixir of the Bible—what Jonson refers to as the "perfect ruby, which we call elixir"—was thought by Mammon to explain the potency of the patriarchal seed. (Richard Harp, *Ben Jonson's Plays and Masques*, 223).

[54] Dekker and Middleton, *News from Gravesend*, lines 621–2.

[55] Ibid., lines 643–5.

[56] Ibid., lines 690–6.

[57] William Rowley, Thomas Dekker, and John Ford, *The witch of Edmonton*. Edited by Peter Corbin and Douglas Sedge, 98.

She is "poor, deformed, and ignorant"—the object of "scandalous malice" and "like a bow buckled and bent together/By some more strong in mischief." The analogy of the bow echoes Hamlet's admission that "They foole me to the top of my bent" and "beyond my bend, beyond my power."[58] As a scapegoat for the ills and hypocrisy of the patriarchal system Mother Sawyer is a: "common sink/For all the filth and rubbish of men's tongues/To fall and run into". As noted by David Stymeist the "common sink" is a "physical manifestation of contagion, the malevolent source of inexplicable disease and death."[59]

Prosecutions against women peaked from 1580 to 1630 under the ruse of moral reform and the desire to eliminate society of sources of evil and trickery. The suspicion surrounding sorcery intersected with the suspicion associated with a deadly disease, not least because both were seen to be products of contaminated morals and association with the devil. In her discussion of Jacobean pamphlets, Deborah Willis draws attention to the witch family unit as a "new kind of threat" wherein witch parents educated their children in the ways of witchcraft. Thus, like a disease, witchcraft was perceived to "spread through communities and across generations."[60] The black dog that only Mother Sawyer can see was designed to symbolize her collusion with the devil. Yet, the black dog has a long history as a symbol of melancholy; a meaning that would have resonated with the audience of the time who understood that negative emotions, in particular fear and melancholy, were believed to make infection by the plague more likely and survival less likely.[61]

The echoing of these references between Shakespeare, Dekker, and Jonson suggests the presence of an implicit dialogue between the playwrights—in plays produced during plague outbreaks—concerning the material and non-material causes of the disease. Although plague may not have been represented visually on stage or through many literal references, its presence is implicit in the dynamic web of historical, medical, and religious associations endemic to its wider cultural understanding. While plague could not be separated from the corruption that the disease was considered to be atonement for, playwrights nevertheless mined the rich seams of vocabulary afforded by the inescapability of its presence. Yet, in our interpretation of the linguistic ambivalence of Renaissance drama and considering the centuries of cultural change that have passed, as modern day readers, we can only be guided by the written word as a "token" of its former meaning. How we read and interpret the meaning of that which is endemic—then and now—remains inevitably shaped by the beliefs, knowledge, and understanding relevant to the culture of our time.

---

[58] Shakespeare, *Hamlet* Act 2, scene 2, 413.

[59] David Stymeist, "Must I be... made a common sink?": Witchcraft and the theatre in "The Witch of Edmonton", 40.

[60] Deborah Willis, "The witch-family in Elizabethan and Jacobean print culture", 21.

[61] See plague accounts by medical writers summarised by DeWall, 138–139.

Bibliography

Acosta, F. J. de. (1604). *The natural and moral history of the Indies* (Vol. 1). Hakluyt Society.

Barroll, J. L. (1991). *Politics, plague, and Shakespeare's theater: The Stuart years.* Cornell University Press.

Berggren, P. S. (2010). Shakespeare's dual lexicons of plague: Infections in speech and space. In Totaro, R., & Gilman, E. B. (Eds.), *Representing the plague in early modern England* (pp. 150–168). Routledge.

Bernard, J. F. (2019). Hamlet's story/stories of Hamlet: Shakespeare's theater, the plague, and contagious storytelling. In D. Chalk & M. Floyd-Wilson (Eds.), *Contagion and the Shakespearian stage* (pp. 213–232). Palgrave.

Black, Jo., et al. (2010). *British literature: A historical overview* (Vol. A). Broadview Press.

Byrne, J. P. (2012). *Encyclopedia of the black death.* ABC-CLIO.

Cawdry, R. (1609). *A Treasury, or storehouse of similes* (Vol. 2). Thomas Creede.

Chalk, D., & Floyd-Wilson, M. (2019). Introduction: Beyond the plague. In D. Chalk & M. Floyd-Wilson (Eds.), *Contagion and the Shakespearian stage* (pp. 1–21). Palgrave.

Dekker, T. (1999). *The shoemaker's holiday.* In Smallwood, R., & Wells, S. (Eds)., Manchester University Press.

DeWall, N. (2012). "Sweet recreation barred": The case for playgoing. In R. Totaro & E. B. Gilman (Eds.), *Representing the plague in early modern England* (pp. 133–149). Routledge.

Florio, J. (2013). *A worlde of words.* Haller, H. W. (Ed.), University of Toronto Press.

Foley, C. D. (2018). "Breathe less, and farther off": The hazardous proximity of other bodies in Jonson's The Alchemist. *Studies in Philology, 115*, 505–523.

Folger Shakespeare. (2017). A digital anthology of early modern English drama. https://emed.folger.edu/fmb Accessed August 31, 2021.

Great Britain, A. (1690). *The perfection of military discipline after the newest method, as practised in England and Ireland.* Nicholas Boddington.

Hexham, H. (1647). *A copious English and Netherduytch Dictionarie.* Rotterdam.

Jonson, B. (2001). *Ben Jonson's plays and masques.* In Harp, R. (Ed.), Norton.

Jonson, B. (1999). *Five plays.* In Wilkes, G.A. (Ed.). Oxford University Press.

Kim, J. (2017). The plague and immunity in "Othello." *Comparative Drama, 51*, 23–42.

Kohn, G. C. (Ed.). (2007). *Encyclopedia of plague and pestilence: From ancient times to the present.* Infobase.

Middleton, T. (2007). *The collected works.* In Taylor G., (Ed.), and J. Lavagnino (Trans.). Oxford University Press.

Montaigne, M. (2021). *The essayes, or morall, politike, and militarie discourses of Lord Michael Montaigne.* In Florio, J. (Ed. and Trans.). University of Florida digital collections. Accessed August 30, 2021. https://ufdc.ufl.edu/AA00004304/00001

Munro, I. (2000). The city and its double: Plague time in early modern London. *English Literary Renaissance, 30*, 241–261.

Novy, M. (2013). *Shakespeare and outsiders.* Oxford University Press.

Nutton, V. (1990). The reception of Fracastoro's theory of contagion: The seed that fell among the thorns?". *Renaissance Medical Learning: Evolution of a Tradition, 6*, 196–234.

Pernick, M. S. (2002). Contagion and culture. *American Literary History, 14*, 858–865.

Parkinson, J. (1640). *Theatrum botanicum: The theater of plants.* The Cotes.

Ridge, H. (2013). Medicine and medical practice in the works of Thomas Middleton and his contemporaries. Doctoral thesis: https://etheses.whiterose.ac.uk/5830/1/HannahRidgePhDThesisFINAL.pdf. Accessed August 30, 2021.

Rowley, W., Dekker, T., & Ford, J. (1999). *The witch of Edmonton.* In Corbin, P., & Sedge, D. (Eds.), Manchester University Press.

Rutkin, D. H. (2019). *Sapientia astrologica: Astrology, magic, and natural knowledge, Ca. 1250–1800.* Switzerland.

Shanahan, J. (2008). Ben Jonson's "Alchemist" and early modern laboratory space. *Journal for Early Modern Cultural Studies, 8*, 35–66.
Simpson, W. (1665). *Zenexton Ante-Pestilentiale. or, A Short Discourse of the Plague: Its Antidotes and Cure*. Printed for George Sawbridge.
Skeat, W. W. (1891). *Principles of English etymology: The foreign element*. Clarendon Press.
Spencer, B. (1636). *Vox Civitatis, or London's complaint against her children in the countrie*. E.P.
Stymeist, D. (2001). Must I be … made a common sink? Witchcraft and the theatre in "The Witch of Edmonton." *Renaissance Reformation/renaissance Et Réforme, 25*, 33–53.
Sullivan, E. (2012). Physical and spiritual illness: Narrative appropriations of the bills of mortality. In R. Totaro & E. B. Gilman (Eds.), *Representing the plague in early modern England* (pp. 76–94). Routledge.
Thomas, P. (2014). *Trees: Their natural history*. Cambridge University Press.
Totaro, R. (2012). Introduction. In R. Totaro & E. B. Gilman (Eds.), *Representing the plague in early modern England* (pp. 1–33). Routledge.
Traister, B. H. (2012). "A plague on both your houses": Sites of comfort and terror in early modern drama. In R. Totaro & E. B. Gilman (Eds.), *Representing the plague in early modern England* (pp. 169–182). Routledge.
Willis, D. (2013). The witch-family in Elizabethan and Jacobean print culture. *Journal for Early Modern Cultural Studies, 13*, 4–31.
Withals, J. (1616). *A dictionarie in English and Latine*. Thomas Purfoot.

Part III
Nineteenth and Twentieth Century Pandemics: Rupture and (Re)configurations of Community

Chapter 5
Three Facets of the Literary Imagination of Cholera: Hysteria, Ridicule, and the Rise of Bacteriology

Aureo Lustosa Guerios

Abstract This chapter studies three different stages of the literary imagination of cholera. It starts by considering the cultural shock caused by the first two cholera pandemics and weighs how that influenced apocalyptic discourses in literature (1826 to the 1860s). Then, it discusses the opposite tendency of denial which is linked to comic texts produced in the 1830s. Finally, it analyzes how these trends were transformed after the remarkable advances in bacteriology from the 1880s onward. By confronting these three facets of cholera's representation, it is possible to identify its influence over the literature of the early twentieth century.

Introduction

Cholera was the great pandemic disease of the 1800s. Although it was not the century's biggest killer, cholera has influenced culture and science to such an extent that its impact "probably exceeded that of any epidemic since the fourteenth-century plague."[1] In the last two hundred years, it has toured the globe no less than seven times in pandemic waves that lasted for years and even decades. As expected from health crises of this magnitude, its spread has impinged upon economics, politics, science, urbanism, and, naturally, culture and literature.

This chapter studies three aspects of the literary response to cholera. It starts with a brief overview of the infection: its cause, symptoms, and history. Then, it analyzes two opposing tendencies which emerged simultaneously in literature: Some accounts were alarming and foretold the death of millions, while others ridiculed the situation and even denied cholera's existence. This chapter studies how these discourses formed in opposition to one another and were related to deep-rooted anxieties linked to previous plague outbreaks. After familiarity emerged, however, both tendencies subsided, though to different degrees: The apocalyptic predictions

[1] Hays (2005).

A. L. Guerios (✉)
University of Padua, Padua, Italy
e-mail: aureolgneto@gmail.com

S. Venkatesan et al. (eds.), *Pandemics and Epidemics in Cultural Representation*,
https://doi.org/10.1007/978-981-19-1296-2_5

became less excessive, while the humorous accounts disappeared almost completely. That would change a few decades later, after the consolidation of the germ theory of disease as the dominant scientific and medical dogma. At this stage, cholera started to be seen as an overcome problem, so comic depictions of it became acceptable once again. Finally, the text argues that these aspects of the literary imagination continued to be used after the menace receded, thus establishing a continuum between the cholera texts of the nineteenth century and the plague texts of the twentieth century.

Cholera Symptoms

Cholera is an infection caused by *Vibrio cholerae*, a bacterium that spreads through contaminated food or water. Once it reaches the small intestines, it induces abundant fluid loss through uncontrollable bouts of diarrhea and, to a lesser degree, vomiting. That leads to severe dehydration and electrolyte imbalances, which, in turn, compromise physical and neurological functions. Patients become weak, confused, and lethargic, have difficulties reasoning, and experience painful cramps. Their appearance is transformed: Patients look thinner and have sunken eyes and dry lips; their skin become wrinkled and inelastic; their face, hands, and feet turn bluish (*cyanosis*). If the patient is not rehydrated, death results in up to 50% of cases, due to heart or kidney failure.

Today, cholera is understood in the minutest details. Its effects in the body can be reversed by therapy that is inexpensive, highly effective, and easy to administer. Prevention strategies are so successful and widespread that they are taken for granted. However, in the nineteenth century, cholera was dreaded as a terrifying scourge that costs the lives of millions around the globe. Its spread was not understood, which meant that prevention strategies were inefficient. Existing therapies consisted above all in bleeding and purging, which only aggravated the situation. Victim's bodies got deformed up to the point of unrecognition—eyewitness accounts often mention "a mummified appearance." Moreover, symptoms erupted suddenly and progressed swiftly. One in every two cases could result in death, often in the shockingly brief timespan of six to forty-eight hours. If that was not enough, cholera's symptoms were repulsive and profoundly humiliating.

Even if the *Vibrio cholerae* is often assumed to have emerged in India, its origins are still a matter of debate among specialists.[2] Regardless of its provenance, the bacterium can be found at the Bay of Bengal, where, in 1817, an outbreak in Kolkata quickly escalated out of control. The region was then ruled by the British East India Company, and it suffered from a combination of factors that facilitated the spread of disease: political instability, large displacements of peoples due to war or pilgrimage, ecological imbalances, and undernourished populations after crops failed in 1816.

Cholera then spread steadily in Asia for seven years, an event that constitutes the first cholera pandemic (1817–1824). By following commercial routes and the

[2] Hamlin (2009).

displacement of troops, it extended from the Arabian Peninsula to Japan and from Indonesia to the Black Sea. In 1824, cholera has also reached the Russian port of Astrakan—which would become a key step in the itinerary of later pandemics—but was prevented from spreading further by a particularly severe winter. As a result, Europe was not visited by cholera at this point.

Hysteria: A Plague Worse Than Plague

Even if Europe escaped the first pandemic, cholera still exerted influence in European thought at this early stage. Governments and entrepreneurs had many interests in Asia, so any potentially destabilizing event drew considerable attention. In 1820, for instance, James Jameson's *Report on the Epidemick Cholera Morbus* was published in Bengal. The treatise is the first of a prolific tradition of medical writing on the disease. It amounts to circa four hundred pages of dense statistics, tables of mortality, weather patterns, and maps. It is of central importance for later developments for two reasons.

Firstly, Jameson concludes that the ultimate causes of the epidemic were anomalous environmental factors (temperature, rains, winds), although lack of hygiene was considered to play a role as well.[3] If the fundamental problem relied on the Indian tropical climate, Europeans clearly had little to fear, for they were beyond the disease's reach. That created a feeling of safety and even invulnerability which would be invalidated a decade later when the second cholera pandemic (1827–1837) reached both Europe and the USA. This unexpected reversal would incite two opposing cultural tendencies: A part of society would continue to feel safe, dismissing cholera's seriousness or even denying its existence, while another would dread it like the plague itself and react with despair.

Moreover, the report employs dismal statements which would be progressively exaggerated as the years passed and the pandemic approached. Jameson declares, for example, that "[cholera] surpassed the plague in the width of its range; and outstripped the most fatal diseases, hitherto known, in the destructive rapidity of its progress."[4] In the passage, he underscores cholera's novelty and draws attention to its sudden eruption. Indeed, cholera's speed is remarkable since even bubonic plague requires a few days to kill. He also asserts that, on this occasion, its morbidity rate (the amount of the population infected) was higher than that usually achieved by the plague. However, throughout the century, it was uncommon for cholera to affect more than 5% of the population, and even very serious outbreaks hardly exceeded a 10% morbidity rate[5]; figures which the plague surpassed manifold in previous centuries. It is also important to notice that Jameson is not claiming that cholera mortality was

[3] Harrison (2020).

[4] Jameson (1820), p. 13.

[5] Guerios (2021), p. 100; Hays, *Epidemics and pandemics*, 229–331.

higher than that of the plague, yet that is precisely how many would understand his remarks.

In 1824, only four years later, Joseph Mather Smith published a treatise proclaiming that "[t]he Cholera Morbus of India is one of the most fatal diseases in the catalogue of epidemics."[6] The assertion was made at a point when the first pandemic was coming to a close and the threat of the second did not yet loom on the horizon. The remark is not necessarily untrue, but it is already noticeable how the focus shifts from promptness of development and morbidity rate to put stress on lethality. The tendency for amplification becomes more evident in 1831 when cholera had already reached Europe, and its importation into the USA was only a matter of time. At this point, Dr. Smith offered a lecture to the College of Physicians and Surgeons in New York, which was subsequently published as *A Discourse on the Epidemic Cholera Morbus of Europe and Asia* (1831). On it, he intensifies his previous declarations and avows that:

> [cholera] has opened a broader outlet of human life than any other epidemic disease. With energies *more terrific than plague* or yellow fever, it admits of no equal in the celerity and fatality of its attacks. In its devastations too, *it has no competitor but reigns supreme over every other destroyer of the human race.*[7]

Now that its arrival was imminent, cholera suddenly became the worst disease of all times, eclipsing even the plague. The conclusion may seem far-fetched, but it is not restricted to Dr. Smith; myriad others shared his alarm. In 1831, when cholera was only a few months away from Paris, the French government commissioned the statistician Moreau de Jonnès to write a report on the first pandemic. The result was an extensive study charged with apocalyptic predictions. He asserts, for instance, that 18 million people died during the first pandemic in India alone—the real number is evaluated to have been one or two million.[8] Given that the population of Europe is estimated to have been about 224 million in 1820,[9] if the same figures envisaged by Jonnès applied, about 18% of Europeans would die. These kinds of frightening predictions were also validated in a smaller scale: In the Persian city of Bassorah, for example, the dead allegedly amounted to "beyond the 18.000 [out of 60.000]," a mortality rate of 33%.[10]

If that was not enough, newspapers augmented the numbers even further. According to R. J. Morris, the "normally calm" *Quarterly Review* proclaimed that cholera was "one of the most terrible pestilences which have ever desolated the earth" and raised the already exaggerated number of casualties to a mind-blowing 50 million.[11] Still another remarkable case is found in a public lecture delivered by François Broussais at the height of the Paris Outbreak of 1832. In his address, the

---

[6] Smith (1824), p. 119.

[7] My emphasis. Smith (1831), p. 9.

[8] Arnold (1993), pp. 161–163.

[9] Inklaar et al. (2018).

[10] de Jonnès (1831), p. 78.

[11] Morris (1976), p. 14.

celebrated physician proclaimed to a terrified audience that the Black Death, "which extinguished almost a third of men [in the fourteenth century]," was likely caused by cholera.[12]

Such apocalyptic discourses were not restricted to physicians, statisticians, and journalists. Literary authors also engaged with the problem, often magnifying symptoms for narrative effect. When, in the popular novel *The Mysteries of Lisbon* (1854), a large number perish during the Lisbon Outbreak of 1833, the Portuguese author Camilo Castelo Branco pledges in a footnote that he is not exaggerating. To prove his point, he tells an autobiographical anecdote: He was a child at that time, and upon arriving at his tutor's house for a lesson one day, he discovered that his teacher and his entire family—seven people in total—had contracted cholera and died in the previous day. The anecdote is hard to believe, yet the very fact that it is used to enhance the novel's trustworthiness reveals how apocalyptic discourses of the kind circulated freely.

Fictional texts also helped to strengthen the cultural overlap between cholera and the plague. Alexander Pushkin is a case in point. After finding himself isolated in the countryside due to a cholera quarantine, the poet worked on a short theater play opportunely entitled *A Feast in Time of Plague* (1830). Instead of discussing the immediate situation of the cholera in Moscow, Pushkin preferred to sketch the plague in London, thus creating a correspondence that would be obvious for his public. Similarly, once the epidemic appears in *The Wandering Jew* (1844), Eugène Sue declares in a footnote that "[i]n 1346, the famous black plague devastated the globe; its symptoms were the same as cholera, and the same inexplicable phenomenon of its gradual progress in stages and along a given route occurred. In 1660 another similar epidemic decimated the world again."[13] Since, according to the author, cholera and plague have the same effects and spread in the same way, they are, to all intents and purposes, one and the same.

Other authors engage with the topic by bringing the disastrous predictions to their logical conclusion. What if the entire human race perished, instead of only a third? That is the essence of the poem *Choléra Morbus*, published in a popular magazine by Léon Gozlan just months before the disease's arrival:

> Stronger than all, it will be known
> who I am. They already say my name
> with fear. In every person,
> my name resounds: Cholera!!
> Make way for this king of kings! […]
> One day, he will rule the universe all alone;
> His scepter is just a bone; his royal mantle is a shroud…
> He kills in grand scale. With his murderous arm,

[12] Broussais (1832), p. 2.

[13] My translation. Sue (1844), p. 145.

He kills an entire people as if a single man.[14]

The poem reenacts the tradition of the *Triumphs of Death*. Petrarch had used the very same motif in his *Triumphus Mortis* (1374), where the plague—personified as an old hag—exterminates thousands, including his beloved Laura. That type of representation was specially employed in the fifteenth- and sixteenth-century paintings, always to illustrate the ravages of the plague.[15] However, Guzman's poem modernizes this tradition by personifying cholera as the Angel of Death. It explicitly declares that the epidemic will eventually annihilate life on earth, effortlessly wiping out whole populations.

Flaubert imagined the disease in a similar light in his short story *La Femme du Monde* (1836), in which a woman speaks directly with the public in a series of aphorisms. Her portrayal is built little by little in accordance with her own declarations, and the reader is expected to anticipate her identity before the end. To be sure, the woman is none other than death itself who brags for having unleashed "cholera, this monstrous cholera, which, with its black nails, its green complexion, its yellow teeth, its convulsing limbs, carries man to the grave more rapidly than an arrow crosses the air or a lightning split the skies."[16] In this case, the idea of humanity's extinction is present only indirectly, but the infection's swiftness is magnified.

Likewise, at the end of *Crime and Punishment* (1866), Raskolnikov lies delirious with fever at the hospital, when he has a vision of an impending cataclysm:

> He dreamt that the whole world was condemned to a terrible new strange plague that had come to Europe from the depths of Asia. All were to be destroyed except a very few chosen. [...] Only a few men could be saved in the whole world. They were a pure chosen people, destined to found a new race and a new life, to renew and purify the earth [...][17]

As noted by Susan Sontag, despite the word "plague," this vision is built upon the traditional imagery of cholera as the "monster from Asia."[18] In fact, the condition was ordinarily called *Asiatic cholera* until well into the twentieth century. At the heyday of nineteenth-century orientalism, caricatures would regularly sketch it as a sort of mummified monster displaying attributes alluding all at once to India, China, and the Ottoman Empire.[19]

---

[14] Leon Gozlan, "Cholera morbus," *L'Artiste* 1 (1831): 53–56. Cited in Nina Athanassoglou-Kallmyer, "Blemished physiologies: Delacroix, Paganini, and the cholera epidemic of 1832," *The Art Bulletin* 83.4 (2001): 694 and 708.

[15] Some famous examples are: the anonymous fresco*Il Trionfo della Morte* (c. 1446) kept at *Palazzo Abatellis*, in Palermo;*Il Trionfo della Morte e la Danza Macabra* (1485) by Giacomo Borlone at the external facade of the *Oratorio dei Disciplini* in Clusone, Italy; or the *The Triumph of Death* (c. 1562) by Pieter Bruegel, currently at the *Museo del Prado*, inMadrid.

[16] My translation. Flaubert (1910), pp. 109–114.

[17] Dostoyevsky (1917), pp. 555–56.

[18] Sontag (1990), pp. 137–38.

[19] To mention only a few: the anonymous coloured lithograph *John Bull between two humbugs*(1831); the anonymous caricature *Plagas de que pronto, á España—quiere ver libre La ARAÑA* (1885) published in the Spanish magazine *La Araña*; the cartoon *V'la le Choléra* (1892) by Adolphe Willette presented by Arsène Alexandre in the anthology *L'art du rire et de la caricature* (1892).

The combination of human extinction and orientalism is also fully formulated in Mary Shelley's *The Last Man* (1826)—probably the first European text to indirectly portray cholera. The novel was written as the first pandemic waned in Asia, and it was published slightly before the start of the second. As effectively revealed by Alan Bewell, the Shelleys had ties to physicians who practiced in India and were likely well informed about its public health situation.[20] In the novel, a new plague "raise[s] its serpent-head on the shores of the Nile" and devastates Asia before moving, via the Ottoman Empire, into Europe and, finally, England. As in previous examples, this "plague" is endowed with the attributes of literary cholera: abrupt emergence, rapid development, changes in color, expiration in hours or even minutes, inflated morbidity rate, and unprecedented mortality. It spreads via miasmas, although on at least one occasion, there is evidence for person-to-person contagion. This confusing pattern of transmission is yet another parallel to cholera since the fecal-oral route by which the bacterium is conveyed does not create any obvious epidemiological patterns. Hence, for decades, the question of cholera's transmissibility remained unresolved. As in the novel, many advocated the infection was non-contagious; others argued it was somehow transmissible from person-to-person—even if it was difficult to pinpoint exactly how—while some even defended the two opposing views at once—as exemplified by a certain Dr. Holland who affirms in 1831: "I have no doubt that the disease is both contagious and non-contagious."[21] Moreover, the plague emerges in *The Last Man* during a British military campaign that fights for Greek independence at the gates of Istanbul—a conquest that is clearly a synecdoche for the entire Asia. This situation matches with exactitude the first encounter of the British with cholera in India in 1818, during the *Third Anglo-Maratha War*.

Beyond these correspondences, *The Last Man* brings about a fictional apocalyptic scenario that is only hinted at by Gozman or Dostoevsky. The novel presents itself as a prophecy by the Cumaean Sybil found in fragments by an unnamed narrator—who Mary Shelley tries to purposely confuse with herself. The frame-narrative of the found manuscript occurs in 1818, exactly the year in which the first pandemic was gathering momentum in India. Plus, it is made in Naples, a city renowned for its plague outbreaks and whose name would later become synonymous with cholera in the literary imagination. The prophecy turns out to be an autobiography of the only person to recover from the coming plague, perhaps becoming the last human alive. To confirm if that is the case, the narrator decides to roam the earth in search of a new Eve with whom to reestablish humankind. The biblical overtones with Adam and Noah serve as a counterpoint to other references to the apocalypse or the plagues of Egypt which materialize throughout the text. It also serves to create a sense of purification and renewal which works as a happy ending of sorts. The tragedy is still immense, but there is at least some hope on the horizon.

That viewpoint contrasts sharply with the resolution of Edgar Allan Poe's T*he Mask of the Red Death* (1842), which is equally informed by descriptions of cholera.[22]

[20] Bewell (2003), pp. 205–242 and pp. 296–314.

[21] Stokes (1921), p. 131.

[22] Guerios (2020), pp. 115–130.

The short story is set in a "country" that was "long devastated" by the fictional Red Death. The disease is a hybrid that combines old and new ideas about pestilence—as noticeable in the double evocation of its name, *Black Death* and *Blue cholera.* Besides, Poe engages in active dialog with the *Decameron* (1348–1352), the plague text *par excellence*; his choices of stage (a medieval abbey), characters (a thousand nobles), and actions (isolation, hedonism, feasting), all follow Boccaccio. Nevertheless, in the context of publication—less than a decade after the arrival of the second pandemic in Philadelphia—any "new disease" would invariably recall cholera in the public's mind. Plus, the diseases' symptoms coincide to a large extent: both cause "sharp pains" and "sudden dizziness"; the "profuse bleeding at the pores" and "dissolution" can be interpreted as an echo of the copious evacuations induced by cholera.[23] Moreover, nearly, all cholera texts focus on the blue coloring of the face as a means to avoid mentioning vomit and diarrhea. The same focus can be observed here since the red stains appear "especially upon the face of the victim." Finally, their abruptness and rapidity are equivalent. Cholera may kill in a few hours, but in popular and medical discourses, this lapse was often shortened, as proved by Dr. Smith, who announces that, in Asia, "[d]eath frequently followed in a few minutes after the attack [of cholera]."[24] Similarly, the Red Death's "seizure, progress, and termination" required solely "half an hour."[25]

These parallels also include the extinction of humanity. Although all the texts discussed so far play with total annihilation, none carries it until the end. Instead, they prefer to conjecture visions of future catastrophes which, overall, point to some sort of rebirth. Yet, that is not the case with *The Mask of the Red Death.* On it, only instants separate the emergence of the disease's personification at the ball and the death of all individuals without exception. Even the energy of inanimate objects is drained off: The clock stops to tic and the torches are magically extinguished. This resolution brings the dire cholera predictions to their logical conclusion. Ultimately, nothing remains, so "darkness and decay and the Red Death" may exert "illimitable dominion over all."[26]

Ridicule: A Laughable Crisis

The promises of doom and gloom were not, however, the only cultural reaction to cholera. As soon as the catastrophic visions started to appear in the months prior to the disease's arrival, they were swiftly met with responses moving in the opposite direction. If some anticipated a reenactment of the Black Death, others simultaneously condemned these views as exaggeration, alarmism, and fearmongering. In his journalistic coverage describing the Paris Outbreak of 1832, Heinrich Heine asserts

[23] Poe (1850), p. 339.

[24] Smith, *A discourse on the epidemic*, 13.

[25] Poe, *Red Death*, 339.

[26] Poe, *Red Death*, 345.

that "all minds were seized with vague terror, and the extent of the malady was magnified beyond limits."[27] The same complaint is found in the first paragraph of a chronicle written during the Naples Outbreak of 1836: "[cholera's] desolations were told, and they seemed invented by those who take pleasure in frightening us"; however, after first-hand experience, it became clear "that its desolations are less frightening than we were led to believe. This is an epidemic like any other […]."[28]

Before populations had direct contact with the infection, it was not unusual for individuals to dismiss its seriousness or to deny its existence altogether. There was a tangible scientific basis for that, since, according to the Miasma Theory of disease, epidemics were caused by noxious vapors in the atmosphere. As such, they were understood in relationship to climate, as well as factors like temperature, humidity, or altitude. Given that cholera was seen as a product of tropical climates, it was reasonable to think it would not reach Europe. So much so that, in typical orientalist fashion, its alleged geographical reach was imprinted on the disease's name: *Asiatic cholera.*

Hence, numerous scientists and government officials saw the prospects of annihilation as ludicrous—in Europe at least. In a letter to his sister, the Italian poet Giacomo Leopardi asserts: "The other evening I spoke with the medical commission sent from Rome to greet the Cholera in Paris, which promises us the coming of the disease in Italy: a prediction that the physicians laugh at in here because they don't believe it."[29] The simultaneous existence of the two opposite views (inevitable infection *versus* humorous rejection) is clear in the comment. Similarly, skepticism and above all denialism often appear in literary texts at this early stage in the 1830s. Authors would use it to interpret the outbreak as supernatural punishment, but also to obtain humorous effect—the German author Heinrich Laube went as far as saying that "[cholera] is the most popular and most effective humorous writer we have ever had in Germany."[30] These two tendencies are combined in Heine's *Cholera in Paris* which employs abundant irony to great comic effect, even if it delivers lyrical reflections on mortality and politics too. According to him, the outbreak erupts at the carnival, where "one could even see maskers, who in caricatures of livid color and sickly mien, mocked the fear of the cholera and the disease itself."[31] The jest is short-lived, though, since immediately after "the merriest of the harlequins felt that his legs were becoming much too cold" and took off his mask to show "a violet-blue face."[32] There is evident sarcasm in the way diarrhea is described here, and the overlap between the first evacuation and the bluish face—a signal of severe dehydration—is highly implausible.

Very similar remarks are made in the memoirs of Chateaubriand, who pledges to have seen "drunkards […] saying as they raised their glasses, "To your health,

---

27 Heine (1893), p. 182.

28 My translation. De Sterlich (1837), pp. 11–16.

29 My translation. Foschi (1983), p. 182.

30 My translation. Heinrich (1834), pp. 66–67.

31 Heine, *The Works*, 167.

32 Heine, *The Works*, 166–167.

Morbus!" Morbus, out of gratitude, rushed up, and they fell dead under the table."[33] The setting is pivotal here. The carnival and the tavern serve to underscore the carelessness of these individuals, and hence, their instantaneous punishment is presented as deserved. These condemnations were usually directed against the lower classes which were portrayed as out of touch with reality.

The depiction was motivated by the popular disturbances which followed the footsteps of cholera. Since mortality among the poor was overwhelmingly higher, many among the marginalized groups became convinced that the epidemic was caused by poisoning, rather than disease. In line with Malthusian perspectives—which saw pestilence and war as efficient measures of population control—and an overall suspicion of the government, conspiracy theories would blame governments for seeking to exterminate the poor. Riots would consequently flare up with regularity: There were significant tumults in Russia, Spain, or Italy,[34] while in the UK alone, Samuel Cohn has counted at least seventy-two riots in only fourteen months.[35]

In literature, denialism usually serves to illustrate the alienation of the lower classes. Such ideas commonly appear in texts featuring turmoil, so much so that in a work dedicated to the plague of the 1630s, Alessandro Manzoni observes that "[u]pon observing this firm persuasion, this insane fear of a chimerical attack, one cannot help but remember what happened likewise in various parts of Europe, a few years ago, in the time of cholera."[36]

Concomitantly, denial is found in comic texts which laugh at the creativity and coarseness of the locution of the lower classes. That is the case of the Italian poet G. G. Belli, who composed a cycle of thirty-four sonnets about cholera while it raged in Rome. The poems depict a few commoners discussing the subject in a tavern. They contain grammatical errors and were written in the Roman dialect, which is perceived as mundane and lacking in prestige. Many words are mispronounced, which creates comicity but also serves to attest the ignorance of those speaking. Thus, *cholera morbus* is ironically transformed into *colla-morbida* (literally "soft glue") by one of the characters. The pun combines the verb *colare* (to pour a liquid) and the noun *colla* (glue) and is an amusing way to describe diarrhea.

The first poem in the cycle also struggles with the disease's name and challenges the infection's existence: "Enough if it is called *cholera* or *choleric*,/I bet […]/that this epidemic, in my view,/is not among us, if it exists at all."[37] The character goes as far as calling those who fear it "stupid" and "damn lunatics" who should "whip their own ass." In contrast, the next poem reinforces the apocalyptic predictions. The speaker believes "Rome has sinned: that is the reason/why plague comes our way/not to leave any laic alive"; in other words, all will die with the sole exception

---

[33] Chateaubriand (1900), p. 486.

[34] Hamlin, *Cholera: the biography*, 81–86.

[35] Jr Cohn (2017), p. 164.

[36] Manzoni (1843), p. 21.

[37] My translation. Giuseppe Gioachino Belli, "Er còllera mòribbus, Converzazzione a l'osteria de la ggènzola indisposta e ariccontata co ttrentaquattro sonetti, e tutti de grinza." in *Tutti i Sonetti Romaneschi* (Roma: Newton, 1998), 841.

of the Church's personnel. In the ninth poem, a drunkard explains how the wells and fountains are being systematically poisoned but concludes with a joke: "to water, they can make whatever they please/on condition that they don't poison my wine."[38] In another reversal, a priest avows in poem twenty-two that "Cholera is coming, my dear brothers/prepare to die, my brethren."[39] Thus, different sonnets in the cycle take different positions within the two extremes of denialism and total annihilation; if one voice asserts cholera is a farce, another assures that few will survive.

For a brief period, comicity served yet another purpose: to help keep good spirits and drive away fear. Insofar as to laugh about something means to be immunized against it, literature could become a prevention strategy of sorts. A common medical recommendation was to seek books as a way to keep entertained and, thus, avoid the fear of cholera. Readers should look for light-hearted works on topics unrelated to disease (love stories, historical novels, etc.), yet still, there are examples of cholera texts—and caricatures as well—produced for the sake of protection. None other than Coleridge has published *Cholera Cured Before Hand*, a poem which claims on the very title to have some power of inoculation. It was "promulgated gratis for the use of the useful classes" in 1832. Once again, it pokes fun at the "incorrect" vernacular of the lower classes, yet it also engages vivid imagery, onomatopoeia, and humorous comparisons to list the precautions that should be taken to evade the disease. Thus, the poem sarcastically pretends to be an agent of prevention, immunization, and perhaps even healing. It is a mechanism to release stress which may contribute to overall psychological health in turbulent times.

Relief: The Protection of Bacteriology

After the initial shock of the new in the 1830s, the population and governments quickly realized that cholera would not bring about the end of civilization, yet it was no laughing stock either. As familiarity arose, the two perspectives abated. Apocalyptic discourses surfaced with lesser frequency, without disappearing completely though—as made manifest by the Gothic explorations in Marie Corelli's *Vendetta: the Story of One Forgotten* (1886), Rodolfo Teófilo's *A Fome* (1890) or Marcel Schwob's *Le Train 081* (1891). However, humorous intakes on the issue became culturally unacceptable soon after the first contact. In general terms, for the next five decades or so, to laugh at cholera was considered as unsympathetic bad taste. As one character puts it in a popular novel: "I play around with fire and with bullets, but not with cholera."[40]

That reluctance can be observed in the criticism aimed at the German scientist and philosopher Gustav Fechner, who wrote a number of satirical works under the pseudonym of Dr. Mises. One of his pieces is *Protective Means Against Cholera*

[38] My translation. Belli, "Er còllera mòribbus," 844.

[39] My translation. Belli, "Er còllera mòribbus," 851.

[40] My translation. Salgari (1973), p. 411

(1832), a lengthy mock treatise that describes the epidemic as a force for good: It had more positive effects than many government policies, and it offered individuals much-needed motivation to behave and improve themselves. The sarcasm was met with disapproval, however. Even the author's biographer would ponder decades later that the enormous suffering caused by the epidemic "directly challenged satire."[41]

Once the menace was validated, humorous intakes on the theme subsided between the 1840s and the 1870s. A considerable number of works tackle the theme during this period—Hans Christian Andersen's *To Be or Not to Be* (1857), Zola's *Les Mystères de Marseille* (1867), or George Eliot's *Middlemarch* (1871-2), to mention just a few, yet none explores the subject comically. I am aware of only one exception, Mark Twain's *The Innocent Abroad* (1869), in which cholera features briefly twice. Nonetheless, given the general playfulness and irony of Twain's text, this narrative could be considered the anomaly which confirms the rule.

The cultural interdiction to laugh about the infection would come to a close only in the 1880s, after considerable developments in science and medicine. Starting in the 1850s, various reforms movements pressed for large projects of sanitation which, along the 1860s and 1870s, resulted in the construction of sewage and water filtration systems. As a result, the urbanization of several cities—Paris, Barcelona, Budapest, and the like—changed drastically. In parallel, the field of bacteriology was blossoming and scientific breakthroughs were being made in swift succession: In the late 1850s and early 1860s, Pasteur conducted his founding experiments on fermentation, which, from 1865 onward, were used by Joseph Lister to revolutionize surgery; in 1873, Gerhard Hansen identified the causing agent of leprosy; between 1879 and 1885, Pasteur and Émile Roux developed the first human-made vaccines against rabies and anthrax; in 1882, Robert Koch discovered the pathogen responsible for tuberculosis and, one year later, the one responsible for cholera.

Koch's identifications of the *Vibrio cholerae*, in 1883, and of its means of transmission, in 1884, have completely rewritten societal responses to cholera. Now that the pathogen and its life cycle were understood, prevention strategies could be designed and implemented with efficiency. This led to a feeling of safety, and with time, the epidemic was increasingly understood as a subjugated opponent, a thing of the past. Its devastations had indeed been steadily declining after the sanitary reforms of the 1860s; nevertheless, even if on a minor scale, outbreaks would still erupt periodically throughout the continent. Regardless of that, from a cultural point of view, cholera was deemed vanquished for the most part after 1884. The serious outbreak in Hamburg in 1892 was regarded with shock precisely for contradicting these discourses of triumph.

Now that the mystery was gone, many of the anxieties built around it rapidly diminished as well. From this point onward, the humorous explorations of the subject would be immediately resumed and explored in myriad texts. In 1886, for example, Maupassant published *Voyage de Santé*, a short story that laughs at the vain attempts of a germophobe to evade cholera microbes. In the following year, the Italian author Federico De Roberto published *San Placido*, in which the inhabitants of a provincial

[41] My translation. Lasswitz (1902), pp. 28–29.

town rebel after the city's long-waited festivities are canceled due to an outbreak. Still, in 1887, Kipling circulated *A Germ-Destroyer*, in which a self-declared scientist invents a method to dispose of bacteria but, during a demonstration, creates panic and a false fire alarm in the residence of India's Viceroy. Around 1889, the popular songwriter Aristide Bruant composed *Voilà the Cholera!*, whose lyrics and scores were sold together to be performed at house parties. The piano accompaniment is lively and cheerful, while the chorus revisits the apocalyptic promises of the 1830s:

It seems we wait the cholera,
the fact is positive.
We are not sure when it will arrive,
But we know it will.
Voilà the cholera! Voilà the cholera!
Voilà the cholera arriving!
On both sides of the river,
everyone will die! [...][42]

The tendency continued into the first decades of the twentieth century, with numerous authors continuing to explore cholera in farcical ways. The classic novel *The Good Soldier Švejk* (1921–23) by the Czech writer Jaroslav Hašek uses a cholera misdiagnosis to deride the incompetent Austro-Hungarian army—in scenes that are highly picaresque and employ abundant scatological humor. Incidentally, these episodes in Hašek's work likely inspired the "diarrhea epidemic" which hinders the activity of the flying squadron in Joseph Heller's *Catch-22* (1961). The theme was also employed by the French author Joseph Delteil who, in the surrealist novel *Choléra* (1923), turns the disease into a person which is engaged in a love triangle. Still, another example is found in the futurist novel *I Burn Paris* by the Polish writer Bruno Jasieński. The work was initially serialized in French in 1928 and, then, published in Polish as *Palę Paryż* in the following year. The plot revolves around a terrorist attack that disseminates "a plague' in Paris" water station. Despite the name used, the fictional outbreak is informed by cholera rather than bubonic plague; nearly, all of the cholera's *topoi* are present in the narrative, including transmission by contaminated water. Interestingly, it portrays a kind of apocalypse, since all the inhabitants of Paris eventually perish with the only exception of prisoners, who survived unscathed since they were naturally isolated from the rest. After they evade the prison, the inmates discover a deserted city and decide to use this opportunity to build a new egalitarian society from scratch. Almost exactly a hundred years after cholera's literary debut in *The Last Man*, Jasieński reuses the same trope of purification and renewal to conclude his narrative, thus combining the trends of annihilation, derision, and scientific triumph.

[42] Bruant (1889–1895), pp. 71–78.

Conclusion

During pandemics, literature fulfills an important role by providing individuals with an opportunity to reflect on the crisis or to escape through entertainment. At the early stages of the European experience with cholera, literary texts have reacted to these needs by adopting strategies that fit in a spectrum: On one end, there are texts such as *The Last Man* or *The Mask of the Red Death* which promise doom and gloom and even contemplate humankind's extinction; on the other, there are popular discourses or texts such as *Er Còllera Mòribbus* which downplay and even deny the existence of the crisis; in between, numerous works tend in degrees to one or the other extreme.

However, both tendencies dwindled after familiarity with cholera developed. The apocalyptic predictions were proved wrong, for cholera was by no means a reenactment of the Black Death. The tendency did not vanish, but later texts tended to be less extravagant and a little more realistic. Moreover, the pandemic also proved it could engender enormous suffering and loss of life. So, as it became clear that cholera was no laughing matter, humorous explorations were abandoned.

That changed when the achievements of the sanitary reforms became increasingly evident and the advancements of the germ theory of disease offered hope and opportunity for commemoration—even if little had been accomplished in therapy and pharmacology. The literary imagination reacted almost immediately, so cholera grew as a theme and became once again a source for humor. In addition, this treatment of the subject was continued by several twentieth-century texts featuring epidemics.

Bibliography

Arnold, D. (1993). *Colonizing the body: State medicine and epidemic disease in nineteenth-century India*. University of California Press.

Athanassoglou-Kallmyer, N. (2001). Blemished physiologies: Delacroix, Paganini, and the cholera epidemic of 1832. *The Art Bulletin, 83*(4), 686–710. https://doi.org/10.2307/3177228

Belli, G. G. (1998). *Tutti i Sonetti Romaneschi*. Newton.

Bewell, A. (2003). *Romanticism and Colonial Disease*. John Hopkins UP.

Broussais, F. J. V. (1832). *Du Choléra-morbus, leçons faites au Val-de-Grâce par le Dr Broussais*. Imprimerie de Fores.

Bruant, A. (1889–1895). *Dans la Rue: Chansons et monologues* (Vol. 1). Aristide Bruant.

de Jonnès, A. (1831). *Rapport au Conseil supérieur de santé sur le choléra-morbus pestilentiel*. Imprimerie de Cosson. BNF Gallica.https://gallica.bnf.fr/ark:/12148/bpt6k5424131n/f85.image.texteImage. Accessed November 25, 2020.

de Sterlich, C. (1837). *Quadri Storici del Cholera di Napoli*. Tipografia Flautina.

de Chateaubriand, F.-R. (1900). *Mémoires d'outre-tombe* (Vol. 5). Garnier Frères.

Dostoyevsky, F. c. (1917). *Crime and punishment*. Trans. Constance Black Garnett. P. F. Collier and Son.

Flaubert, G. (1910). *Œuvres de jeunesse* (Vol. 1). Conard.

Foschi, F. (1983). *Epidemie nella terra di Leopardi*. Bulzoni.

Guerios, A. L. (2020). Blue death: The masking of cholera as plague in Poe and Pushkin. In K. Tampakis & G. Vlahakis (Eds.), *Science and literature: Imagination, medicine and space* (pp. 115–130). Institute of Historical Research.

Guerios, A. L. (2021). *Cholera and the literary imagination in Europe, 1830–1930*. University of Padua.

Hamlin, C. (2009). *Cholera: The biography*. Oxford UP.

Harrison, M. (2020). A dreadful scourge: Cholera in early nineteenth-century India. *Modern Asian Studies, 54*(2), 502–553. https://doi.org/10.1017/S0026749X17001032

Hays, J. N. (2005). *Epidemics and pandemics: Their impacts on human history*. Abc-clio.

Heine, H. (1893). *The works of Heinrich Heine* (Vol. 7). Trans. Charles Godfrey Leland. William Heinemann.

Inklaar, R., de Jong, H., Bolt,J., & van Zanden, J. (2018). Rebasing 'Maddison': new income comparisons and the shape of long-run economic development. Maddison Project Working Paper 10. https://www.rug.nl/ggdc/historicaldevelopment/maddison/releases/maddison-project-database-2018. Accessed July 25, 2018.

Jameson, J. (1820). *Report on the Epidemick Cholera Morbus, as it visited the territories subject to the Presidency of Bengal, in the years 1817, 1818, and 1819*. Government Gazette Press.

Jr Cohn, S. K. (2017). Cholera revolts: A class struggle we may not like. *Social History, 42*(2), 162–180.https://doi.org/10.1080/03071022.2017.1290365

Lasswitz, K. (1902). *Gustav Theodor Fechner*. Frommanns Verlag.

Laube, H. (1834). *Reisenovellen* (Vol. 1). Otto Wigand.

Manzoni, A. (1843). *Storia della colonna infame*. Gaetano Nobile.

Morris, R. J. (1976). *Cholera 1832: The social response to an epidemic*. Croom Helm.

Poe, E. A. (1850). *The works of the late Edgar Allan Poe* (Vol. 1). J. S. Redfield.

Salgari, E. (1973). *La Rivincita di Yanez*. Mursia.

Smith, J. M. (1824). *Elements of the etiology and philosophy of epidemics*. J. and J. Harper.

Smith, J. M. (1831). *A discourse on the epidemic cholera morbus of Europe and Asia*. J. Seymour.

Sontag, S. (1990). *Illness as metaphor/AIDS and its metaphors*. Anchor Books.

Stokes, J. (1921). *The history of the Cholera epidemic of 1832 in sheffield*. J. W. Northkind.

Sue, E. (1844). *Le Juif errant* (Vol. 7–8). Cans et compagnie.

Chapter 6
The Blue Death: Cholera and Reimagined Community in Nineteenth-Century Havana

Bethany M. Wade

Abstract The first case of cholera was diagnosed in Havana on February 25, 1833. Two months later 8000 inhabitants of the city were dead. This article considers how contemporary literature interpreted the social and economic trauma of living through quarantine and lockdown. With a focus on texts by Cuban intellectual José Antonio Saco, physician Dr. Juan Francisco Calcagno Monti, and authors Ramón de Palma and Cirilo Villaverde, I argue that the social disruption that followed cholera created space for authors to reimagine the social order in Havana. Long-simmering social rifts came to the foreground. Economic factions who put their own interests above the population were critiqued. Meanwhile, the rhetoric of contagion was conflated with race, poverty, and immorality. Blaming their own vulnerability on the perceived moral and physical sickness of the city's marginalized populations, Havana's urban elite used cholera to justify imposing modern technologies of security and control. This historical case offers insight into how cultural texts produced by artists and authors in the response to the social, economic, and political disruptions that accompany pandemics radically reimagine the limits of community.

> …(cholera) is as deadly and destructive as the plague;
> that it is spread in the same way as the plague…
> and that consequently, it requires the same
> precautionary measures as the plague.
> Dr. Francisco Calcagno (1832, 14)

> No, our people have suffered in silence the horrors of the most destructive epidemic found in their annals; and when the number of victims who expired daily, took to the heart of all the sad conviction that cholera had invaded us, raised their eyes with resignation to heaven, and adoring the decrees of Providence, some sought asylum in neighboring towns and fields, and others waited for death sitting in their homes.
> José Antonio Saco (1833, 89)

B. M. Wade (✉)
Emory University, Fox Center for Humanistic Inquiry, Havana, USA
e-mail: bethany.wade@emory.edu

S. Venkatesan et al. (eds.), *Pandemics and Epidemics in Cultural Representation*,
https://doi.org/10.1007/978-981-19-1296-2_6

Today I raise my eyes to Heaven
Sad Havana — now free from crying:
and the scene of death and horror.
No longer the mourning cars resound
Through your streets and squares and walls,
God wanted to grant us safety
Oh! Solace in so much pain.
JoséSeverino Boloña, *Al Te Deum* (1833, 24–25)

Angélica sat in her sumptuously appointed living room, reading as the death carts rattled by. The book she read was Alessandro Manzoni's *I promessi sposi*, an Italian novel that described the devastation of the 1630 plague outbreak in Milan. Driven into isolation as cholera ravaged Havana, her only escape from boredom was reading and watching the wagons pass with their macabre contents (Palma, 1838, reprinted in Alvarez Garcia, 1974). Angélica was herself a fictional character in a Cuban novel describing the devastation of the 1833 cholera epidemic in Havana. The author, Ramón de Palma, used this scene to illustrate the global and historical awareness of the city's inhabitants as they endured the terror of the first local cholera outbreak.

The first case of cholera was diagnosed in Havana on February 25, 1833. It was an event the inhabitants of the city had anxiously anticipated. Newspapers routinely reported on the progress of the disease. The people of Havana tracked cholera as it entered Europe in 1831; their fear grew when it crossed the Atlantic. Havana's intellectual and medical community published work about this new plague that transfixed even as it terrorized the port's inhabitants. They were not alone in their preoccupation. Across the globe the arrival of cholera—indeed, even the threat of its arrival—was accompanied by chaos and panic. It was a painful and messy way to die. There was no treatment or cure. Terrified of the blue, twisted faces that heralded imminent death, vulnerable populations whispered in horror of the Blue Death. Desperate state and local officials implemented inconsistent and ineffective measures to stave it off. Failing to stop cholera's progress, the resulting trade disruptions instead exacerbated already existing economic crises in many countries. In response, disparate narrative patterns emerged as citizens attempted to understand this global phenomenon on a local level.[1]

This chapter considers how Cuban authors interpreted the social and economic trauma of living through quarantine and lockdown. The spread of pandemic cholera overlapped the co-emergence of biomedical discourse, a creole intellectual class, and Cuban narrative fiction. These fields embodied new, untested ways of representing the world. The language of biomedicine did not yet have fixed meanings. Concepts such

[1] The literature on cholera is extensive and offers perspectives on the disease in numerous geographic contexts. For major works on the subject, see Myron Echenberg, *Africa in the Time of Cholera: A History of Pandemics from 1817 to the Present*; Richard Evans, *Death in Hamburg: Society and Politics in the Cholera Years, 1830–1910*; Catherine Jean Kudlick, *Cholera in Post-Revolutionary Paris: a Cultural History*; Charles E. Rosenberg, *The Cholera Years: The United States in 1832, 1849 and 1866*; Christopher Hamlin, *Cholera: The Biography*; Michael Zeheter, *Epidemics, Empire, and Environments: Cholera in Madras and Quebec City, 1818–1910.*

as contagion, contamination, and infection were especially fluid, drawing on past connotations and imagined future usage. This very fluidity made cholera a fruitful metaphor through which intellectuals in Havana defined competing political and racial ideologies. The nascent vocabulary of biomedicine infiltrated medical, political, and intellectual writing; collectively these constituted a discursive net through which concepts of race and class were sifted and refined. The Cuban novel—emerging in the aftermath of the 1833 cholera outbreak—was a potent didactic tool that made scientific and medical texts accessible to a populace facing changes in its social, political, and economic structure. Together, I argue, this body of cholera literature shaped collective consciousness in that era.

This chapter develops in three parts. In the first, I explore the emergence of pandemic cholera and the rhetoric attached to it on a global scale. Competing models of epidemic disease clashed in medical depictions of cholera, contributing to new popular understandings of susceptibility, health, and illness. I next follow these ideas into early-nineteenth-century Havana to show how different people understood health and illness in this tropical port city. Physicians, adventurers, and intellectuals documented an urban environment that was seen as the font of modernity and future prosperity for the island, and conversely as a colonial holdover that was detrimental to human health. The stakes of these conceptualizations were high, as the 1830s were a defining period in the early development of anti-colonial sentiments in Cuba. Finally, I conclude by showing how a diverse collective of intellectuals in nineteenth-century Havana mobilized the discourses around cholera, the urban environment, and the susceptibility of certain bodies to illustrate the need to reform the existing social and political order in the island. With a focus on texts by Cuban intellectual José Antonio Saco, physician Dr. Juan Francisco Calcagno Monti, and authors Ramón de Palma and Cirilo Villaverde, this chapter explores how cholera became a potent metaphor for authors to reimagine the social order in Havana. This historical case offers insight into how cultural texts produced in response to the social, economic, and political disruptions that accompany pandemics radically reimagine the limits of community.

Representing Pandemic Cholera

The first global cholera pandemic began in 1817 with an outbreak in Jessore, India. Cholera-like diseases had been documented since ancient times and, by the nineteenth century, the term cholera was used as a catchall for any diarrheal illness. The 1817 outbreak was something new: an epidemic form of the disease. Initially, it spread across the Indian subcontinent. Over the next five years, the virulent disease moved along trade and military routes to Central Asia and the Middle East. Cholera entered China in 1821 and Japan the following year. Then, six years after it started, the pandemic unexpectedly died out. The farthest west it reached was the Caspian Sea (Hamlin, 2009).

This new cholera was terrifying in its virulence and unpredictability. The idea of microscopic, living organisms as the vector of epidemic disease had yet to gain acceptance. Since the late-eighteenth century, two opposing theories dominated the scientific community's explanation of epidemics. Contagionists (germ theory), who believed that disease was passed between individuals, faced off against anti-contagionists (miasmatic theory) who argued disease was the result of corrupted environments. The miasmatic view shaped medical understandings of epidemics until the late-nineteenth century. The spread of epidemics was attributed to "miasmas," or bad air, caused by poisonous emanations from decomposing biological matter: rotting vegetation, decaying carcasses, human feces, even molds inside dwellings (Barnes, 2006; Halliday, 1999; Lees & Lees, 2007). These deadly clouds could be recognized by their foul smells. It was widely accepted by doctors and scientists that illness was caused by the patient's environment.

It was also accepted that certain patients—through their behaviors and lifestyles—were predisposed to epidemic disease. According to this view, immoral behavior, crowded living conditions, poor diets, and a lack of cleanliness left the poor susceptible to miasmatic infection. Predating modern systems of sanitation, nineteenth-century cities were cesspits of filth and human garbage. Marginalized areas were in the worst condition. More than anywhere else in the city, these neighborhoods—often the home of people of color, immigrants, and the poor—were ripe with the stench of decay. It made perfect sense within the logic of miasmatic theory that the inhabitants of these environments would be vulnerable to the disease (Althammer, 2012; Kudlick, 1996; Rosenberg, 1987).

For much of its early history, cholera was understood as a miasmatic disease. Western observers—unscathed by the first pandemic—quickly characterized cholera as an Eastern illness. Dovetailing with Orientalist discourses which imagined the East as a site of corruption, the environment of India, along with the living conditions and character of its inhabitants, were used to explain the emergence of pandemic cholera. And, if India was the source of this new contagion, Western intellectuals also had an explanation for its spread. Rather than implicating colonial commercial and military networks, both the press and scientific publications commonly ascribed cholera's dispersal to Hindu pilgrimages (Arnold, 1986; Boyle, 2015; Hamlin, 2009). As late as 1892, American William G. Eggleston opined that "so long as the pilgrimages continue Europe and this country will be endangered and will be visited by cholera, despite the millions of dollars spent in precautionary measures (Eggleston, 1892, 128)." Nations and populations that were less civilized and advanced than the Christian West were susceptible. This virulent form became known as Asiatic cholera.

The Western narrative of cholera as an Eastern disease was tested when a second pandemic began in 1829. Once again it started in India and spread along military and trade routes to Asia, Russia, and the Middle East. This time the West saw no inexplicable reprieve. By 1831, cholera had entered Western Europe. From here, it crossed the Atlantic to the Americas. The complacency of the West was quickly replaced with fear.

> The far distant storm, which startled our countrymen on the banks of the Ganges fifteen years ago, and has since ravaged, with devious but too fatal course, every territory from the Straits of Malacca to the Pas de Calais, has at length burst on our shores! In Asia, the fiend was contemplated by us with curiosity—in the wilds of Russia, with suspicion—in Germany, with alarm—but on English soil, with terror! (Anon, 1832, 163)

European states imposed strict quarantines; a model used since the plague. These measures were intended to keep goods and persons from purportedly less salubrious regions from entering uninfected cities (Chase-Levenson, 2016; Salas-Vives & Pujadas-Mora, 2018). Absent medical or scientific understanding of the cause of cholera, these efforts failed to halt the progress of the disease.

Chaos followed in the wake of cholera. Local officials scrambled to manage outbreaks on an ad hoc basis. In many cities, decisions were guided by two imperatives: protect trade and keep the population from panicking. Public health boards—cobbled together to deal with the emergency—often suppressed the news of the disease's arrival to achieve these goals. Local officials denied that an outbreak was the dreaded Asiatic cholera, insisting instead it was a more benign "local cholera." Authorities released deliberately low numbers to keep public panic under control. Newspapers printed false information, contributing to the confusion. And, even as they denied the extent of an outbreak, city officials imposed armed *cordons sanitaires* to isolate diseased neighborhoods.

Conspiracy theories and misinformation abounded. Many working-class populations—hit hardest by cholera—saw outbreaks as a plot to repress the poor. Popular riots broke out in St. Petersburgh in June of 1831. Crowds targeted sanitary inspectors and doctors who they accused of spreading cholera by poisoning the wells of the proletariat (Barabanova & Kraikovski, 2018; Cohn, 2017). In Madrid, rumors circulated that members of the regular orders—wealthy clergy aligned with reactionary political factions—were poisoning the well-water of working-class neighborhoods. On July 17th, 1834, mobs of Madrileños attacked the monasteries; by the end of the day some one hundred monks were dead (Pérez Ledesma, 2001; Sanabria, 2009). In cities around the world, popular unrest simmered in the wake of cholera outbreaks. In Havana, cholera became integrated into growing concerns about race, class, and power in the existing colonial system.

Representing the Commercial City

Havana, in the early 1800s, was already a large and prosperous port, even if it had yet to reach the heights it would achieve later in the century. Late-eighteenth-century officials had begun reforming the old colonial city to reflect its growing wealth and prestige. In 1818, Richard Henry Bonnycastle described the city's elegant stone houses, the great square, and the magnificently ornamented convents and churches. He remarked that the inhabitants of Havana were "more polished than in most of the American cities" and had founded numerous "societies for the encouragement of the arts and sciences (Bonnycastle, 1819, 189)." In his opinion, Cuba was "the most important of the West India islands" with Havana the "key" of Spain's American holdings (Bonnycastle, 1819, 189).

This prosperity would grow rapidly as the century progressed, fueled by a thriving sugar economy. After the revolution in Haiti led to the founding of the first Black republic in 1804, racist, slaveholding nations refused to trade with the new state. Trade embargos, French reparations, and diplomatic isolation—imposed on the fledgling nation—contributed to the collapse of Haiti's previously dominant sugar industry. Cuban planters stepped in to fill this lucrative position, and the island's economy took off.[2] John Howison, upon entering Havana harbor in 1825, encountered "a forest of masts, surmounted with the flags of all nations, and vessels of every description, from the ship of war to the coasting-sloop (1825, 106)." The port was full of raucous energy. Vessels lined the wharfs "so close together that a small boat [could] hardly find room to make a landing (108)." A motley array of shipmasters, seamen, Spanish merchants, and black laborers jostled for position on the quay. A new arrival seeking to escape this cacophony must wend past "the piles of boxes, bales, and casks that everywhere obstruct the way (108)." The city beyond was only slightly less chaotic.

The streets of Havana bustled with activity. Crowds of carriages and pedestrians moved about at all hours of the day and night. The city's elite flitted about in *volantes* and *quitrines*—distinctly Cuban carriages that were as much about status as they were about transportation. In the afternoons, the wide Alameda would be packed with carriages as the city's wealthy gathered to promenade. In the evenings they attended dances, the theater, and the opera (Niell, 2013; Pérez Jr., 2017). This upper class of Havana was divided between peninsular and creole elites. The peninsular elite were Spaniards who worked in the colonial government to enforce Spanish rule and authority on the island. The Creole elite were descendants of Europeans who had stayed and prospered in Cuba. The Creole-owned plantations were the economic engines of the island in the nineteenth century (McGillivray, 2009; Paquette, 1988). Resentful of trade and taxation policies that were designed to benefit Spanish coffers, the Creole plantocracy engaged in continuous maneuvering in an attempt to gain leverage and influence over commercial policy (Pérez Jr., 2003). For all their differences, they were united in their determination to keep the sugar money flowing.

The prosperity of Cuba's sugar industry was reliant on the exploitation of enslaved labor. Even as the carriages of the wealthy passed through the city streets, the enslaved persons who made that wealth possible moved about on foot. The rich western provinces, where Havana is located, were the principal importers of enslaved labor. Still, only two in ten of the enslaved lived in the city; the vast majority were sent to toil on the plantations (Simons, 1996). And, even as the revolution in Haiti created Cuba's booming sugar industry, it left the white population of Cuba with a fear of enslaved uprisings (Reid-Vazquez, 2011; Simons, 1996).

---

[2] For discussions of the impact of the Haitian Revolution on the Atlantic World, see: Laurent Dubois, *Avengers of the New World: the Story of the Haitian Revolution*; Ada Ferrer, *Freedom's Mirror: Cuba and Haiti in the Age of Revolution*; Julia Gaffield, *Haitian Connections in the Atlantic World: Recognition after Revolution*; Karen Salt, *Unfinished Revolution: Haiti, Black Sovereignty and Power in the Nineteenth-Century Atlantic World.*

Between the white elite and the enslaved Africans existed large middling- and lower-class populations. Free people of color made up a large portion of this population. Progressive manumission policies and immigration meant a growing population of free people of color in Havana (Knight, 1970; Paquette, 1988; Simons, 1996). While legally barred from many professions, this population worked as artisans, skilled tradesmen, and small business owners. The free population of color—with expanding economic and social networks in the city—was seen as a threat by both Spanish and Creole elites (Castellanos & Castellanos, 1988; Knight, 1970). Many white Cubans considered the proximity of the enslaved and free population of color as a potential site of racial insurrection. A second, no less potent, fear was the potential for racial mixing. Authorities attempted to restrict the expansion of the free population of color while encouraging the growth of the lower and middling class white population (Childs, 2006; Reid-Vazquez, 2011). This vast middling population—riven with class and racial divides—filled the city's markets and streets with the steady hum of industry and commerce.

Representing the Miasmatic City

Havana's economic growth was accompanied by an expanding population, and the city's inhabitants overflowed the walls to establish thriving *extramuros* neighborhoods. The population grew from 76,000 in 1774 to 160,000 in 1841. New inhabitants meant new neighborhoods, *extramuros* barrios like Regla, Horcón, and Cerro. These were composed largely of the poor and people of color. The 1841 census was the first in which census takers counted these areas as a continuation of the capital. When included, these residents swelled the population of the newly coined *Gran Habana* to over 184,000. The city was a tapestry of extravagant wealth and abject poverty. Three-quarters of Havana's residents lived outside the old walled city (Garcia, 2016; Pérez Jr., 2001). Representations of Havana captured its dual nature. It was the font of wealth and civilization from which Cuba would move forward into prosperity. It was also a dangerous environment which threatened the very life of its inhabitants.

Havana's infrastructure was ill prepared to house such a large population. It was, at its heart, an old colonial city. The always-crowded streets were narrow, for the most part unpaved, and encrusted with a film of filth and decay. Even affluent neighborhoods were ripe with the detritus of human life and industry; poorer and *extramuros* areas were in worse condition. The medical community saw the city's ever-expanding population as a factor exacerbating an already dangerous miasmatic environment.

Popular and medical opinion held that something in Havana's environs promoted epidemic disease. Howison, who described the city's chaotic port, tells a more somber tale of the peril new arrivals faced in Havana's "pestilential atmosphere (1825, 154)."

> The city is indeed filled and surrounded with sources of disease. The streets are badly aired and odiously dirty; the water is obnoxious to the eye and to the taste, and the harbor forms a receptacle for the innumerable impurities which are daily thrown from four or five hundred vessels of all descriptions and sizes. The miasmata arising from such a quantity

> of putrescent materials, conjoined with the scorching heat of the sun, soon operate upon a European constitution, and produce the most fatal consequences (Howison, 1825, 128)

It was not just outsiders who considered the city's atmosphere unsafe. Dr. Tomas Romay, a leader of public health reform in the city, outlined the peril in the scientific language of miasmatic theory. He, and medical theorists more broadly, considered air to be composed of different gases. Pyrogenic gases supported life and were consumed by the respiration of living organisms. The proximity of large numbers of people rapidly consumed these life-giving pyrogenic gases. Mephitic gases, on the other hand, were toxic and deadly. These were emitted by the decay of biological matter—sewage, refuse, and rotting bodies (Arlidge, 1892; Bichat & François, 1827; Corbin, 1986). To Romay, Havana's climate, urban design, and overcrowding made the very air dangerous, if not deadly (Romay y Chacón, 1806).

Romay considered the unique geography of the city to be a factor that increased the threat to the population. In Havana "the heat of the climate, the lack of ventilation, and the vapors of many gathered bodies" accelerated the air's degradation (Romay y Chacón, 1806, 19–20). Further, narrow streets, the elevation of *la Cabaña* (a large fort guarding the harbor), and the surrounding mountains precluded wind from dissipating the corruption. These conditions—a hot climate, stifling buildings, and the proximity and respiration of so many people—rapidly exhausted the pyrogenic gases in Havana. Romay was certain that the growing population, crowded together in the city, was leaching vitality out of the air.

The city's medical profession saw environmental factors as making Havana a space especially vulnerable to epidemic disease. Even as the conditions of Havana depleted the life-giving gases, they produced dangerous mephitic gas. The city was ringed with swamps and stagnant waters. The corpses of animals, along with other filth "unworthy of naming," were thrown into the streets and squares. Pigsties, candle factories, and meat warehouses generated waste, while forges and factories increased the heat and humidity. These, combined with the excrement of the population, exuded toxic gases (Romay y Chacón, 1806, 25). When cholera was first diagnosed in the city's poor *extramuros* neighborhoods, it fueled the idea that these "unhygienic" spaces were dangerous to the city as a whole.

Morality, Predisposition, and Medicine

The social and economic upheaval wrought by the surging sugar economy led to a divided intellectual community in Havana. The foremost intellectual collective in Cuba—the *Real Sociedad Patriótica de la Habana*—was formed by conservative elements of Havana society in 1792 to promote agriculture, commerce, education, and industry. The membership was split when reformists challenged the influence of conservative colleagues. The polarization of the intellectual community in the 1830s marked the beginning of a period of intense creole cultural production in the fields of education, the social sciences, natural sciences, music, and literature (Benítez-Rojo, 1986; Pía Bruno, 2016).[3] Reform-minded thinkers experimented with medical, intellectual, and literary forms as they articulated their evolving vision of a future Cuba.

Contagionist and anti-contagionist theories of disease were an active part of the intellectual debate around cholera in Havana. There was a significant overlap and fluidity in various author's use of terms and concepts; these ideas were evolving and therefore open to interpretation. Havana's preeminent expert on cholera, Dr. Francisco Calcagno, synthesized government reports, books, medical treatises, and the records of physicians from across the Atlantic World in his studies on the origins of cholera and the threat it poised to the island. Based on his analysis, Calcagno asserted that:

> Too evident and well-proved its (cholera's) property was eminently contagious. It spread both against the wind and downwind. It followed the course of navigable waters and that of great roads: it always manifested itself in a country immediately after the arrival of people from a country where the epidemic reigned. (Calcagno, 1832, 8)

While Calcagno described cholera as contagious, he was convinced that the disease spread with people, not through them.

Calcagno posited it was goods, not people, that were the primary vector when it came to the spread of epidemics. This aligns with nineteenth-century, pre-germ theory beliefs about infection, which held that disease-causing miasmas could infect susceptible objects. These objects could then carry the infection across long distances (Barnes, country where the epidem).[4] Calcagno described how contagious diseases moved "by means of imperceptible emanations or miasmas (1832, 16)." While it was possible an infected person could pass them directly to a healthy person, this was rare and happened "at a very short distance, about a few inches from the patient (1832, 16)." The danger, Calcagno explained, came from infected objects.

> …contagious miasmas are fixed on the blankets, the canvases, the clothes, the furniture and other effects touched by the patient; and these effects, thus infected and transported far away, communicate the disease through their contact, which is even more dangerous than that of the patient himself. (Calcagno, 1832, 16)

According to the model of infection, contaminated objects were the primary medium through which epidemic disease was brought into new, uninfected environments.

Calcagno argued that the situation was even more dire. This infectious potential could lay dormant for long periods, emerging when the object entered a suitable environment.

---

[3] This creole intellectual flourishing marked the emergence of nascent visions of *cubanidad*. These ideas would grow in influence, contributing to nationalist movements mid-century.

[4] This was a "loosely articulated but firmly held conviction that foul or contaminated air could be imported from overseas in vessels and goods, and under certain conditions could spark deadly disease outbreaks." Some goods were understood to be extremely vulnerable to infection: those subject to decay, anything permeable, and items of animal origin (Barnes, 2014, 76).

> These miasmas are not fixed indistinctly on all bodies: they do not stick to smooth bodies, such as metals, glass, & c. and they have a particular affinity, or are united in preference to porous bodies, such as hair, wool, silk, cotton, thread, and in general all fabrics, feathers, paper, leather, sponges, & c. These substances, packed or stored in dark places, can preserve the miasmas in all their efficacy for months and years, and transported by means of trade, they can spread the disease at any distance. (Calcagno, 1832, 17)

Calcagno maintained that this external threat could only be prevented by strict and inviolable quarantine. Violators of the quarantine protocols he outlined should face capital punishment, for if cholera spread to Cuba it would be catastrophic. Calcagno reiterated that hot and humid climates like Cuba "favor and increase" the spread of epidemic disease. As a result, "if cholera-morbus unfortunately comes to Havana, it will wreak havoc even more fierce than in Europe (Calcagno, 1832, 3)."

Havana faced both internal and external risks. The first came from the threat of infected cargo entering the busy port and introducing cholera. The second was an environment predisposed to spread the disease. There was, however, an additional internal threat. Extending the rationale that the nature of certain goods and environments predisposed them to infection, the medical community of Havana considered certain bodies more likely to contract cholera. The link between morality and disease was imported to Havana from the prevailing medical opinion in France, which held that physical and moral discipline were the foundation of health. According to liberal thinkers, the educated and the enlightened understood the need to control their baser natures and had the self-discipline to follow a moderate diet and avoid excess in behavior (López Denis, 2003; Selene Zander, 2015). In Havana, the Public Health board noted that "disease was less frequent among well-to-do individuals who practiced a stricter code of cleanliness in their persons…" (Minutes of Superior Board of Health, February 1, 1832, quoted in Ramos, 1994, 183). Mirroring the pattern observed by Foucault in eighteenth-century France, the white creole community pathologized the so-called deviant segment of the population of Havana in contrast to an imagined "normal" code of conduct.

Because cholera outbreaks clustered within the poverty-stricken sectors of cities, this moral deviance was associated with those surviving on the margins. In Paris, liberal elites targeted peasants immigrating from the countryside. In Prussia, it was the proletariat class. In New York and London, the cities' working classes were the focus of liberal outrage (Althammer, 2012; Hamlin, 2009; Kudlick, 1996; Rosenburg, 1987). Calcagno captured this in Havana when he informed his readership that cholera attacked "[t]he lowest class of society, the poorest class, the most ignorant, the most poorly fed and poorly sheltered (Calcagno, 1833, 35)." The connection between poverty and contagion resonated with deepening liberal notions that there coexisted both the "worthy poor" and an unredeemable class of persons at the bottom of the social body. In the second case, it was more than just their poverty which made them susceptible to cholera: It was their behavior. Conflating physical and moral weakness, creole intellectuals believed disease would strike "those careless individuals given to excesses in food and drink, those who have strayed from virtue's path and have allowed vices to degrade their morals, those who are tormented by remorse and dishonest passions (Calcagno, 1833, 55–56; 66)." In Havana, where deep-seated fears about racial insurrection were fueling racism in the white community, this idea of physical and moral deviance was mapped onto black bodies.

For the creole intellectual class of Havana, poverty and blackness were increasingly interchangeable. When cholera struck Havana, it initially flourished in the poverty-riddled areas outside the walls that were home to many people of color. This was taken as evidence that certain bodies were predisposed to the disease. Dr. Diego Manuel Govantes, in his observations of the *extramuros* neighborhood of Jesús María, attributed the areas rampaging outbreak to the high number of "very poor people and blacks" who lived there (Govantes, 1833, 8–9). Dr. Don Ramón de Coloma y Garcés acknowledged that the "miserable slavery in which blacks live" contributed to the high rate of cholera among the population of color. Still, while living conditions were a factor, it was the race's essential "nature" to which he ascribed their vulnerability: "indolent and incapable of observing the slightest hygienic principle (Coloma y Garcés, 1834, 10–11)." José Antonio Saco echoed this belief when he asserted that "in the constitution of the African race there seems to be a predisposing principle for cholera (1833, 21)." This inherent weakness, he explained, was exacerbated by "the lack resources and immorality in which a great number of them live (1833, 21)." This moment in intellectual production, which linked African bodies with a predisposition to cholera, contributed to discourses that would evolve into scientific racism later in the century (Sartorius et al., 2014).

In light of the perceived internal vulnerabilities in Havana, there was initially support for strict measures to prevent the importation of this disease. A public health board was established to manage the city's response. In August of 1831, the board implemented a stringent quarantine on the port. Despite the danger, by 1833 the wealthy and powerful Count of Villanueva and a faction of the sugar elite successfully lobbied public health officials to loosen the quarantine. Cholera entered Havana shortly after this decision (Aguilera Manzano, 2007; Beldarraín & Espinosa, 2014). The city, and later the island, was devastated by the disease. In response, both camps in Havana's divided intellectual community appropriated the malleable concepts of contagion, contamination, and infection to bolster competing ideologies.

Medicine, Politics, and Racial Insecurities

Cuba's creole intellectual movement was led by notable thinkers José Cipriano de la Luz y Caballero, Domingo del Monte, and José Antonio Saco. By the 1830s, Saco began to self-consciously identify as a historian, philosopher, and social critic. Accused of spreading subversive ideas, he countered that this collective pursued the "considerations of a very high lineage" as they followed "the noble mission of writers (Saco, 1858, 85–86)." He and his compatriots determined to advance their reform project through intellectual production rather than in social, economic, or political action.

The group found a platform when its members took over the *Revista Bimestre Cubana*, a publication of the *Sociedad Patriótica*. It was in the *Revista* that Antonio Saco published his analysis of Robert Walsh's "Notices of Brazil in 1828 and 1829," a text concerned with the future of slavery in Brazil. In his analysis, Saco articulated an anti-slavery position that he would develop over time. Even in this early iteration, the dissonance at the core of the reform agenda was apparent. While the group advocated for the modernization and democratization of the sugar economy, they simultaneously held deeply racist beliefs. The system of chattel slavery—especially since the sugar boom—had sharply increased the population of color on the island. The reformers' solution to this problem was *blanqueamiento*; the whitening of the population. Replacing the labor of enslaved Africans with white colonists would stop this demographic trend (Benítez-Rojo, 1986; Gomariz, 2014; Opatrny, 1994; Raminelli, 2021). Then, over time, the procreation of white men with women of color would lead to a gradual whitening of the population. In the words of Saco:

> If mestizos are born from the union of a white woman and a black man, this would be regrettable. This would diminish our white population and weaken it in every conceivable aspect; but since the opposite happens, far from considering it a danger, I consider it positive. (Saco, 1847, 39–40)

These men envisioned a Cuban whiteness which would assimilate what they considered a problematic blackness. However, articulating anti-slavery sentiment was not tolerated by the colonial state and the creole elite who sought to maintain the system which generated their obscene wealth.

A series of contentious debates between José Antonio Saco and Ramón de la Sagra were published in the *Revista* over a period of years. Both prominent liberal thinkers and members of the *Sociedad Patriótica*, Saco and Sagra had oppositional visions for the future of the island. Born in Galicia, Sagra was a proponent of centrist liberal policies. He believed Cuba should remain a colony of Spain and subservient to the metropolis. Saco, a native of the Cuban city of Bayamon, was part of the creole reform movement. He envisioned a future where Cuba was a province of Spain—equal in status to peninsular provinces and with an independent government. The two men frequently clashed over their ideological differences; by the 1830s they legitimized their political views through a debate over the origin and nature of cholera (Aguilera Manzano, 2009; López Denis, 2000). Their debate over the origins of cholera was a very public example of the weaponization of intellectual production in the city.

Sagra, a botanist with an international reputation, asserted that epidemic disease was caused by corrupted environments. He bolstered his political position with observations of the spread of cholera. He noted that some cities and towns remained free of outbreaks despite close contact with infected populations. In other cases, areas with strong quarantines still experienced outbreaks. He argued that if cholera were spread through contact, this would not be the case. Instead, he held that environmental conditions caused epidemics. Cuba, like the rest of the Antilles, was hot and humid: the ideal environment for the propagation of disease. The island's vulnerability to cholera stemmed from the "natural inferiority" of Cuba's climate, geography, and, by correlation, inhabitants. To Sagra, a strong colonial system was necessary to guide and manage the "inferior" Cubans (Aguilera Manzano, 2009; López Denis, 2000).

Saco disputed these claims, arguing that disease did not arise from the environment but was brought into it by contaminated objects. In building this argument, he considered the many different theories of cholera's origins. These included electricity, poisonous insects, emanations from the earth, the air and its temperature, humidity, winds, and chemistry, and also the influence of heavenly bodies (Saco, 1833). However, it was the question of the contagious nature of the disease which most interested him. Drawing on medical literature, he contended that contagion involved "contact, or close or immediate touching of a morbid product that spreads certain diseases (1833, 42)." The morbid product could be either miasmas or viruses (a substance whose nature was still a mystery to medicine). Different diseases had different manners of transmission: Direct contact (syphilis); inoculation (smallpox); and through the air and other substances (scarlet fever). After reviewing all the theories, Saco opined that cholera was contagious, "a disease that is transmitted, that is communicated from the infested to the non-infested, whatever the way this transmission or communication may be (1833, 42)." While the mechanism of its transmission remained unclear, Saco argued that protecting Cuba involved quarantine to keep the "morbid product" out. When the Spanish colonial officials loosened quarantine protocols—a decision that ultimately allowed cholera to arrive on the island—Saco pointed to this as a failure of Spanish colonial authority. The Spanish state, in league with the merchants and plantation owners who profited from the sugar industry, sacrificed the health and well-being of Cubans in order to protect peninsular economic interests. Saco and Sagra used the nascent language of biomedicine to debate opposing political views. Since censorship limited critical political writing in this period, reformers advocating for change began to use cholera as an allegory for the corruption and incapacity of the colonial social order.

Racial Insecurity, Narrative Fiction, and The Boundaries of Community

A subset of reformers began experimenting with literature as a genre to express their political ideals. Domingo del Monte, a contemporary of Saco and another leader in the reform intellectual movement, left the restrictive environment of Cuba 1827. He toured Europe and the United States, where he connected with liberal thinkers who were using literature as a space for political critique. Upon his return to Cuba, he formed a *tertulia* (a literary salon) which would promote the development of a Cuban literary community. His tertulia connected progressive thinkers from Havana, from Cuba more broadly, and from abroad. Its members circulated, read, critiqued each other's work alongside an array of international thinkers (Aguilera Manzano, 2009). As members of the larger reform intellectual community, they were similarly concerned with the policies of the colonial state and Cuba's racial composition. However, the censorship of the era meant they could not publish political works with direct condemnations of the state (Benítez-Rojo, 1986; García-Barrio, 1977; Opatrny, 1994; Sartorius, 2013).

Del Monte's tertulia was the incubator of Cuba's first prose fiction. Just as Saco had self-consciously claimed his authority as a scholar and writer, its members identified as literary critics, poets, and, increasingly, as authors of narrative prose. In an 1832 essay, Del Monte wrote that to be a novelist was a noble calling, requiring the "qualities of poet, philosopher and antiquarian (Del Monte, 1929, 217)." The Cuban creators endeavored to "vulgarize the most notable periods of national history in the form of novels" by offering detailed accounts of local places, people, and events (Del Monte, 1929, 216–217). Ramón de Palma —a member of Del Monte's tertulia and one of the first to publish narrative prose—reinforced this perspective when he pronounced that novelists were "the flower and cream of human ingenuity (Palma, 1838, reprinted in 1964, 106)." According to Palma, novels were on par with the works of science and history (1964, 105–6). Writing at a liminal moment between the era of romanticism and realism, these author's mimetic representations of Cuban life blended the tradition of *costumbrismo* with the romantic novel.[5] Privileging the need to convey a truthful and objective vision of contemporary life, this group simultaneously sought to show the corruption in the political and social systems of colonial Cuba. Dedicated to the belief that creating literature should be "a mission of morality and improvement," this collective produced the foundational texts of the Cuban literary canon (Del Monte, 1929, 243).[6]

Members of the tertulia employed a series of metaphors to explore their fears about racial and class miscegenation, the corruption of the upper classes who benefited from the colonial order, and their imagined future of a disciplined, rational society run by white creole men. The authors drew on the characterization of cholera in contemporary fields of medicine and the social sciences, which had established a number of "truths" about bodies, disease, and predisposition. First, illness was caused by an infection—contagionists and anti-contagionists were united in this belief, even if they debated the mechanism of how it spread. Second, that some bodies were predisposed to infection. In Havana, these were the bodies of the poor and people of color. A correlated belief was that infection could lie dormant: a seemingly healthy person or object could unknowingly carry the seeds of future disaster. And, finally, that predisposition was not, in and of itself, enough to cause disease. Disease manifested in conjunction with external factors, such as environmental conditions and individual behavior. This model, which emerged to explain cholera, was extended by these authors as a metaphor for the Cuban social body. The bodies of almost-white women were the potential carriers of corruption. The disordered and chaotic

---

[5] *Costumbrismo* was an artistic movement (both in literature and visual arts) that depicted the everyday manners and customs unique to a city or region. The form featured detailed representations of regional social behavior and people. *Costumbrista* stories were frequently satirical and critical in nature.

[6] In Havana, members of the tertulia included Jose Zacarias Gonzalez del Valle, Ramon de Palma, Jose Jacinto Milanes, Gabriel de la ConceptionValdes, Juan Francisco Manzano, Cirilo Villaverde; Anselmo Suarez Romero, Bias Oses, Jose de la Luz, and Jose Luis Alfonso.

urban landscape symbolized the policy failures of the colonial state. And the wealthy characters peopling this landscape embodied the dissolute creole elite.

The bodies of almost-white women became a trope through which the authors' fears and desires for the Cuban nation were played out.[7] This figure carried both the potential for Cuban greatness and the risk of degeneration. We see this interplay in Ramón de Palma's novella *El cólera en la Habana* and Villaverde's novel *Cecilia Valdés,* both of which center on two beautiful creole women. *El cólera en la Habana* detailed Angélica's life in the city during the outbreak.[8] *Cecilia Valdés* was set in Havana immediately before the cholera outbreak.[9] Both authors depicted women who were, superficially, representative of an idealized creole standard of beauty. Still, their whiteness was destabilized by markers denoting a rift between the surface appearance and what lies beneath.

Palma and Villaverde described these young women as paragons of Creole feminine beauty. Both are portrayed as having dark hair, fair skin, bright eyes, and carrying themselves with sensuality and a lightness of movement. Palma's Angelica, with her curly hair "blacker than jet," a "delicate complexion," and striking eyes as "dark as the veil of night" and shining with the "lights of life" physically matched Villaverde's Cecilia, with her "high forehead…crowned with abundant naturally wavy black hair," "mother of pearl" skin, and "black almond eyes…all animation and fire (Palma, 1974, 49–50; Villaverde, 2005, 12; 227)." Palma's narrator had long "dreamed of an ideal beauty;" in Angélica he encountered this ideal (1974, 49–50). Villaverde's narrator was equally inspired, stating that Cecilia's beauty was reminiscent of "the virgins of the most renowned painters (2005, 12)." Yet even as the authors catalogued their characters' beauty, they disrupted this aesthetic perfection by insinuating markers that point to potential corruption below the surface.

Angelica was presented as an outsider in Havana society, having only recently arrived in the city. The other characters in the novel, while curious, knew nothing of her background or genealogy. The mystery of her origins deepened when the narrator stated that "she was not white like the tepid daughters of the North, because the sun of the tropics had shed some slight tints on that smooth and delicate complexion (Palma, 1974, 50)." Was this darkness a sign of a mixed bloodline? Palma offered no further clarification. He left it ambiguous, undermining the appearance of whiteness

[7] For a discussion of the racialization of the feminine in Cuban cultural production, see: Fraunhar, *Mulata Nation: Visualizing Race and Gender in Cuba*; Guevara, "Inexacting Whiteness: Blanqueamiento as a Gender-Specific Trope in the Nineteenth Century"; Pilar Egüez Guevara, "Dangerous Encounters, Ambiguous Frontiers: Dance, Sex, and Intimacy in Nineteenth-Century Cuba."

[8] This was the second novel published by Palma in 1838. In the first, *Una pascua en San Marcos*, Palma was openly critical of the corruption and frivolity of Cuba's upper class. Its publication was met with censure and Palma faced arrest. His subsequent works took a subtler approach, masking his critical observations within a sentimental romantic aesthetic (Benítez-Rojo, 1986).

[9] Short treatments of Cecilia Valdés were published in Havana 1839. However, the canonical novelization of Cecilia Valdés was published by the exiled Villaverde in New York in 1886. While cholera was only mentioned in the text, Villaverde—free from Cuban censors and radicalized by his exile—offered a more direct criticism of colonialism and slavery in his later version (Genova, 2016).

by alluding to an impure interior. Cecilia, meanwhile, had been born with "skin as white as coconut meat (Villaverde, 2005, 387)." Despite this, throughout the novel characters referred to her as the "bronze virgin," indicating her skin, like Angelica's, was a warmer shade than "pure" whiteness. This was reinforced when the narrator stated that "a knowing eye could not help noticing that her red lips had a dark border or edging, and that the bright glow of her face ended in a sort of half-shadow near her hairline. Her blood was not pure and it could be stated with assurance that three or four generations back it had been mixed with Ethiopian blood (Villaverde, 2005, 13)."

The similarities in these young women's appearance were not happenstance. Members of the tertulia—intimately familiar with each other's writings—deployed a shared reference of images with recognizable connotations. By using similar if not near-identical descriptions of Angelica and Cecilia's physical appearances, Palma and Villaverde point to a common background. Both could be interpreted as white-passing women with mixed blood somewhere in their ancestry.[10] Still, the characters' shared physical characteristics were the extent of their commonalities. In Palma's novel, Angélica was legally and socially white: an heiress from the countryside whose story ends in marriage. Villaverde left no doubt concerning Cecilia's racial heritage: she was a *mulata* (a woman of mixed race). Her story ended in tragedy, for herself and everyone around her. If we return to the metaphor of infection, the similarity in the two women's physical appearance pointed to them both carrying the contamination of mixed blood. Their disparate outcomes were not the result of this inner flaw; rather, the predisposition they carried in their blood was activated by external triggers.

For all that they espoused the whitening of the Cuban population, these authors feared the impacts of miscegenation. Through trajectories of the almost-white Angélica and Cecilia the authors examined how this embodied weakness could manifest under the right stimuli. It was the women's interactions with the streets of Havana and its "deviant" inhabitants that triggered their own potential for corruption. In the novels, the colonial city—dirty and overpopulated—symbolized the failure of colonial rule. Both Palma and Villaverde considered that the unregulated mixing of people across racial and class barriers in Havana fostered aberrant, if not immoral, behavior. In the city, Angélica encountered harmful elements she had been isolated from in the countryside. She was often depicted at her window, fascinated by the goings-on in the street outside. After cholera struck, she and her mother were isolated in their house. Bored and irritated, she was overpoweringly drawn to the window. From her perch she encountered a body collector; a disreputable character who handled the infected bodies of dead cholera victims. "More of a beast than a human," the

[10] This invoking of the almost-white creole woman was familiar in the literature emerging from the tertulia. In José Z. González del Valle's *Carmen y Adela* (1838/39), Carmen was a physically beautiful creole girl whose delicate complexion was marred by purplish shading around her eyes. This subtle imperfection marked her as having an inner impurity which manifested in her emotional excesses. In Anselmo Suárez's *Carlota Valdés* (1838), the eponymous character was never able to overcome "the seal of misfortune imprinted on her forehead." She, like Cecilia, was stigmatized by her origin as a foundling left at the Casa Cuna.

narrator described how "his eyes were red from excessive drinking," and his expression "resembled the look of a hungry hyena when it gazes at a corpse beneath its claws (Palma, 1974, 69–71)." Thoughts of this figure exerted "diabolical influence" over Angélica's mind, inducing an unsettled state that led her to leave the safety of her isolation.

Palma suggested that the corrupted city infected Angelica's mind through her contact with deviancy. Spurred by her newly disordered thinking, the young woman cajoled her mother into a carriage ride through the streets. The choleric city Palma described was a strange and terrifying landscape. The abandoned city was illuminated by the melancholic half-light of twilight, which "gave things a fantastic and sepulchral aspect, which together with the silence that reigned, made the objects seem like visions (Palma, 1974, 75–76)." Occasionally, this eerie half-light was

> …illuminated by the red flame of a tar barrel of the kind that burned at the doors of hospitals, and if this sinister and funerary clarity was mixed with the moans and curses of the cholera victims, the noise of the cars, and the sight of the corpses and their drivers, then the spirit was impressed so that everything was represented as a scene from hell (76).

Horrified, the women fled back to their home. Too late, as corruption returned with them; Angélica's mother quickly succumbed to cholera. Cholera, Palma showed, stripped away the illusion of the prosperous colonial city to reveal the decay that lay underneath.

The colonial city, and, by extension, the colonial state, offered little protection to the vulnerable. Even Angélica, largely shielded from the urban squalor, fell victim to the influence of the city streets. What, then, happened to Cecilia? Cecilia, so similar in appearance she might have been Angelica's sister, had grown up in Havana with no one to curb her exploits. Her first appearance in Villaverde's novel was as a young child. The narrator described this young Cecilia as "so pure that a person was tempted to believe that she would never cease to be what she was, an innocent young maiden (Villaverde, 2005, 13)." This was not to be, as:

> The streets of the city, the squares, the public establishments, as was noted previously, were her school, and in such places, one may surmise, her tender heart, meant perhaps to harbor virtues, the loveliest of feminine charms, drank instead in torrents the poisoned waters of vice, and nourished itself from her earliest years on the lewd scenes staged each day by an indecent and depraved people. (Villaverde, 2005, 13–14)

Angélica, guarded from the hazard of the colonial city, experienced only a brief fall into disordered thinking and behavior. Cecilia, raised in this environment her entire life, was irredeemably damaged before she reached adolescence. Colonial rule had failed to impose the order required to protect those who lacked the discipline to control themselves.

To the reformers, the creole plantocracy who collaborated with the colonial state were equally complicit in the failure to protect the vulnerable. In the novels, the city was the playground for the debased elite who benefited from the existing political and social structure. Through the experience of these young women, the author's presented the venality and degeneration of this class as another threat to a vulnerable Cuba. Angélica was from a respectable, wealthy family—albeit an outsider not entrenched in Havana's social scene. While beautiful, she was portrayed as self-centered, vain, and only interested in her own entertainments. Her suitor, Jacinto de Leiva, was the scion of a prominent Havana family and seemed content to wile away his days at parties and social gatherings. Palma painted an unflattering portrait of the panic that ensued when the dissolute elite recognized their vulnerability to cholera. Some—like Jacinto's family—fled the city to the relative safety of the countryside (Palma, 1974, 73). Those left behind barricaded themselves inside, waiting in fear as the city suffered. In the face of the disease, the elite took care of themselves and left the city to die. Palma impressed upon his reader that this class, interested only in their profligate lifestyle, had abdicated their moral authority to lead.

Villaverde was even more scathing in his treatment of Havana's elite in *Cecilia Valdés*. Villaverde created a series of characters who represented the worst of this class. Cecilia's father was a Spanish merchant, a plantation owner, and a slave trader. He displayed few—if any—redeeming qualities.[11] His determination to spare himself shame for having fathered a mixed-race daughter outside of his marriage was at the heart of Cecilia's tumultuous life. His wife, the daughter of an aristocratic Havana family, spoilt her children while jealously policing her husband's behavior. Their son, Leonardo, was a wastrel, a womanizer, and a manipulator. The doomed love story of the novel was that of Cecilia and Leonardo—unknowingly siblings embarking on an illicit affair.[12] Fully aware of his children's budding romance, Cecilia's father was still unwilling to admit to his indiscretion. When he found himself unable to control his son's lust, Cecilia's father had her arrested for seducing Leonardo (Villaverde, 2005, 450–461). The wealthy characters in Villaverde's novel were a collection of unprincipled, immoral, self-entitled individuals. Cecilia, metaphorically the mixed-race Cuban nation, was betrayed by the class responsible for her care.

Angélica and Cecilia—both symbols of a mixed-race Cuban nation—were in peril in these novels. The colonial political order and the immoral social elite at the top of the system were incapable of guiding Cuba into a prosperous future. In each work, the discourses circulating around cholera, disease, and racialized bodies informed the author's account of the flawed social and political order. It also shaped the manner in which degeneration could be avoided. Medical opinion dictated that the key to preventing cholera was discipline—of both the mind and the body. In the novels, the young women lacked discipline. Their parents, peers, and society failed to impose discipline upon them. In *Cecilia Valdés*, the characters continued on an escalating path of destruction. Cecilia's father never repented. Cecilia herself became pregnant with Leonardo's illegitimate daughter. In a fit of jealousy when he was to marry a

---

[11] In one of the few direct mentions of cholera in the text, Cecilia's father noted that cholera was rampaging through Poland. He expressed his lack of concern, as "Señor Cholera will have to keep rolling right along to reach us, and by that time … who knows where you and I will be!" This correlates with Palmas' critique of the elite's selfish response to the threat of cholera (Villaverde, 2005, 56).

[12] The themes of miscegenation and incest frequently overlapped. The novella *Petrona y Rosalia* (1838), written by fellow tertulia member Felix Tanco y Bosmeniel, also featured intergenerational rape, the secrecy around an illegitimate mixed-race daughter, and the incestual sexual relations between siblings.

girl of his own class and race, she had him killed on his wedding day. There was no redemption in the colonial system.

However, in submitting to discipline Angélica changed her fate. Palma's novella ends with a series of marriages. Jacinto belatedly returned to Havana, shamed for having abandoned it in fear. Angélica, the almost-white upper-class creole married Jacinto, her white, upper-class creole equivalent. They turn away from their indolent youth to pursue a mature and respectable future (Palma, 1974, 99–100). Still, it was the marriages around them that are even more pointed. A doctor who had fallen in love with Angélica during the pandemic married a young woman who was "worthy of him, although not as beautiful as Angélica (Palma, 1974, 100)." Jacinto, noting his service and sacrifice during the outbreak, stated that "happy is the country lucky enough to birth many men like you, for the glory of your profession and the consolation of humanity (Palma, 1974, 98)." Even a criminal who helped Angélica during the epidemic—a Spanish deserter serving his sentence in Cuba—left the island to marry his Spanish love. Palma ended his novel with a tentative vision for Cuba that would be expounded on throughout the century. This was a disciplined society ruled by rational, liberal philosophies. Cubans married and bred with Cubans, keeping to their class. The *patria* would be born from disciplined creole wombs like Angélica's.

Conclusion

Cholera was a global catastrophe that played out in local spaces. In Havana, the cholera outbreak of 1833 was a foundational moment in creating discourses of that would expand over the nineteenth century into a Cuban nationalist movement. Doctors, intellectuals, and artists used the language of biomedicine to illustrate what they saw as the ineptitude of the ruling class, the danger of the colonial city, and the risk of social miscegenation. Cholera, which had devastated the population and disrupted social and economic life on the island, was used as a metaphor through which reformers defined a vision of reform that would include new boundaries of race and class. In the cultural production of the era—from Dr. Calcagno's treatises on cholera, to Saco's political arguments, to the narrative fiction of Palma and Villaverde, evolving understandings of infection, disease, and discipline—both physical, moral, and on a larger-scale social—became interwoven with reformers' hopes and insecurities about the future of Cuban society. While in 1838 these ideas were tentative and fluid, over time they were incorporated into the political changes taking place in the Cuban social and economic systems.

These ideas would also impact urban and social reforms. In the immediate aftermath of the outbreak, newly appointed governor Miguel Tacón began extensive modernization projects in Havana. Using the cholera outbreak as evidence of the dangerous nature of the unregulated colonial city, he began transforming the urban landscape into a hygienic, modern metropolis. Hygiene, linked through the outbreak with ideas of the deviant and worthy poor, became a touchstone in the introduction of modern social institutions. *Juntas de Sanidad*, or public health boards, took on the responsibility of regulating and policing public spaces to ensure orderly and hygienic behavior. The reform of the institutions of *beneficencia*, or charity, became the mechanism through which the worthy poor were transformed into productive members of liberal society. *Policía*, both the police and policies of good governance, defined acceptable behavior while punishing the deviant. Fueled by the racist discourses that emerged from the cholera outbreak, these institutions increasingly targeted populations of color as dangerous to the social body. Intellectual and literary works about cholera were the medium through which these ideas were introduced into the public consciousness. Once planted, they became the debates which defined the direction of Cuban society for the next one hundred years.

Bibliography

Aguilera Manzano, J. M. (2007). El uso político de la epidemia de Cólera Morbo en la Habana. *Canadian Journal of Latin American and Caribbean Studies, 32*(64), 181–208. https://doi.org/10.1080/08263663.2007.10816931

Aguilera Manzano, J. M. (2009). The informal communication network built by Domingo Del Monte from Havana between 1824 and 1845. *Caribbean Studies, 37*(1), 67–96.https://doi.org/10.1353/crb.0.0101

Althammer, B. (2012). Poverty and epidemics: Perceptions of the poor at times of cholera in Germany and Spain, 1830s–1860s. In Althammer, B., et al. (Eds.), *Poverty and Sickness in Modern Europe: Narratives of the Sick Poor, 1780–1938* (pp. 93–116). Bloomsbury Publishing.

Anon. (1832). Epidemic cholera. *The Medico-chirurgical Review, 16*(31), 163–180.

Arlidge, J. T. (1892). *The hygiene, diseases and mortality of occupations*. Percival.

Arnold, D. (1986). Cholera and colonialism in British India. *Past & Present, 113*(1), 118–511. https://doi.org/10.1093/past/113.1.118

Barabanova, K., & Kraikovski, A. (2018). The management of cholera epidemics and the Neva River in St. Petersburg in the nineteenth century. *Water History, 10*(2), 163–181. https://doi.org/10.1007/s12685-018-0219-4

Barnes, D. (2006). *The great stink of Paris and the nineteenth-Century struggle against filth and germs*. Johns Hopkins University Press.

Barnes, D. (2014). Cargo, 'infection', and the logic of quarantine in the Nineteenth Century. *Bulletin of the History of Medicine, 88*(1), 75–101. https://doi.org/10.1353/bhm.2014.0018

Beldarraín, E., & Espinosa, L. M. (2014). El cólera en La Habana en 1833: Su impacto demográfico. *Diálogos Revista Electrónica, 15*(1), 155–173. https://doi.org/10.15517/DRE.V15I1.8372

Benítez-Rojo, A. (1986). Power/sugar/literature: Toward a reinterpretation of Cubanness. *Cuban Studies, 16*, 9–31. http://www.jstor.org/stable/24485974

Bichat, X., & François, M. (1827). *Investigaciones fisiológicas sobre la vida y la muerte*. García.

Bonnycastle, R. H. (1819). *Spanish America: Or, a descriptive, historical, and geographical account of the dominions of Spain in the Western Hemisphere, continental & insular*. A. Small.

Boyle, S. A. (2015). Cholera, Colonialism, and Pilgrimage: Exploring global/local exchange in the Central Egyptian Delta, 1848–1907. *Journal of World History, 26*(3), 581–604. https://doi.org/10.1353/jwh.2015.0039

Calcagno, J. F. (1832). *Aviso sobre el cólera-morbus y modo de preservarse de su invasión*. D. Juan Roquero.

Calcagno, J. F. (1833). *Tratado completo del cólera-morbus pestilencial*. Gobierno y Capitanía General.

Castellanos, J., & Castellanos, I. (1988). *Cultura afrocubana: el negro en Cuba, 1492–1844*. Ediciones Universal.

Chase-Levenson, A. (2016). Early nineteenth-century Mediterranean quarantine as a European system. In A. Bashford (Ed.), *Quarantine: Local & Global Histories* (pp. 35–53). Palgrave Macmillan.

Childs, M. D. (2006). *The 1812 Aponte rebellion in Cuba and the struggle against Atlantic slavery*. University of North Carolina Press.

Cohn, S. K. (2017). Cholera revolts: A class struggle we may not like. *Social History, 42*(2), 162–180. https://doi.org/10.1080/03071022.2017.1290365

Coloma y Garcés, R. de. (1834). *Colera-morbo observado y tratado en las ciudades de la Habana y San Carlos de Matanzas en la isla de Cuba*. R. Howe.

Corbin, A. (1986). *The foul and the fragrant: Odor and the French social imagination*. Harvard UniversityPress.

Del Monte, D. (1929). Sobre la novela histórica. In *Escritos de Domingo del Monte, Tomo II*. Cultural, S.A.

Dubois, L. (2005). *Avengers of the new world: The story of the Haitian revolution*. Harvard University Press.

Echenberg, M. (2011). *Africa in the time of cholera: A history of pandemics from 1817 to the present*. Cambridge University Press.

Eggleston, W. G. (1892). Oriental pilgrimages and cholera. *The North American Review, 155*(428), 126–128. https://www.jstor.org/stable/25102418

Evans, R. (1987). *Death in Hamburg: Society and politics in the cholera years, 1830–1910*. Oxford University Press.

Ferrer, A. (2015). *Freedom's mirror: Cuba and Haiti in the age of revolution*. Cambridge University Press.

Fraunhar, A. (2018). *Mulata nation: Visualizing race and gender in Cuba*. University Press of Mississippi.

Gaffield, J. (2015). *Haitian connections in the Atlantic World: Recognition after revolution*. University of North Carolina Press.

Garcia, G. (2016). *Beyond the walled city: Colonial exclusion in Havana*. University of California Press.

García-Barrio, C. (1977). The abolitionist novel in nineteenth century Cuba. *CLA Journal, 21*(2), 224–237. https://www.jstor.org/stable/44329348

Genova, T. (2016). Foundational frustrations: Incest and incompletion in Cirilo Villaverde's Cecilia Valdés. *Decimonónica: revista de producción cultural hispánica decimonónica, 13*(1), 66–86.

Gomariz, J. (2014). Esclavitud, blanqueamiento y modernidad periférica en Cuba: Gaspar Betancourt Cisneros El Lugareño. *América Sin Nombre, 19*, 73–81. https://doi.org/10.14198/AMESN.2014.19.08

Govantes, D. M. (1833). *Exposición histórica de algunas observaciones sobre el cólera-morbo-espasmódico*. Gobierno y Capitanía General.

Guevara, G. R. (2005). Inexacting whiteness: Blanqueamiento as a gender-specific trope in the nineteenth century. *Cuban Studies, 36*, 105–128. https://doi.org/10.1353/cub.2005.0033

Guevara, P. E. (2016). Dangerous encounters, ambiguous frontiers: Dance, sex, and intimacy in nineteenth-century Cuba. *New West Indian Guide, 90*(3–4), 225–256. http://dx.doi.org.proxy.library.emory.edu/10.1163/22134360-09003001

Halliday, S. (1999). *The great stink of London: Sir Joseph Bazalgette and the cleansing of the Victorian metropolis*. Sutton Publishing.

Hamlin, C. (2009). *Cholera: The biography*. Oxford University Press.

Howison, J. (1825). *Foreign scenes and travelling recreations*. Oliver & Boyd.

Knight, F. W. (1970). *Slave society in Cuba during the nineteenth century*. University of Wisconsin Press.

Kudlick, C. J. (1996). *Cholera in post-revolutionary Paris: A cultural history*. University of California Press.

Lees, A., & Lees, L. (2007). *Cities and the making of modern Europe, 1750–1914*. Cambridge University Press.

López Denis, A. (2000). *Saco, Sagra y el cólera morbo*. Pinar del Rio. Ediciones Vitral.
López Denis, A. (2003). Higiene pública contra higiene privada: cólera, limpieza y poder en La Habana colonial. *Estudios Interdisciplinarios de América Latina y el Caribe, 14*(1), 11–33.
McGillivray, G. (2009). *Blazing cane: Sugar communities, class and state formation in Cuba, 1868–1959*. Duke University Press.
Niell, P. B. (2013). Rhetorics of place and empire in the fountain sculpture of 1830s Havana. *The Art Bulletin, 95*(3), 440–464. https://doi.org/10.1080/00043079.2013.10786083
Opatrny, J. (1994). José Antonio Saco's path toward the idea of 'Cubanidad.' In *Cuban Studies 24*, ed. Enrico Mario Santí (pp. 39–56). Pittsburgh University Press.
Palma, R. de. (1964). La novela. In *Cuba en la UNESCO: Homenaje a Cirilo Villaverde* (pp. 105–113). Biblioteca de la Sociedad Económica de Amigos del País.
Palma, R. de. (1974). El cólera en la Habana. In Alvarez Garcia, I. (Ed.), *Noveletas Cubanas*,. Editorial de arte y literatura.
Paquette, R. L. (1988). *Sugar is made with blood: The conspiracy of La Escalera and the conflict between empires over slavery in Cuba*. Wesleyan University Press.
Pérez Ledesma, M. (2001) Studies on anticlericalism in contemporary Spain. *International Review of Social History, 46*(2), 227–255.https://doi.org/10.1017/S0020859001000128
Pérez Jr, L. A. (2001). *Winds of change: Hurricanes & the transformation of nineteenth-century*. Cuba University of North Carolina Press.
Pérez Jr, L. A. (2003). *Cuba and the United States: Ties of singular intimacy*, 3rd edn. The University of Georgia Press.
Pérez Jr, L. A. (2017). *Intimations of modernity: Civil culture in nineteenth-century Cuba*. University of North Carolina Press.
Pía Bruno, M. (2016). La cubanidad de Domingo del Monte: Entre la insularidad y el hispanismo Americano. *Anales De Literatura Hispanoamericana, 45*, 75–98.https://doi.org/10.5209/ALHI.55115
Raminelli, R. (2021). Reformers of slavery: Brazil and Cuba c. 1790 and 1840. *Varia História, 37*(73), 119–54. https://doi.org/10.1590/0104-87752021000100005
Ramos, J. (1994). A citizen body: Cholera in Havana (1833). *Subaltern Studies in the Americas, 19*(46), 179–195. https://www.jstor.org/stable/41491512
Reid-Vazquez, M. (2011). *The year of the lash: Free people of color in Cuba and the nineteenth-century Atlantic World*. University of Georgia Press.
Romay y Chacón, Tomás. (1806). *Discurso sobre las sepulturas fuera de los pueblos*. Don Estevan JoséphBoloña.
Rosenberg, C. E. (1987). *The cholera years: The United States in 1832, 1849, and 1866*. Universityof Chicago Press.
Saco, J. A. (1833). *Carta sobre el cólera morbo-asiático escrita por el editor de la revista cubana a un amigo suyo residente en la isla de Cuba*. Gobierno por S. M..
Saco, J. A. (1847). *Carta de un cubano á un amigo suyo; en que se hacen algunas observaciones al informe fiscal sobre fomento de la poblacion blanca en la isla de Cuba, etc*. Impr. de J. Gomez.
Saco, J. A. (1858). *Colección de papeles científicos, históricos, políticos y de otros ramos sobre la isla de Cuba ya publicados, ya inéditos*. Tomo Segundo. D'Abusson y Kugelmann.
Salas-Vives, P., & Pujadas-Mora, J. -M. (2018). Cordons sanitaires and the rationalisation process in southern Europe (nineteenth-century Majorca). *Medical History, 62*(3), 314–332. https://doi.org/10.1017/mdh.2018.25
Salt, K. (2018). *Unfinished revolution: Haiti, Black sovereignty and power in the nineteenth-century Atlantic World*. Liverpool University Press.
Sanabria, E. (2009). *Republicanism and anticlerical nationalism in Spain*. Palgrave Macmillan.
Sartorius, D., et al. (2014). Colonial transfusions: Cuban bodies and Spanish loyalty in the nineteenth century. In K. A. Coles (Ed.), *The cultural politics of blood, 1500–1900* (pp. 229–250). Palgrave Macmillan.
Sartorius, D. (2013). *Ever faithful: Race, loyalty, and the ends of empire in Spanish Cuba*. Duke University Press.

Selene Zander, J. (2015). Contagious invasions: The 1833 cholera epidemic in Havana. *Revista De Estudios Hispánicos, 49*(1), 3–23. https://doi.org/10.1353/rvs.2015.0026

Severino Boloña, J. (1833). Al Te Deum. *Colección de poesías arreglada por un aficionado a las musas*. José Boloña.

Simons, G. (1996). *Cuba: From conquistador to Castro*. St. Martin's Press.

Villaverde, C. (2005). *Cecilia Valdés Or El angel hill*. (H. Lane, trans.) Oxford University Press (original work published 1882).

Zeheter, M. (2015). *Epidemics, empire, and environments: Cholera in Madras and Quebec City, 1818–1910*. University of Pittsburgh Press.

Part IV
The Spanish Flu and Its Afterlives

Chapter 7
Reading Toronto's Response to Spanish Influenza: *The Globe* and *Daily Star* Report on the 1918–20 Pandemic's Second Wave

Cheryl Thompson and Lucy Wowk

Abstract Through a content and discourse analysis of Canada's prominent English-language newspapers during the Spanish Influenza pandemic, this article explores how newspapers, as cultural texts, negotiated the pandemic. Specifically considering not only the ways in which issues of community and contagion were not singularly reported on, but also how the pandemic and those who fell victim to it were sensationalized through editorials. Where *The Globe*, founded in 1844 was, by 1900, required reading for the educated and business community in Toronto, the *Daily Star*, founded in 1900, was recognized by progressive businessmen as a labor newspaper, becoming a voice for the federal Liberal government. This article examines how each newspaper reported on the pandemic, and the overlapping social issues and business interests of the period (e.g., World War I, smallpox, eugenics, immigration, women's rights, and blackface theater) that compounded the virus. How did newspapers negotiate the virus? What does the newspaper coverage reveal in terms of complex intersections between pandemics and media culture, representation and myth, cultural memory and forgetting?

Introduction

In fall 1918, the first reports of an influenza outbreak in Toronto, Canada, the largest city in the province of Ontario, appeared in *The Globe*. On 24 September, the newspaper reported on a dispatch from a Niagara (region) military camp, located 130 *km* southwest of Toronto, that described 188 cases of "Spanish influenza in the Polish military base" (*The Globe* September 24, 1918). The newspaper claimed that there was no outbreak at the Canadian camp, but it did have reported cases of "la grippe."

C. Thompson (✉) · L. Wowk
Ryerson University, Toronto, Canada
e-mail: cheryl.thompson@ryerson.ca

L. Wowk
e-mail: lucy.wowk@ryerson.ca

S. Venkatesan et al. (eds.), *Pandemics and Epidemics in Cultural Representation*,
https://doi.org/10.1007/978-981-19-1296-2_7

The terms "grip" or "grippe" were frequently used as synonyms for influenza.[1] Though there were four suspected cases in the Canadian camp, *The Globe* reported it as the "common" influenza, not a potentially more contagious and deadly communicable disease. In his examination of what became known as "Spanish influenza" in Canada, Mark Osborne Humphries found that as the disease circled the globe it hit in three waves: "the first in the spring of 1918, the second in the autumn, and the third in the winter of 1918–19, extending in some places into 1920. The first wave is said to have caused few deaths but much sickness; it would probably have gone unnoticed in history were it not for the second and more deadly autumn wave between August and December 1918 that caused the majority of deaths" (Humphries, 2013).While *The Globe* was first to report on influenza, by October, the city's other popular newspaper, the *Daily Star*, also reported on the considerable rates of sickness in young people.

The first reports of influenza in the *Daily Star*, appearing on 4 October, included the death of a twenty-year-old cadet named Geoffrey Scott, who died within a week of having contracted influenza (*Daily Star* October 4, 1918). On the same day, the first reports counting the deaths from influenza across hospitals appeared, with the most cases (540) occurring at a military base hospital. The *Daily Star* also named the illness when it reported, "Last Friday little ten-year-old Edna Willis [...] was taken ill and died within a few days from what is believed to be the Spanish influenza" (*Daily Star* October 4, 1918). Where the *Daily Star*'s article personalized the disease, *The Globe*'s early reports took a broader perspective. Contrary to *The Globe*'s early reporting on the second wave, which emphasized the location and source of the influenza, the *Daily Star*'s early reporting on influenza cases in Toronto during the autumn of 1918 started with death counts from hospitals and details of individual afflictions. The newspaper focused on specific stories about families who had become sick, the neighborhoods in the city with outbreaks, and the military soldiers who died because of influenza. For example, the neighborhood of Parkdale, about six kilometers west of downtown, was one area of focus because it was "a lively activity centre for young people" (Whitzman & Slater, 2006). St. John's Ward (also known as "the Ward") was also flagged by the city for public health concerns long before the pandemic hit. In 1911, for example, a report by Toronto medical officer of health Dr. Charles Hastings stated that the neighborhood had an immigrant population of over 11,000, which increased to 17,000 by 1918 (Lorinc, 2015). As a result, it was a tightly packed community with sanitation and public health issues that was also considered a "slum" by the city's elites. The demographics of Toronto in the first decades of the twentieth century not only contributed to the spread of influenza, but it also impacted the city's disparate response to outbreaks.

This article examines the role that newspapers played in the representation of Spanish influenza, including the city of Toronto's public health response. Through a content and discourse analysis of *The Globe* and the *Daily Star*, the most read English-language newspapers, this article explores how, as cultural texts, both negotiated the pandemic while remaining in dialogue with broader socio-cultural shifts occurring at

[1] According to etymologists at Merriam-Webster, "grippe" means "seizure" in French.

the time. Based on a sample of 252 articles from *The Globe* and 204 articles from the *Daily Star*, we examined how pandemic reporting overlapped with the two largest concerns of the era: World War I and women's rights. We focused primarily on the second wave (autumn 1918 through spring 1919) for two reasons. First, the newspaper reporting was at its height during this time; and second, most of the deaths from the pandemic occurred during these months. We chose to study *The Globe* and the *Daily Star* because while scholars of Spanish influenza have studied Canada's public health response, community spreads and quarantines, city/town preparedness, virus transmissibility, and the initial sources of the outbreak (e.g., World War I maritime travel, and military camps), there has yet to be a study that strictly isolates the media's reporting on the pandemic.

As media and communication scholars, we believe it is vital to unpack the historical role the media has played in public health crises. Further, *The Globe* and the *Daily Star* were considered a bastion of Canadian Liberalism. Liberalism is defined as a particular way of thinking about human needs and the political good, and while it does not belong to a single political party, in Canada Liberalism developed as an important strand in the ideology of both Conservative and Liberal parties (Axworthy, 2015). Liberalism came to Canada with the United Empire Loyalists who settled primarily in Nova Scotia, Quebec, and later Ontario (Axworthy, 2015). While devoted to British institutions (especially to the monarchy), overtime, Canada's ruling governments opposed American republicanism; but they were equally passionate about individual liberty. As such Canadian Liberalism, as expressed through dominant media, often wrestled over how to report on the country's ethnic pluralism on the one hand while espousing national narratives on the other hand.

At the National Liberal Conference in 1919 the party, led by Mackenzie King, defined its politics as "thoroughly democratic" and more broadly as progressive and open minded with an aim to establish responsible self-government that was not directly tied to the British Empire ("A message of Liberalism" 1919). At the same time, the province of Ontario in the later years of the 1910s was forced to redefine itself. As Peter Olivier writes, "The strains created by the Great War undeniably played havoc with the existing party structure, and the rising tide of farm and labour unrest, which catapulted two new political movements to the fore, presented a challenge which would have sorely tested any government" (Olivier, 1972). Thus, the political climate in Canada—nationally and in Ontario—was unstable and unpredictable at the time of Spanish influenza.

Importantly, there were nuanced differences between how dominant media reported on the news. The newspapers broadly targeted the same readers—white, Anglo-Saxon Protestant, English-speaking men—but there are key distinctions that emerged in terms of the substance of their respective reporting. *The Globe* was a morning paper that oriented toward both Liberal and Conservative audiences, targeting metropolitan readerships without resorting to "sensationalist headlines" (Jenkins, 2010, p. 59). Comparatively, the *Daily Star* was an evening paper that competed for a local audience of middle- and lower-income readers, often relying on denationalization (Jenkins, 2010, p. 50). How did *The Globe* and the *Daily Star* negotiate the virus by catering to their audiences? What does the newspaper coverage reveal in terms of complex intersections between pandemics, media culture, and representation? To understand the cultural context of the Spanish influenza spread throughout Toronto in autumn 1918, we first discuss the demographics of Toronto at the time; then, we report on *The Globe* and the *Daily Star*'s coverage of war heroes and athletes in comparison to women's rights during the pandemic's second wave; and finally, we discuss the shifts in the city's public health response to the subsequent smallpox outbreak of 1920.

Toronto in the 1910s

In the first decades of the twentieth century, Toronto began taking steps to modernize its public health institutions to address concerns around housing and sanitation that were fueled by an influx of immigrants. In the Ward, people were doubling up in crumbling rental dwellings because they could not find affordable housing (Monsebraaten, 2015). In a December 1918 report by the Bureau of Municipal Research, a non-profit advocacy group, the Ward was said to have "slumlike living conditions" (Monsebraaten, 2015, p. 222). As Laurie Monsebraaten writes, "the seventy-four page research document was both a warning and a call to action" (Monsebraaten, 2015, p. 223). Inadequate settlement services for immigrants was a major issue; as well, many adults in the city's low-income community had little formal education, and many children worked as messengers, hawked newspapers or toiled in factories to supplement their family's income (Monsebraaten, 2015).

Toronto's newspapers often fueled moral panic about slum conditions in neighborhoods like the Ward and Parkdale. According to Carolyn Whitzman and Tom Slater, Parkdale was not being singled out necessarily as a slum neighborhood but concerns about central city decline in Toronto were starting to permeate its borders (2006, p. 680). They write further that despite this, "newspapers and government reports reacted with horror and indignation to … 'tenements,'" even though they had virtually nothing in common with the substandard housing in New York or London. Throughout the spring of 1912, *The Globe* ran a series of strongly worded editorials condemning the 'plague of disease-breeding tenements' disguised under the name of 'apartment houses,' threats to 'morals as well as health' that would result in 'a city of stunted children and unhappy adults' (Whitzman & Slater, 2006, p. 680). Because Toronto was a staunchly English, Scottish and Irish city, the stain of unsanitary housing conditions was largely attributed to its non-Western European, Black, and Asian immigrants—those individuals whose first language was not English.

Initial Reporting on the Pandemic

Given the historical link between war and disease, Humphries writes that "it is not surprising that at the outbreak of the Great War in 1914 public health officials feared that a global war might bring new diseases home to civilian populations" (2013, p. 56). Before officials in Canada recognized the presence of influenza among the Canadian Expeditionary Force (CEF), the first reports that a virus was present among soldiers of the American Expeditionary Force (AEF) came out of Camp Funsten, Kansas, beginning March 5, 1918 (2013, p.56). Among soldiers in Canada, at least four military hospitals registered heightened numbers of admissions for influenza and pneumonia between February and April 1918, however, while soldiers across Ontario experienced heightened morbidity, there was low mortality during the first wave of the pandemic in the spring and summer months of 1918 (Bogaert, 2017, p. 48). While the public health records for Toronto pointed to heightened admissions for influenza-like illnesses in the first half of 1918 there were still no widespread quarantines or shut downs (Humphries, 2013, pp. 74–76).

Importantly, influenza reports in *The Globe* and *Daily Star* did not appear until the fall when the second wave hit, joining other communicable diseases spreading in the city at the time. For instance, in August 1918, *The Globe* reported that, according to a Provincial Board of Health (PBH) report for July of that year, communicable diseases—smallpox, scarlet fever, diphtheria—were on the decline province-wide, while Toronto's Department of Public Health (DPH) found that cases of measles and scarlet fever had increased during the month, as compared to July 1917 (*The Globe* August 1, 1918). By September, however, "Ontario's soldiers were … among the first Canadians affected by the second, more deadly wave of pandemic influenza that erupted." (Bogaert, 2015, p. 7233). September through October, *The Globe* was the first newspaper to describe the pandemic as "influenza" or "flu." "Two more deaths from influenza were reported here to-day, and the body of a former Hamiltonian who died of the disease at Sudbury was brought to this city," reported *The Globe* on October 14, 1918. "More than five hundred cases of the 'flu' were reported in Hamilton last week. The reports during Saturday showed that the epidemic was not abating," the newspaper added (*The Globe* October 14, 1918).

Toronto's response to the second wave mirrored that of other cities and towns across the province with one exception: because of its size, the city's newspaper coverage of the virus was centered around Medical Officer of Health (MOH) Dr. Hastings, who spearheaded the response. In this regard, we examined how *The Globe* and *Daily Star* reported on Dr. Hastings' briefings, and those of his provincial counterpart, Chief Medical Officer (CMO) Dr. John W.S. McCullough, and the narratives constructed around military deaths, both overseas and locally due to Spanish flu.

Toronto's MOH and Ontario's CMO During the Second Wave

By October 1918, nearly half a million Torontonians had endured four years of war in which 70,000 had enlisted in the CEF and served overseas in the trenches on the Western Front (MacDougall, 2006, p. 61) Under the direction of Dr. Hastings, MOH from 1910 to 1929, Toronto became a showplace for modern methods of public health administration and preventive activity (MacDougall, 2006, p. 61). As Heather MacDougall explains further, "Between 1910 and 1917, Hastings spearheaded the introduction of pasteurized milk, well-baby clinics, municipal housekeepers, and health education by public health nurses based on successful examples found in leading British and American cities." (2006, p. 61) This public health work was intimately tied to sanitation and broader discussions about the hygiene practices of lower-income Torontonians. "From the moment he took office," writes Lorinc, "the snowy-haired obstetrician focused intensively on eradicating slum conditions in … poor areas. Hastings promoted dramatic changes to … infectious disease control, vaccination and housing" (2006, p. 61).

Among his many contributions, Dr. Hastings was instrumental in bringing public health nursing as a profession into the delivery of public health services; that is, he revolutionized "the idea of sending nurses out into the community to connect with people in their own homes, whether it was for the purposes of TB [tuberculosis], communicable disease control or teaching new mums how to adapt to their babies" (Lorinc, 2015, p. 265). Starting with a staff of three public health nurses in 1910, Hastings expanded the health education component of his staff's work and in 1914 created a Division of Public Health Nurses, and then, using a generalized system that stressed health education rather than curative services, he spearheaded Toronto's DPH which sought to forge links with more than 200 local voluntary groups through the Neighbourhood Workers' Association (NWA) (MacDougall, 2006, p. 62). When the pandemic's second wave hit, however, there were disagreements between Dr. Hastings and the province's CMO Dr. McCullough on how to respond. The conflicted and sometimes contradictory responses of the CMO and city's MOH were not unique to Toronto; across the province medical professionals had varied and sometimes contradictory responses to the pandemic.

Writing in the *Canadian Medical Journal* in 1918, McCullough said:

> In Ontario influenza is not required by law to be reported, placarded or quarantined, and does not come under the regulations regarding communicable disease except that, by a recent Order-in-council, land and unoccupied buildings may be taken summarily by a board of health for hospital purposes.
>
> With respect to the closing of schools, churches, theatres and other public assemblages, each Medical Officer of Health or Local Board of Health has power to close such places if it seems desirable. (McCullough, 1918)

When the disease first appeared in Toronto, MacDougall explains that "the MOH and military authorities appealed for calm, provided a detailed description of the symptoms, strongly recommended resting in bed, and exhorted the sick to call for medical assistance" (MacDougall, 2006, pp. 64–65). However, after the first civilian casualty was reported, a schoolgirl who died in Toronto General Hospital on September 29, 1918, in spite of growing public pressure for isolation and quarantine, Dr. Hastings did not issue a stay-at-home order because the bulk of cases remained military men in the base camp located in the city (MacDougall, 2006, p. 65). Both Dr. Hastings and Dr. McCullough agreed, however, that wearing masks in public was not warranted; as a result, Torontonians, and Ontarians in general, were not required to use them (MacDougall, 2006, p. 68). Importantly, *The Globe* and *Daily Star* varied in their reporting on the disagreements between Dr. Hasting's and Dr. McCullough's response to influenza, specifically their approach to placarding and quarantines, hospitalizations, and deaths.

The first reports of Dr. Hasting's declaration of an influenza outbreak in Toronto appeared in *The Globe* on October 3, 1918 when, in an interview with the newspaper, he said, "there was now little doubt that the disease was the same as has been spreading through the United States" (*The Globe* October 3, 1918). Dr. Hastings was unsure whether it was a new disease or just a form of grippe but he conceded that "some bacteriologists [who] claim[ed] to have found the old influenza organism, while others [had]... not located it in the new outbreak" (*The Globe* October 3, 1918). One week later, *The Globe* quoted Dr. Hastings' declaration that influenza was spreading "most rapidly among the poor" (*The Globe* October 10, 1918). There was no discussion of specific poorer neighborhoods such as the Ward or Parkdale. At that press conference, Hastings also issued a warning to the public to take care of themselves, and to not "do anything that [... would] lower their vitality" (*The Globe* October 10, 1918). Emphasizing nourishment, rest, exercise, walking to work, and avoiding overwork, his advice lacked specificity about the different influenza strains and there was no guidance on how to reduce the rates of infection.

On 15 October, reports in the *Daily Star* informed the public that schools and dance halls were closed in Toronto until further notice; parties were also banned (*Daily Star* October 15, 1918). In the same article, the newspaper cited a petition from the Association of Toronto Theatre Managers (ATTM) which asked that theaters not be closed, given that they were "thoroughly well ventilated" and the air quality, the ATTM argued, was superior to "other places where people congregate" (*Daily Star* October 15, 1918). Unlike *The Globe*, which focused its reporting on official statements from Dr. Hastings regarding death tolls at military camps and hospitalization tallies, the *Daily Star*'s reporting positioned Dr. Hastings' opinions on how best to respond to the influenza spread alongside personal narratives—lists of the names, ages, and addresses of those who died—and reactions from politicians, such as a move from city council to amend temperance regulations due to an increased demand for liquor for medical purposes in the city (*Daily Star* October 15, 1918).

Another headline from the *Daily Star*, published on 16 October detailed cases of sickness, including a doctor, and a number of cases at the Toronto Conservatory of Music (*Daily Star* October 16, 1918). In a report at the end of October, the *Daily Star* quoted Dr. Hastings who pointed to the New York Health Board's decision to amend their Sanitary Code as something that should be done in Toronto; the Board made it a misdemeanor for landlords not to have heat in dwellings due to the epidemic.

However, "at the last session of the Legislature we tried to get an amendment to the Public Health Act, which would enable us to control heating and ventilation, but we failed," said Dr. Hastings to the *Daily Star* (*Daily Star* October 21, 1918). Dr. Hastings also gave his opinions on overcrowded street cars. "Until the Ontario Railway Board acts and enforces its order for sufficient cars to carry the passengers, an ordinance forcing the company to carry only seating capacity on its cars would leave the public with only two choices at rush hours," he said, adding "People could stand on the streets until cars were available, or they could walk home. We recommend walking where it is possible, but obviously Toronto is too big for people employed downtown to walk all the way home. Also, if they stood on the street corners until there was a seating room in the cars, it would be ten o'clock at night before they got home" (*Daily Star* October 21, 1918). These reports indicate a lack of coordinated responsiveness to the pandemic, especially related to the most vulnerable citizens of the city, who would not be able to walk home due to the distance from home to work or hours of work.

The *Daily Star* also gave details about people's ability to adapt to the changes brought on by the influenza outbreak, the impact of the disease on families, and Dr. Hastings' professional opinions as the voice of authority not only on infectious disease prevention but also on legislative responses to the pandemic. By comparison, *The Globe* primarily focused on province-wide death tolls, hospitalizations, and quarantines. Each newspaper also responded differently to military and celebrity influenza deaths. Unlike small town newspapers that did not have space for lengthy individual obituaries, Toronto's newspaper did. By the end of October through November 1918, as more soldiers and eventually famous athletes died from influenza complications, the *Daily Star* featured in-depth, individual obituaries that contrasted with *The Globe*'s reports, which remained largely focused on death tolls, closures, and quarantines. The passing of famous people (who tended to be athletes) was eulogized in newspaper features in ways that pointed to the sensationalization of the epidemic.

Reporting on the Deaths of Soldiers and Athletes

According to Kandace Bogaert's analysis of Commonwealth War Graves Commission (CWGC) records, "the presence of influenza among soldiers on outbound troopships, and the high number of deaths among soldiers from pneumonia and influenza in the summer, demonstrates that pandemic influenza was already present in Canadian military camps." (Bogaert, 2017, p. 57). Thus, when soldiers and civilians alike died from influenza the root cause was "secondary pneumonia infections which caused people to turn blue from lack of oxygen and cough up purulent, bloody sputum" (Humphries, 2013, pp. 56–67). Beginning in October, local newspapers across the country began to report on the deaths of soldiers most prominently because many were very young.

In 1918, the act of publishing charts that presented statistics on deaths from influenza by date appeared with regularity in the *Daily Star* for nearly the entire month of October but the reporting in both newspapers on soldiers' deaths in hospitals varied. On 3 October, for example, *The Globe* reported on guidance from Dr. Hastings that there were fifty soldiers with influenza in the city's General Hospital military wing (*The Globe* October 3, 1918, p. 6). Soldiers were said to be in quarantine, and visitors were refused admittance as a precautionary measure against spread (*The Globe* October 3, 1918, p. 6). Five days later, in addition to reporting on civilians who were falling ill from influenza, *The Globe* noted, "most of the local cases so far have been in the camps of the Royal Air Force, but a marked improvement has been noted there. The improvement is noticeable in so far as new cases are concerned" (*The Globe* October 8, 1918). As the hospitals began to reach capacity, *The Globe* focused its reporting on the mounting pressures on the healthcare system, outbreaks at military camps not only in Toronto but also in Hamilton, Niagara Falls, and across the province, including outbreaks in Quebec. Like *The Globe*, the *Daily Star* reported on civilian deaths, which included details on their names, where they lived, and how they died, but it also created a heroic narrative out of these soldiers, which highlighted the important sacrifice they had made in fighting a war on one front, and influenza on the other.

Some of the reporting sensationalized the deaths in ways that would have generated more public interest in the *Daily Star* than the pandemic itself. We are not suggesting that the contents were embellished or exaggerated as there was emphasis placed on the particularities of the soldiers' lives, but as compared to the lives of civilians impacted by influenza, it is important to observe how the coverage on young white men whose deaths during World War I were described as causing "concern about the stability of society" reflected editorial bias (Nicholas, 2009, p. 254).

In a *Daily Star* article appearing on 16 October with a headline, "Soldier Victims of 'Flu,'" the newspaper reported that fourteen soldiers had died of influenza at the Base Hospital; it then listed the names of all the soldiers and their hometowns, and for those who resided in Toronto, their exact home address was printed (*Daily Star* October 19, 1918). It also reported on funerals for soldiers who died overseas. In one article, appearing on 19 October, the tone of the obituary made it clear that the loss of such a soldier at the height of a war due to influenza was taking the lives of not only young men but also brave soldiers: "Deceased, who was 22 *years* of age, was invalided home early in the present year after having been overseas three years. His death occurred on Wednesday from pneumonia, following influenza, after a few days' illness. His home was at 11 Bartlett avenue [sic]. The remains were deposited in the soldier's plot at Prospect Cemetery" (*Daily Star* October 19, 1918). Unlike *The Globe*, which primarily reported on Dr. Hastings' briefings about deceased soldiers due to influenza, the *Daily Star* published obituaries directly from military associations and families while emphasizing not only their lives lived but when/how they served in the war, and when/how they died from influenza. When hockey players began to die from influenza, however, both newspapers treated them as if they were fallen war heroes. In this regard, newspapers performed the role of converging the soldier with the athlete—men who were white, young, and as strong and virile as soldiers.

"Sarnia Star Dies of 'Flu,'" read a headline in *The Globe* on November 2, 1918 (*The Globe* November 2, 1918). The story, while not a frontpage lead, paid respects to Ontario Hockey Association (OHA) player Frank Montgomery who had died at his home in Oshawa, Ontario. Born on January 5, 1899 in Peterborough, the death certificate confirms that he was only nineteen when he died on 28 October; the cause of death listed as "pneumonia following flu" (*The Globe* November 2, 1918). There were no other reports of hockey player deaths in *The Globe* until 1920 when on 3 February, the newspaper printed a notice of the passing of James P. Grannary, one of Canada's foremost hockey players at the time, who had succumbed to Spanish flu in Detroit, Michigan (*The Globe* February 3, 1920).

Unlike *The Globe*, which focused its reporting on the deaths of OHA players, the *Daily Star* set its focus not only on hockey players but also the deaths of baseball and football players, as well. On October 18, 1918, the newspaper told readers that "The 'flu' has carried off another well-known Toronto sporting figure" (*The Globe* February 3, 1920). Harry ("Chuck") Jackman, owner of the Hotel Widmer, who died from pneumonia at the age of forty-eight was, in his younger days, said to be "captain and manager of the famous Athletic baseball team" (*Daily Star* October 18, 1918). At the same time that soldiers and sports heroes were eulogized, Toronto's newspapers reported on women's rights, which was as debated as the responses to the pandemic. The pandemic coverage of women, especially those who labored outside the home, overlapped with the fight for women's suffrage.

Reporting on Women's Roles and Rights

The war had brought about many social and cultural changes, the role of women was one of them. "Women's wartime efforts in politics and the workforce seemed to undermine traditional ideas about gender roles," writes Jane Nicholas, adding that "anxieties about a younger generation that seemed so different from its predecessor brought to the fore issues" (2009, pp. 245–255). Rising rates of unmarried women and "illegitimacy" also contributed to concerns over family life as did the number of physically and psychologically wounded soldiers who returned from the war (Comacchio, 1997, p. 7). Significantly, discourse about women's roles as nurses during the pandemic appeared in *The Globe* throughout October 1918, such as a report from Hamilton where efforts were being made to "train all women who are willing to serve as nurses in the rudiments of nursing that they may minister to those stricken with the disease" (*The Globe* October 17, 1918, p. 2).

Conversely, the *Daily Star*'s editorials on women highlighted their evolving roles as nurses during the pandemic, and those who had succumbed to the flu at the city's hospitals. "Grace Hospital reported three more cases since yesterday among the nurses," the newspaper reported on October 10, 1918, adding, "This hospital is still refusing patients on account of the shortage in the nursing staff, and is allowing no visitors in the public wards except in serious cases, and only the immediate relatives in the private wards" (*Daily Star* October 10, 1918, p. 21). Given the healthcare worker shortage, women were also said to be part of a recruitment plan administered by the Home Bureau of Nursing to act as untrained helpers at Toronto's Western Hospital

(*Daily Star* October 10, 191, p. 21). Beyond hospitals, the *Daily Star* reported on an outbreak of flu among women who worked at banks in the city. "The two head offices which are hardest hit are the Dominion Bank with 15 girls absent from a staff of 120, and the Canadian Bank of Commerce, also with 15 absent from a staff of 60 girls," the newspaper stated, adding "Factories which employ large numbers of men and women are taking extreme measures, and are using liquid disinfectants in the rooms to prevent the spread of the disease" (*Daily Star* 12 October 1918). Where, in the pages of the *Daily Star*, women were essential to the pandemic response, in *The Globe*, women's fight for the franchise, even during the pandemic's second wave (October 1918 through March 1919), was just as essential to its editorial coverage.

"There seem to be no statistics compiled to show just how many names of women have been added to the voters' lists in Toronto this year," reported *The Globe* on New Year's Day 1918 (*The Globe* January 1, 1918). "Next year we're going to right that point," declared one determined woman described by the newspaper as "active in all women's work." "Who works longer hours, or more honestly and laboriously earns her living than the mother and housekeeper in the home?" (*The Globe* January 1, 1918). By the spring, a Federal bill for the enfranchisement of women passed the committee stage and was set for a third reading when it faced opposition from the majority of the Quebec members on the grounds that each province had the right to determine the question for itself, not the federal government. "French-Canadian opinion is hostile to women's suffrage, but there will be no compulsion upon the women of Quebec to go to the polls. They will have the privilege of refusing to exercise that right," reported *The Globe* (April 12, 1918). By October, the British House of Common adopted a resolution in favor of granting women the right to sit in Parliament. "It is to be hoped that the position of British Parliamentarians, which has been promptly and heartily commended by an influential portion of the Conservative Press of Ontario, will do something to bestir the Government of this Province," reported *The Globe*, adding that "It may be that the action of the British House of Commons and its commendation by the newspaper spokesmen of the Conservative party in Ontario will serve to awaken the laggard legislators at Queen's Park. It is notorious that they have to be prodded into progressive policy" (*The Globe* October 26, 1918).

Smallpox Leads to Changes

As Toronto returned to "normalcy" in 1919, the mayor and Board of Control (BOC) recommended budget cuts to municipal services, including the Health Department (MacDougall, 2006, p. 72). However, despite effective organizing and the yeoman services that DPH staff had performed during the pandemic, all "were forgotten or ignored when a mild form of smallpox appeared in October 1919. Anti-vaccination groups organized rallies attended by some city council members who objected to

Hastings's dynamic leadership and his demand that mandatory vaccination be instituted" (MacDougall, 2006, pp. 72–73). Just after the city appeared to have won the fight against influenza in fall 1919, smallpox was becoming a more general concern throughout Ontario. Unlike influenza, when smallpox grew to epidemic levels, the public health response was immediate, and a vaccine/vaccination campaign followed suit. But like Spanish flu, convincing the public to adhere to smallpox public health recommendations led to much controversy.

In 1887, the Ontario government passed the first *Vaccination Act* which stipulated parents had to have their children vaccinated within three months of birth and re-vaccinated when necessary, every seven years (Arnup, 1992, p. 160). In 1914, however, the Ontario government passed a second *Vaccination Act*, which made a general vaccination order mandatory in the face of an outbreak of smallpox; as Katherine Arnup asserts, "The Act also empowered the local Medical Officer of Health to require vaccination certificates of 'all pupils' within their jurisdiction. While the passage of the act went virtually unnoticed at the time, it was to become the centre of a storm of controversy some five years later" (p. 165). Following the 1919 outbreak, Dr. Hastings, who was still Toronto's MOH, sought a proclamation demanding the general vaccination of the population, and when presented to the City Council for approval, Mayor Thomas Church, a staunch Conservative who was backed by the city's independent-minded Conservative newspaper, the *Telegram*, refused to back the vote, which split down the middle (p. 165). In response, Dr. Hastings ordered the vaccination of all school children, and he even threatened to exclude from attending school any children who could not produce a certificate of vaccination (p. 165). The backlash against Dr. Hasting was swift.

In a Letter to *The Globe*'s Editor, one Torontonian said, "is it not the right of each individual citizen to decide a matter of such vital importance for himself or herself? Irrespective of the merits or demerits of vaccination as a preventive of disease, when such a majority is opposed, is not the action of the Provincial Medical Health Board a tyrannical one at best, and entirely at variance with the principles of equality and human liberty upon which our Provincial Government is founded? Toronto, Ont.–One Who Knows" (*The Globe* December 26, 1919). With the support of Dr. McCullough, the city proceeded with its vaccination rollout.

Between the outbreak of smallpox in October 1919 and its decline in March 1920, 2864 cases of the disease were reported in Toronto; and despite anti-vaccination campaigns, some 200,000 people received vaccinations against smallpox (Arnup, 1992, pp. 166–167). This number, while significantly less than Spanish influenza, still meant that the city's public health system in the last years of the 1910s had been tested like never before and while it was challenged, it survived and would undergo further changes such as the Ontario government's creation in 1921 of the Division of Public Health Education, the first of its kind in Canada.

Conclusion

If there is one conclusion that can be drawn from reading *The Globe* and *Daily Star*, it is that both held editorial biases that reflected their partisanship, appealing to both Conservative and Liberal audiences during the pandemic. While they remained true to medical data as it related to deaths and did not underreport or ignore the severity of the pandemic, the *Daily Star*'s reporting was more sensational in its tone, while *The Globe* stayed so true to facts and figures it failed to humanize the deaths. What is clear, however, is that both newspapers supported the overlapping women's rights movement that was made even more visible through women's involvement in both World War I efforts and nursing, but *The Globe* politicized the fight for suffrage turning women's rights into a wedge issue with English and French Canada.

One gap apparent in our examination of Toronto's newspapers is the lack of reporting on non-English-speaking immigrants, who were targeted by eugenicists and labeled "feeble-minded" and unsuitable for Canada. *The Globe* and *Daily Star* represented the pandemic as a communicable virus; they did not represent the other virus–xenophobia. The only reason the Ontario government did not pass eugenic legislation, for example, was because of voracious opposition from Catholic leaders, and the Ontario Liberal party's unwillingness to jeopardize their support among Catholic voters (McLaren, 1990, pp. 122–125). Ultimately, reading Toronto's response to Spanish influenza reveals a complex milieu of medical opinion, political posturing, and social change.

Acknowledgements This article was supported through a Social Sciences and Humanities Research Council Insight Development Grant.

References

Arnup, K. (1992). Victims of vaccination? Opposition to compulsory immunization in Ontario, 1900–90. *Canadian Bulletin of Medical History, 9*(2), 160.

Axworthy, T. S. (2015). Liberalism. In *The Canadian Encyclopedia.* https://www.thecanadianencyclopedia.ca/en/article/liberalism

Bogaert, K. (2015). Cross protection between the first and second waves of the 1918 influenza pandemic among soldiers of the Canadian Expeditionary Force (CEF) in Ontario. *Vaccine, 33*, 7232–7238.

Bogaert, K. (2017). Military and maritime evidence of pandemic influenza in Canada during the summer of 1918. *War & Society, 36*(1), 44–63.

Comacchio, C. (1997). Dancing to perdition: adolescence and leisure in interwar English–Canada. *Journal of Canadian Studies, 32*(3), 7.

Hixon, L. (2021). 'The Dread Influenza': Milford in the Grip of the 1918 Pandemic. *Historical Journal of Massachusetts, 49*(1), 12.

Humphries, M. O. (2013). Paths of infection: The first world war and the origins of the 1918 influenza pandemic. *War in History, 21*(1), 56.

Jenkins, W. (2010). Homeland crisis and local ethnicity: The Toronto Irish and the cartoons of the 'Evening Telegram' 1910–1914. Urban history review/Revue d'histoire urbaine. *Special Issue:*

Encounters, Contests, and Communities: New Histories of Race and Ethnicity in the Canadian City, 38(2):59.
Lorinc, J., et al. (2015). Introduction. In J. Lorinc (Ed.), *The ward: The life and loss of Toronto's first immigrant neighbourhood* (p. 15). Coach House.
MacDougall, H. (2006). Toronto's health department in action: Influenza in 1918 and SARS in 2003. *Journal of the History of Medicine and Allied Sciences, 62*(1), 61.
McCullough, J. W. S. (1918). The control of influenza in Ontario. *Canadian Medical Journal*, 1084.
McLaren, A. (1990). *Our own master race: Eugenics in Canada, 1885–1945* (pp. 122–125). McClelland and Stewart.
Monsebraaten, L., et al. (2015). The BRM's wake-up call. In J. Lorinc (Ed.), *The ward: The life and loss of Toronto's first immigrant neighbourhood* (p. 222). Coach House.
Nicholas, J. (2009). Gendering the Jubilee: Gender and modernity in the diamond Jubilee of confederation celebrations, 1927. *The Canadian Historical Review, 90*(2), 254.
Olivier, P. (1972). Sir William hearst and the collapse of the Ontario conservative party. *The Canadian Historical Review*, 21.
Ottawa: National Liberal Committee (1919). *A message of Liberalism, including the resolutions adopted at the National Liberal Convention of August*, 1919, pp. 21–23.
Whitzman, C., & Slater, T. (2006). Village Ghetto land: Myth, social conditions, and housing policy in Parkdale, Toronto, 1879–2000. *Urban Affairs Review, 41*(5), 677.

Chapter 8
Outbreak Narratives of the 1918 Influenza Pandemic: Contagion, Community, and Politics in Myla Goldberg's *Wickett's Remedy* and Thomas Mullen's *The Last Town on Earth*

Pei-chen Liao

Abstract This chapter investigates the possible reasons and underlying significance of the 1918 influenza narratives' absence and outbreak. It first of all examines the historical amnesia related to the 1918 influenza in the 1920s and the later return of the repressed memories in personal and fictional flu narratives published in the 1930s. It then turns to two recent American historical novels as examples, Myla Goldberg's *Wickett's Remedy* and Thomas Mullen's *The Last Town on Earth*, to explore how the outbreak narratives by writers who did not have any personal contact with the 1918 influenza pandemic transcend the predominant discourse of trauma and illness that dominates early flu stories while configuring the contours of contagion, community, and politics that are more relevant to contemporary readers than the influenza itself. At first glance, through the public imagination of both the local and national communities as a population that are susceptible to the contagious influenza, Goldberg's and Mullen's outbreak narratives seem to construct what Priscilla Wald in *Contagious* terms the "epidemiology of belonging" (Wald in Contagious: Cultures, Carriers, and the Outbreak Narrative. Duke University Press, p. 29, 2008). Nonetheless, subtle details about the scapegoating of foreigners and strangers as potential disease carriers and the exposure of marginalized communities to infectious situations lay bare the extension of biopolitics to necropolitics and the precarious conditions under which lives become disposable and deaths acceptable.

Introduction

Dubbed the "mother" of all influenza pandemics (Taubenberger & Morens, 2006, 15), the 1918 influenza pandemic first broke out in the early spring of 1918, quickly spread out within a year and mysteriously disappeared in 1919. Young adults were astoundingly more vulnerable to the highly infectious and lethal influenza than the

P. Liao (✉)
National Cheng Kung University, Tainan City, Taiwan
e-mail: pcliao@ncku.edu.tw

S. Venkatesan et al. (eds.), *Pandemics and Epidemics in Cultural Representation*,
https://doi.org/10.1007/978-981-19-1296-2_8

elderly and children, who are usually considered weaker and more fragile. Despite its severity, many European countries and the USA censored wartime coverage of the flu to prevent media coverage from revealing casualties to Germany. The neutrality of Spain during WWI allowed journalists to report more freely on the wide spread of the influenza and the high number of casualties. This thus created the false impression that the influenza originated in Spain, causing people to misleadingly call it the "Spanish flu" or the "Spanish Lady" (Fisher, 2012, 13; Price-Smith, 2009, 78–79). The origin of the 1918 influenza is still under debate, but Barry (2004) suspects that it very likely appeared first in an army base in Haskell County, Kansas (para. 2). Overlooking the contagion of the disease, infected American soldiers boarded ships to France and spread the virus across Europe. If, as Barry suggests, the 1918 influenza pandemic originated in an American army base, the influenza could be considered WWI's "lethal twin" (Fisher, 2012, 1). It also arguably facilitated the end of WWI.

Having caused more casualties than WWI due to infecting over one-third of the general population and killing approximately 50 million people, the 1918 influenza pandemic was, however, mysteriously absent from literary narratives in the 1920s and most history books until it was rediscovered in the 1970s by medical historians like Alfred W. Crosby. Fisher (2012) contends that Crosby's book (1976), *Epidemic and Peace, 1918*, "deserves credit for doing the first systematic study of the pandemic" (5). Just as there were at least three waves of influenza outbreak during a 12-month period in 1918 and 1919,[1] literary narratives as well as cultural and medical historical studies of the influenza also exploded in waves in the late twentieth century and early twenty-first century. Phillips's (2014) research finds that, from the mid-1970s to the early 1990s, the first wave of flu historians "was largely pre-occupied with probing in the dark, just to establish what had happened in 1918 and whether this was important historically" (800). From 1999 to 2014, the year in which Phillips's article was published, the second wave of the flu historiography, stimulated by outbreaks of avian flu, swine flu, and severe acute respiratory syndrome (SARS), attends to the flu's relevance to the present "for comprehending and addressing current epidemics" (Phillips, 2014, 796). This period also witnessed the publication of more than a few American historical novels such as Myla Goldberg's *Wickett's Remedy*, Thomas Mullen's *The Last Town on Earth*, and many others that are set against the backdrop of the 1918 influenza pandemic.[2] Expectedly, the centenary of the pandemic in 2018 triggered another wave of the flu historiography and literary representation, which is reaching its peak with the outbreak waves of the present coronavirus disease-2019 (COVID-19).

---

[1] The first wave lasted approximately from February 1918 to June 1918, the second from August 1918 to December 1918, and the third from December 1918 to April 1919. For more detailed discussions on the three waves of the 1918 influenza pandemic, see Taubenberger and Morens (2006), Fisher (2012), Price-Smith (2009), and Bristow (2012).

[2] For a sampling of the twenty-first-century historical novels about the 1918 influenza pandemic, see Reina James's *This Time of Dying* (2006), Charles Todd's *An Unmarked Grave* (2012), David Cornish's *1918: The Great Pandemic* (2013), Cat Winters's *In the Shadow of Black Birds* (2013), Robert Masello's *The Romanov Cross* (2013), Makiia Lucier's *A Death-Struck Year* (2014), and James Rada's *October Mourning* (2016).

While memories of the 1918 influenza pandemic were repressed in its immediate aftermath, its scholarly historiography and literary narratives have exploded over the past two decades of the twenty-first century, like the emerging and reemerging of disease outbreaks. Wald (2008) coins the term "outbreak narratives" to "designate those epidemiological stories" that "influence how both scientists and the lay public understand the nature and consequences of infection, how they imagine the threat, and why they react so fearfully to some disease outbreaks" (3). This chapter, first of all, examines the historical amnesia related to the 1918 influenza in the 1920s, as well as the return of the repressed memories in personal and fictional flu narratives published in the 1930s. It then turns to two recent American historical novels as examples, Goldberg's *Wickett's Remedy* and Mullen's *The Last Town on Earth*, to examine how the outbreak narratives by writers who did not have any personal contact with the 1918 influenza pandemic transcend the predominant discourse of trauma and illness that dominates early flu stories while configuring the contours of contagion, community, and politics that are more relevant to contemporary readers than the influenza itself. The two novels, along with other contemporary outbreak narratives that revisit the historical past of the 1918 influenza, collectively reveal "a fascination not just with the novelty and danger of the microbes but also with the changing social formations of a shrinking world" (Wald, 2008, 2). In the time of COVID-19, the "epidemiology of belonging" (Wald, 2008, 29) that Goldberg's and Mullen's outbreak narratives construct through both imagined communities and imagined immunity against human contact provides guidance to contemporary readers that leads to an understanding of not only the social interdependence network but the disciplinary mechanisms of biopolitics and its extension to necropolitics that renders lives disposable and deaths acceptable.

Amnesia, Trauma, and Memory

In order to better capture and cope with the current COVID-19 pandemic, increasing numbers of scholars are looking back to history to learn lessons from the 1918 influenza pandemic. Medical historians are comparing COVID-19 and the 1918 influenza pandemic.[3] COVID-19, as well as the centenary of the influenza pandemic in 2018, has given rise to the most recent wave of flu historiography and literary studies. As Arnold (2020) points out, "COVID-19 might be a 'novel' coronavirus to the epidemiologist, but to the historian, there is much about the behavior of the disease and the socio-political responses it has elicited that appears decidedly lacking in novelty" (569). Like COVID-19, the 1918 influenza was deemed a pandemic because of its global spread, but despite its heavy death toll, the influenza was quickly forgotten. How memories of the 1918 influenza were repressed or lost should deliver a dire warning for the present.

[3] See, for example, Daihai He et al.(2020), Scarpa et al. (2020), and Arnold (2020).

In the wake of the influenza pandemic, very little was written in the USA about it, far less than was written about WWI. Willa Cather's *One of Ours* (1922) is one of the earliest flu novels, followed by Thomas Wolfe's highly autobiographical novel, *Look Homeward, Angel* (1929). Surprisingly, none of the writers of "that golden literary age," such as Ernest Hemingway, William Faulkner, and Francis Scott Fitzgerald, wrote about the 1918 influenza. Many of them, however, "found war a weighty topic to expound upon, allowing exploration of such themes as manhood, patriotism, and courage" (Mullen, 2007b, 17). Crosby's book (1976) noted that the pandemic "never inspired awe, not in 1918 and not since" (311). It was "America's forgotten pandemic," as the new title of his book highlighted when it was reissued in 1989. In the USA, influenza was cast in the shadow of the Great War and "subsumed within a more compelling story of patriotic suffering and aspiring nationhood," just as "imperial violence supplanting epidemic violence, the Amritsar massacre of April 1919, with its nearly 400 deaths, put the twelve million influenza deaths in the shade" (Arnold, 2019, 198). Except for sporadic accounts of personal memory and loss in fictional and non-fictional works, American society had been dominated by a sense of amnesia for nearly six decades since the flu's first outbreak in 1918. This lacuna shows that public health and disease did not earn as much public attention as war, although, as my analysis of Mullen's and Goldberg's novels will later demonstrate, the questions of life, health, and death are connected with not merely medical science and epidemiology but also politics.

To some extent, discussion of the 1918 influenza pandemic was blocked by military and medical authorities and was overshadowed by post-war optimism and public efforts to rebuild the nation. Not only did military officers censor wartime coverage but medical authorities also silenced reports on the influenza "so as not to publicize their rout by a virus" (Phillips, 2014, 806). Byerly (2005) asserts that "historians have relegated the worst epidemic in modern history to the sidelines of human history" after taking "cues from hundreds of war memoirs and official medical histories that efface the role of disease in modern warfare" (188). Price-Smith (2009) likewise claims that, "in the aftermath of the Great War[,] military forces sought to diminish the perceived impact of the contagion, particularly since military medical officers had been utterly powerless in the face of such a pathogen" (63). It should be noted that the silence surrounding the 1918 influenza pandemic was not merely a result of a conspiracy on the part of medical and military authorities but also due to "an exhaustion of public attention, compassion and resources" (Fisher, 2012, 6). In the 1920s, with the end of the influenza pandemic, American society in general shared "the upbeat chorus of optimism with expressions of faith in modern science and a conviction that progress was sure to follow" (Bristow, 2012, 157). Journalistic reports and literary narratives about the influenza could be taken as deterrents to societal attempts to recover from illness and to regain the health and strength essential to the rebuilding of the nation in the post-war era.

Even if political suppression and post-war optimism were not taken into account, the 1918 influenza pandemic itself seemed too banal and too ubiquitous to be represented in words in its immediate aftermath. As Kolata (1999) points out, the influenza exerted "no obvious effect," for it "did not usher in a long period in which death from influenza was a new and constant threat," and neither did it "leave behind legions of crippled and maimed or disfigured survivors" (54). Lacking visible damage, it may not have aroused as much awe or dread as WWI. It was so ubiquitous as it became a daily fact of life that it was "difficult to describe and difficult to turn into tangible characters" (Outka, 2019, 26). Compared with the military spectacles and violence of the Great War that produced a state of emergency and crisis, the mundanity and attritional violence of influenza arguably engendered "a state of acceptance" (Sandset, 2021, 1411), wherein death was expected and accepted. What has been overlooked and should be brought into view, as Goldberg's and Mullen's novels manifest, are the "everyday death-worlds created by health inequality" and the preexisting and underlying conditions that "foster ill-health and premature deaths" (Sandset, 2021, 1413), such as socio-economic status and racial inequalities.

In contrast to the state of acceptability, the absence of the influenza pandemic from most history books and interwar literature may also be argued as a reflection of the inability of human language to articulate both the mental wound inflicted on its survivors and the horror of its high death rate. Nonetheless, linguistic inability is an insufficient explanation for this lacuna because, while the American society struggled to forget the past, the aftereffect of the influenza, a decade later, found its way into novels and short stories, such as John O'Hara's "The Doctor's Son" (1935), William Maxwell's *They Came Like Swallows* (1937), and Katherine Anne Porter's novella, *Pale Horse, Pale Rider* (1939), to which Crosby's seminal book is dedicated. Under the interplay between cultural amnesia and retroactive acts interpreting and remembering it, the 1918 influenza pandemic was neglected but not completely forgotten. Modernist novelists' belated response in the 1930s is evidence of the influenza's enduring impact. These writers had varying degrees of personal contact with the pandemic. Centering largely on the focalization of patient characters, their flu stories "preserve details about the pandemic that scientific reports and histories do not convey" and "transmit to us what it was like to experience the pandemic up close" (Ware, 2009, 80, 70). Mediating between loss and recovery, they represented influenza as trauma, a nightmare, or an apocalypse and tried to cope with it through writing. For example, as Bollinger (2013) suggests, Porter's *Pale Horse, Pale Rider* blurs the line between autobiography and fiction while structuring the novella in ways intricately tied to apocalyptic literature that aims "to shed divine light onto human experience, to offer the consolation and reassurance that, regardless of the current overarching trauma, human suffering has meaning" (375). Although trauma is usually perceived as "a response, sometimes delayed, to an overwhelming event or events" like wars (Caruth, 1995, 4), literary representation of the influenza through the lens of trauma captures mundane, everyday experiences of illness which can also cause fear, anxiety, and depression.

It is understandable that modernist writers who either lived through the 1918 influenza pandemic or had indirect contact with those who did may concern themselves with the experience of the influenza and the interior psyche of individuals. Their articulation of traumatic experiences, however, may drive the reader's attention into the private, emotional, and psychological distress of the traumatized self and away from any meaningful contextualization of the influenza in public or historical space. It also raises questions about the effects of narratives as a medium through which the trauma resulting from the influenza can be remembered and worked through. Belling (2009), for example, is of the opinion that, "in its inadequate struggle to contain and express the experience of the influenza, the medium of literary fiction carries traces, the ash and dust, but not the vivid blood, of a trauma that is now almost impossible to apprehend" (59). If, even to the modernist writers who had direct or indirect contact with the pandemic, the traumatic experiences of the influenza were almost impossible to apprehend or articulate, how then do twenty-first-century writers approach it? A paradigm shift is emerging in the literary representation and interpretation of the influenza.

Contagion, Community, and Politics

As exemplified by *Wickett's Remedy* and *The Last Town on Earth*, the twenty-first-century outbreak narratives of the 1918 influenza are less concerned with questions related to narrative modes or articulation of trauma than with the relevance of the 1918 influenza to contemporary society and the challenges that pandemics pose to a shrinking world. Through the lens of contagion, *Wickett's Remedy* and *The Last Town on Earth* contextualize the outbreak of the 1918 influenza against the backdrop of WWI and shed light on local and national policies related to disease control and prevention. At the same time, they bring out the dilemmas faced by medical personnel, frontline essential workers, and policy makers struggling between life and death, between social distancing and contact, and between health and the economy. These are the challenges that our contemporary society is similarly facing while coping with COVID-19. Trauma discourse brings to the forefront a narration of individual loss and illness as well as the repression of memory, and yet, when dealing with the influenza pandemic, it has limits related to a focus on personal and psychological wounds. The discourse on contagion pertains to both the infection of disease and emotions and underscores the intricate relationship between the physical body and the socio-political body.

The notion of contagion bespeaks the contradictions of communicability inherent in the formation of a community. Etymologically from the Latin *con* (together with) and *tangere* (to touch), "contagion" suggests "the primal contact of a touch," and the adjective, "contagious," denotes "as much about the powerful circulation of ideas, emotions, or affects as about disease agents" (Wald, 2016, vi). The transmission and spread of viruses arouses public fear of human contact and hence necessitates the imposition of quarantines and isolation on persons diagnosed with an infectious disease. And yet, as Wald (2008) further notes, paradoxically "[t]he interactions that make us sick also constitute us as a community" (2). Such a constitution of "community" indicates "an epidemiology of belonging through which people might experience their emergence as 'a population'" (Wald, 2008, 18). Ironically, in a shrinking world of interdependence, the protection of the health of the population justifies the growth of what Michel Foucault terms the "biopower" and reinforces national belonging as well as borders. Portraying the importance and danger of both human contact and biopolitics, Goldberg's and Mullen's outbreak narratives illuminate both the tension and the intersection of imagined immunity and imagined community. Their depiction of contagion, on the one hand, "offers a visceral way to imagine communal affiliations in national terms" (Wald, 2008, 51). On the other hand, it calls imagined communal belonging into question by pinpointing the inextricable link of biopolitics with racism, sovereignty, social hierarchy, and capitalism to the extent that it is extended to what Mbembé (2003) calls "necropolitics," namely the "subjugation of life to the power of death" (39).

Instead of keeping the 1918 influenza pandemic overshadowed by WWI, Goldberg's and Mullen's outbreak narratives elaborate the confluence of the two that reflects a combination of multiple powers: sovereign, biopolitical, and necropolitical. In their novels, through waging the Great War and sending people to the battlefields, Germany and the USA exercise sovereign power to "take life or let live" and "the right to kill those who represented a kind of biological danger to others" (Foucault, 1978, 138). It does not matter that some citizens like Lydia, the protagonist of Goldberg's novel (2006), "had not appreciated the war's breadth and scope, the brute numbers of men involved" (75). Mullen's novel (2007a) also depicts the misdeeds of the American Protective League, "deputized by the Department of Justice to make sure their fellow Americans were not fomenting dissent, interfering with the draft, or disparaging the war effort" (118). The supreme power of the state is strengthened, as Goldberg's and Mullen's novels foreground, when the influenza runs rampant during WWI. The authorities, in the name of getting the pandemic under control, are duty-bound to exercise a biopower intended to manage the population, to "foster life or disallow it to the point of death" (Foucault, 1978, 138). As both novels employ military and medical metaphors to underline the intersection of war and disease, they explore the conditions under which the biopower to protect the local and national communities is transformed into the necropower to divide people into "those who must live" and "those who must die" (Mbembé, 2003, 17). Under the double influence of the Great War and the 1918 influenza pandemic, the lives of soldiers, foreigners, and strangers are especially subject to both the state of emergency and the state of acceptance.

In *Wickett's Remedy*, the foreigners are excluded from the imagined community and fictionalized as both the enemies and virus carriers who must die because they are deemed threatening to the safety and well-being of the community. As the USA enters the Great War, Lydia's husband, Henry Wickett, wishes to join the army to serve the country, but he is turned down because he is too small and fragile. Even so, as the official propaganda that promotes the Liberty Loan proclaims, American citizens who are not strong enough to fight in the battlefield can support the Allied cause by purchasing war bonds. The Liberty Loan is strategically placed to "provide the force for an American vaccine against *Germ*any," allowing common people to participate in war effort to fight the "*Germ*-man" enemy and to keep the "infection at bay" (Goldberg, 2006, 70). While the country is striving to contain "the dread disease of war" so as "to reclaim the health of Europe and to defeat the *Germ*-man

Kaiser" (Goldberg, 2006, 71), a flu stealthily strikes Boston. When Henry catches a cold and has a fever, Lydia feels "as violated as though the Hun had invaded Boston itself" (Goldberg, 2006, 78). For Lydia, the flu that claims Henry's and many other Americans' lives is "an enemy that lives right here, not across an ocean" (Goldberg, 2006, 204). Not only is Germany rhetorically compared to germs as a figure of speech or the influenza compared to the Hun to accentuate the dreadful effects, but rumor has it that "a Gerry submarine was spotted in the Harbor releasing gas into the air. Spanish influenza gas" (Goldberg, 2006, 135). As discussed previously, the 1918 influenza did not originate in Spain, neither was there any direct evidence to substantiate rumors of a German virus leaked to attack the Allies. Goldberg's novel shows that, addressing the pandemic in the language of war, or vice versa, signifies a biopower that is even more prevalent in wartime than in peacetime since a virus may be used as a bioweapon to break down the immunological ecosystem in both a medical and a national frame. The racist appellation "*Germ*-man" disease resonates with contemporary American readers hearing President Trump call COVID-19 "the Chinese virus." Furthermore, the imagined community of the nation, as embodied by the second-person plural voice "we" in the novel's fictional official propaganda, is formed in defense against the enemy both outside and inside the USA, namely Germany and the influenza.

Imagining a fictional remote American mill town called Commonwealth under self-imposed quarantine to keep the flu at bay, Mullen's *The Last Town on Earth* takes a step further to portray people's hysteria about the spread of a virus via suspected spies who invade the town and the naïve assumption that descendants of German immigrants would never contract the flu. In the remote forests of the Pacific Northwest, where Mullen's novel is set, Charles Worthy leads a group of lumber workers to build Commonwealth on principles of freedom and democracy. Many of its inhabitants do not fight in Europe because they are either deemed "essential war workers" at the mill or not registered for service due to their conscientious objections to the "crooked" war (Mullen, 2007a, 25). The novel opens with Commonwealth as "the last town on Earth" (Mullen, 2007a, 42) because it is safe from both the Great War and the influenza, which quickly spreads from eastern cities to the state of Washington and inflicts the surrounding towns. In Commonwealth, rumors are going around, as Goldberg's novel similarly depicts, about "a black cloud over the Atlantic," which "the Germans released from a battleship" (Mullen, 2007a, 74), bringing the virus to Boston. Upon hearing of the rumor, Charles's adopted son, Philips, immediately responds, "Then I guess Elsie's family has nothing to worry about" because "they're German" (Mullen, 2007a, 74). Given that Philips is secretly in love with Elsie, his comment is certainly not an ill-intentioned criticism. Nonetheless, singling out the German origin of Elsie's family, Philips's comment reminds readers of the outbreak narrative in which "the stranger/carrier materializes, and amplifies, the disequilibrium that strangers characteristically represent" (Wald, 2008, 57). While the local community befriend Elsie's German family, Philips and the other armed guard posted at the gates shoot two unnamed soldiers who attempt to enter Commonwealth for food and shelter. Philips has doubts about the decision to shoot the soldiers, but with newspapers' "warning of sabotage on the home front," the townspeople suspect that the soldiers may be German agents sneaking into the country and bringing the flu to American army bases (Mullen, 2007a, 199). Their ungrounded suspicion becomes an acceptable reason, if not an excuse, to justify the murder of the two soldiers. By casting immigrants and unknown soldiers as suspects of being German spies or potential virus carriers, Mullen's novel identifies the fundamental instability of the nation as an imagined community and, in the language of nativism, marks the stranger or the estranged familiar figure as a threat to both the security and health of national and local communities. Under the circumstances, the life of the strangers has to be sacrificed.

Public health in an imagined national community cannot be maintained merely by murdering, excluding, or scapegoating the outsider/foreigner as a source of a plague. Behavior of the civilian population must be managed and their space confined, too. In *Wickett's Remedy*, for example, a city ordinance, issued by the Board of Health in Boston, forbids all public meetings, including schools, and all places of amusement (Goldberg, 2006, 311). Also in Seattle, as Elsie tells Philip, "[t]hey've canceled school in most towns, and closed any other places people get together," and inhabitants "can even get arrested" for not wearing a mask when going outside (Mullen, 2007a, 41). Not only are restrictions, bans, and punishment implemented by the state to discipline the population as a whole, but self-responsibility and self-regulation of the body is advised during the pandemic. As Goldberg's novel illustrates, citizens are instructed by the Department of Health to "take the following simple steps to protect themselves":

> Spray nose and throat daily with dichloramine.
>
> Get plenty of rest in bed.
>
> Keep windows wide open.
>
> Eat meals regularly and do not curtail on quantity.
>
> Beware of persons shaking their handkerchiefs.
>
> Don't use common towels, cups, or other articles which come into contact with face.
>
> Don't spit in public places. (172).

These steps are so simple that they can be observed by nearly every individual and practiced repetitively until they become part of a daily routine. Nevertheless, disciplinary and biopolitical power is "most dangerous precisely when we do not notice it," functioning "in an automatic, invisible, and perfectly *ordinary* way" (Lorenzini, 2021, S42). As ordinary and neutral as they may sound, wearing a mask, resting in bed, opening windows, and eating regularly, etc. have become biopolitical strategies that meticulously govern life to the extent that the population may become docile and normalized.

While proactive policies are made, and advice is given, albeit well-intentioned, to protect and sustain life, the deployment of biopolitics evokes worries about the increase in surveillance mechanisms and concurrent doubts about their effectiveness. In *The Last Town on Earth*, even if the quarantine policy is approved of by the majority in a town meeting, some residents are concerned about posting armed guards to block the roads into and out of the town, rendering Commonwealth much like the "work camps" or "prison-like factories" from which they have fled (Mullen, 2007a, 22). As soon as Yolen, one of the residents, begins to develop influenza symptoms, his house is guarded, as his wife complains, "I'm a prisoner in my own home!" (Mullen, 2007a, 237). The residents' calling the town "prison like" or feeling like a "prisoner" reminds readers of Foucault's critique of "panopticism." In *Discipline & Punish*, Foucault traces the birth of the prison to the measures taken in the seventeenth century to confront the outbreak of the plague in a town. The first measure to be taken is "a strict spatial partitioning," namely "the closing of the town and its outlying districts, a prohibition to leave the town . . . . the division of the town into distinct quarters, each governed by an intendant" (Foucault, 2008, 1). Under quarantine and strict surveillance, prison-like Commonwealth manifests an irony that is noted nowadays in many countries, including the USA, which have undergone months of full lockdown with COVID-19. Notwithstanding the strictly monitored quarantine in Commonwealth, some of its townspeople find ways to secretly sneak to Timber Falls, a flu-stricken neighboring town, in order to buy alcohol and "to visit girlfriends, to visit . . . . prostitutes" (Mullen, 2007a, 340). The desire for human contact prevails among some community members. When they return, they carry the virus back from Timber Falls and spread it to others. Finally, in Mullen's novel, the influenza lays siege to Commonwealth, as if it were mocking the futility of state control.

In a similar vein, Goldberg's novel captures the idea of the prison through its portrayal of Gallops Island (then called Gallups Island), a quarantine station at the outer edge of Boston Harbor, while bringing medical research strategies and ethics into question. As Goldberg explains in "Author's Note," her portrayal of Gallups Island is not purely fictional; rather, it is based on historical records. In November and December of 1918, Gallups Island was "the site of a US Public Health Service study designed to determine the cause and mode of spread of influenza," and the test subjects were "inmate volunteers taken from the Deer Island Naval Prison" (Goldberg, 2006, 367). In the novel, after her husband's and brother's deaths from the influenza, Lydia applies for a position as a nurse aide in the Gallups Island project. Upon arrival, Lydia is informed that, in case of a German invasion, staff members at the quarantine station "took shifts posting watch for the ominous rise of a periscope from beneath the surface of the waves" (Goldberg, 2006, 220). Feeling like "a prisoner of war," Lydia is stunned by the view of the inmate volunteers, "clad in rough, gray uniforms," and "hampered by shackles at their wrists" (Goldberg, 2006, 234). Healthy inmates participated in the project in trade for early release from prison. However, as test subjects, they were quarantined alone in a locked room after being given bacterial samples and ran a big risk if they contracted the influenza. After knowing that the second group of volunteers would be given direct "secretions" from influenza patients, Lydia could not help but ask one of the doctors if he hates the convicts for their crimes and feels that, as an assistant, she is "helping to kill people" (Goldberg, 2006, 281, 282). The controversy over the Gallups Island Project, as well as the security measures and self-imposed quarantining in Commonwealth, highlights the dilemma of upholding the right to health or the loss of other human rights, such as the right to freedom of movement. Another question, how to strike a proper balance between majority will and minority rights also arises from the controversy underscored in the novels.

The desire to protect public health gives rise to biopower and generates a sense of communal affiliation, but it is arguable that members of the imagined community are equally protected. Lorenzini (2021) maintains that "biopolitics is always a politics of *differential vulnerability*" (43). Mbembé also notes that "the calculus of life passes through the death of the Other" (2003, 18). Instead of eradicating "social and racial inequalities by reminding us of our common belonging to the same biological species," biopolitics "structurally relies on the establishment of hierarchies in the value of lives, producing and multiplying vulnerability as a means of governing people" (Lorenzini, 2021, 43–44). Goldberg's and Mullen's outbreak narratives precisely exemplify such hierarchies. As discussed previously, strangers and undesirable subjects like the prisoners who can be readily murdered or sacrificed appear to have a life that does not deserve to be lived. The novels also reveal "the obliteration of the very many invisible workers forced to work anyway so that others can continue to hole up in their homes. . . ." (Latour, 2021, S26). Medical personnel are evidently the most noticeable heroes during a pandemic as they risk their lives to save others, like Doc. Banes, "a thin, older man with spectacles that had been dirtied by the wet coughs of countless patients" (Mullen, 2007a, xi), and Acting Assistant Surgeon Percival Cole, who dies of the influenza on Gallups. Heading the Gallups Island project, Dr. Joseph Gold is deeply distressed by Cole's death, regretting that he "cannot transmit flu to a single goddamn Navy deserter but kills off his best assistant surgeon"(Goldberg, 2006, 346). Dr. Gold's inner thoughts reveal that, according to social standards, a convict's life is less valuable than a medical official's although both are dedicated to scientific research that "will prevent future deaths, future suffering" (Goldberg, 2006, 227). Goldberg's and Mullen's novels simultaneously draw the reader's attention to other frontline workers who, despite providing essential services at high risk of acquiring the infection, are less recognized than medical personnel. In Mullen's novel, for example, when the contagion becomes rampant in Commonwealth, Metzger's General Store owned by Elsie's family stays open to provide the townspeople with food, supplies, and daily necessities, eventually causing the infection and death of Elsie and her mother. In the words of Clover (2021), "Horrified humans immediately noted this was a blood sacrifice to capitalism . . ." (S30). Likewise, in Goldberg's novel (2006), before assisting in the Gallups Island project, widowed Lydia has to go out to work when "everyone seemed to be suffering from a cold" (152). In the store where she works as a counter girl, the owner, Mr. Gorin, makes money from the influenza since he "ran a sale on handkerchiefs and instructed his girls to inform customers that silk handkerchiefs—which cost twice as much—were the most hygienic, though Lydia observed that Mr. Gorin used plain cotton" (Goldberg, 2006, 152–53). While Mr. Gorin is thinking of profits, his employees are exposed every day to the flu, and one of them even faints while serving a customer. These examples illustrate the vulnerability of workers under the triple impact of the pandemic, capitalism, and social hierarchies.

Although drastic times call for drastic measures, *Wickett's Remedy* and *The Last Town on Earth* suggest that pandemics require comprehensive measures to protect the health, safety, and well-being of all members of the imagined community, be it local, national, or global. Beyond the psychoanalytical discourse of trauma and the rhetoric of recovery that commonly informs the Spanish flu narratives of the 1920s and 1930s, Goldberg's and Mullen's twenty-first-century outbreak narratives supplement patients' accounts with those of the strangers, policy makers, and essential frontline workers, allowing for a political reading of contagion, communicability, and community. The novels show that biopower is neither good nor bad in essence, but it was dangerous during the 1918 influenza pandemic as it is in the time of COVID-19, given the manipulation of biopolitics to extend and intensify state control and to curb human rights. Illuminating the danger of biopower being turned into necropower, Goldberg's and Mullen's novels further complicate Wald's (2008) contention that "[c]ontagion was the color of belonging, social as well as biological" (12). Ideally, as Wald (2008) expounds, if all people are commonly susceptible to the pandemic, the epidemiology of belonging testifies "the common bonds of humanity, and the idea of a plague as a great equalizer, affecting rich and poor, worldly and devout" (12). Even if human contact and interdependence is natural and inevitable, as Goldberg's and Mullen's outbreak narratives illustrate, reimagining the community based on the common bonds of humanity would entail, first of all, the abolition of racism, nativism, and social hierarchies. Otherwise, such a reimagined community as Wald has imagined would exist only in the imagination.

References

Arnold, D. (2019). Death and the modern empire: The 1918–1919 influenza epidemic in India. *Transactions of the RHS, 29*, 181–200.

Arnold, D. (2020). Pandemic India: Coronavirus and the uses of history. *Journal of Asian Studies, 79*(3), 569–577.

Barry, J. (2004). The site of origin of the 1918 influenza pandemic and its public health implications. *Journal of Translational Medicine*, 2(3). https://doi.org/10.1186/1479-5876-2-3

Belling, C. (2009). Overwhelming the medium: Fiction and the trauma of pandemic influenza in 1918. *Literature and Medicine, 28*(1), 55–81.

Bollinger, L. (2013). Trauma, influenza, and revelation in Katherine Anne Porter's 'pale horse, pale rider.' *Papers on Language and Literature, 49*(4), 364–389.

Bristow, N. K. (2012). *American pandemic: The lost worlds of the 1918 influenza epidemic.* Oxford University Press.

Byerly, C. R. (2005). *Fever of war: The influenza epidemic in the U.S. army during World War I.* New York University Press.

Caruth, C. (1995). Introduction. In C. Caruth (Ed.), *Trauma: Explorations in memory* (pp. 3–12). Johns Hopkins University Press.

Clover, J. (2021). The rise and fall of biopolitics: A response to Bruno Latour. *Critical Inquiry, 47*, S28–S32.

Crosby, A. W. (1976). *Epidemic and peace, 1918.* Greenwood Press.

Crosby, A. W. (1989). *America's forgotten pandemic: The influenza of 1918.* Cambridge University Press.

Fisher, J. E. (2012). *Envisioning disease, gender, and war: Women's narratives of the 1918 influenza pandemic.* Palgrave Macmillan.

Foucault, M. (1978). *History of sexuality, volume 1: An introduction*. Trans. Robert H. Pantheon Books.
Foucault, M. (2008). Panopticism from discipline and punish: The birth of the prison. *Race/Ethnicity: Multidisciplinary Global Contexts, 2*(1), 1–12.
Goldberg, M. (2006). *Wickett's remedy*. Anchor Books.
He, D., Zhao, S., Li, Y., Cao, P., Gao, D., Lou, Y., & Yang, L. (2020). Comparing COVID-19 and the 1918–19 influenza pandemics in the United Kingdom. *International Journal of Infectious Diseases, 98*, 67–70.
Kolata, G. (1999). *Flu: The story of the great influenza pandemic of 1918 and the search for the virus that caused it*. Farrar, Strauss, and Giroux.
Latour, B. (2021). Is this a dress rehearsal? *Critical Inquiry, 47*, S25–S27.
Lorenzini, D. (2021). Biopolitics in the time of coronavirus. *Critical Inquiry, 47*, S40–S45.
Mbembé, A. (2003). Necropolitics. Trans. Libby Meintjes. *Public Culture, 15*(1), 11–40.
Mullen, T. (2007b). P.S. ideas, interviews & features . . . In *The last town on earth* (pp. 1–22). Harper Perennial.
Mullen, T. (2007a). *The last town on earth*. Harper Perennial.
Outka, E. (2019). *Viral modernism: The influenza pandemic and interwar literature*. Columbia University Press.
Phillips, H. (2014). The recent wave of 'Spanish' flu historiography. *Social History of Medicine, 27*(4), 789–808.
Price-Smith, A. T. (2009). *Contagion and chaos: Disease, ecology, and national security in the era of globalization*. MIT Press.
Sandset, T. (2021). The necropolitics of COVID-19: Race, class and slow death in an ongoing pandemic. *Global Public Health, 16*(8–9), 1411–1423.
Scarpa, R., Caso, F., Costa, L., Passavanti, S., Vitale, M. G., Trojaniello, C., Del Puente, A., & Ascierto, P. A. (2020). May the analysis of 1918 influenza pandemic give hints to imagine the possible magnitude of corona virus disease-2019 (COVID-19)? *Journal of Translational Medicine, 18*(489). https://doi.org/10.1186/s12967-020-02673-6
Taubenberger, J. K., & Morens, D. M. (2006). 1918 influenza: The mother of all pandemics. *Emerging Infectious Diseases, 12*(1), 15–22.
Wald, P. (2008). *Contagious: Cultures, carriers, andthe outbreak narrative*. Duke University Press.
Wald, P. (2016). Forward. In K. Nixon & L. Servitje (Eds.), *Endemic: Essays in contagion theory* (pp. 5–8). Palgrave Macmillan.
Ware, R. W. (2009). Thomas Wolfe's 1918 flu story: The death of Ben in the context of other literary narratives of the pandemic. *The Thomas Wolfe Review, 33*(1), 67–82.

Part V
Pandemics/Epidemics in Indian Literature

Chapter 9
Spatial Pathologies: The Biopolitics of Disease, Death, and the Caste-Body in Ananthamurthy's *Samskara: A Rite for a Dead Man*

Gaana Jayagopalan

Abstract This essay explicates the narrative and symbolic value of triangulating caste, disease, and death in enabling a spatially informed sense of the caste-body and the diseased body politic. Through a reading of Ananthamurthy's *Samskara: A Rite for a Dead Man*, this essay argues that the plague functions as a narrative landscape to foreground the body politics of the community. *Samskara* is an important interjection into our present times especially highlighting the intersectionality of disease. The novel, known for its radical perspective of caste and rituals that inscribe and legitimize caste positions, is a pertinent narrative that also foregrounds the community's body politic located in the caste proscriptions of the body. This biopolitics is central to the negotiation of disease, death, and caste in this essay. The production of disease, as well as defilement, alongside the contours of a diseased and outcaste body, become crucial vectors of investigation here. Therefore, Ananthamurthy creates a narrative world that generates a movement away from a single pathological pathway to understanding disease and locates it within a Brahmin space to bring home questions that have been cast(e) away until then. Using the frameworks of understanding caste and disease through the prism of biopolitics, the essays seeks to explore the following questions: Which bodies are legitimized as bodies-in-death? Which of them are cast out of the body politic of the community? How does disease intersect with the cast(e)-out body and the dead body? By closely analyzing spatialization of the diseased dead body in the community through the conceptual lens of "autoimmunity," and the diseased body politic of the community, the essay argues that the inscriptions of caste, disease, and death on the body enable an embodied engagement with the spatial dimensions of life, living, and dying. *Samskara* functions as an illustrative representation that expands the textual geographies of the body-in-disease, the caste-body, and the body-in-death.

G. Jayagopalan (✉)
Associate Professor in English, Department of Language and Literature, Alliance University, Bangalore, India
e-mail: gaanaj.25@gmail.com

S. Venkatesan et al. (eds.), *Pandemics and Epidemics in Cultural Representation*,
https://doi.org/10.1007/978-981-19-1296-2_9

Introduction

This essay explicates the spatial politics of the caste-body and deceased/diseased body alongside its consequential abhorrence of touch in the Brahmin subject as discursively rendered in U. R. Ananthamurthy's novel *Samskara* (1965). Translated into English by A. K. Ramanujan as *Samskara: A Rite for a Dead Man (1975)*, the novel asks crucial questions relating to tradition and modernity in a small Brahmin settlement in Karnataka. How does death and disease manifest itself within the contours of notional Brahmanical purity? How does a dead Brahmin body continue to occupy an interstitial, suspended position of being and non-being? These are some of the questions addressed in this essay.

Originally written in Kannada, through *Samskara*, Ananthamurthy explores how the biopolitics of death management fails in the Brahmin Agrahara (space where Brahmins reside in and around temples) of Durvasapura due to an interesting inversion of the anxieties that a dead body perpetuates in the Agrahara. Naranappa, one of the principal characters in the novel, has passed on due to plague. The novel explores how his death brings the mechanisms and strategies of Brahmanical regulation of *Brahmin subjects* within the space to a standstill. Naranappa has died a death no Brahmin desires, especially because of the highly tenuous position he held within the society. Alive, he was seen as every Brahmin's nemesis, flouting the regulations of the caste space. In his death, too, he escapes the caste sovereignty that will have rooted his social identity inscribed by the community on his body.

What this inevitably points toward is a fascinating reversal of power—turning its locus onto the diseased/deceased Naranappa, the community fears their own ostracization if he is cremated without proper directions from Praneshacharya. This is a morbid reminder of a community losing touch with humanity. In my reading, I explicate how the plague functions as a narrative landscape to foreground the body politics of the community and as a measure of the community's depleting social health and increased distance in social bonding. In using plague as a trope to unravel the community's moral corruption, Ananthamurthy draws our attention to the degeneration of an outcast(e) body, while also simultaneously hinting at the symptomatic degeneration of the community's body politic. I propose that Naranappa's dead body, placed within the physical space of the Agrahara, becomes a site of what I call spatial pathologies (both biomedical and moral diseases) that come to frame his death, as well as the Agrahara's Othering as diseasing.[1]

It is noted that death and dying are inevitably anchored in space and place. In a weighty engagement with spatialities of death, Maddrell and Sidway in *Deathscapes* (2010) remark: "Death and dying draw attention to the meanings that we invest in space and place, in as much as spaces and places offer a lens through which to understand death and dying" (Maddrell and Sidaway, 2010, 1). *Samskara*, among other things, is a spatial and place-based account of death and dying. Naranappa's death stirs the strongest emotional and caste-based anxieties of purity and pollution among the Agrahara's other resident Brahmins. It represents a finitude of who remains a

[1] Etymologically, "disease" ties up with a "lack of ease" or causing inconvenience.

Brahmin if the cremation is carried out, notwithstanding the Brahmin tenets that govern who is a Brahmin and who is not. This anxious turn in failing to recognize Naranappa's death as a human death that warrants a respectful and dignified cremation, and instead, being ridden by the nervousness (of remaining a Brahmin if the rites of this cremation is carried out), is at the center of the novel.

The essay begins with a critical overview of the novel, followed by a delineation of how Naranappa's dead body functions as a significant spatial pivot around which an entire community's precarious social and biological immunity rests. Following this, the essay explores two thematic paradigms of spatialities as a mode of making sense of a diseased/deceased body in the Agrahara: (1) the Agrahara as a physical space homogenizing caste purity, and (2) the caste-body as a spatio-pathological realm.

Samskara: An Overview

Famed and infamous for its representation of an insulated and decadent Brahmin community in the fictional Durvasapura, *Samskara* (meaning "rites of passage for the dead") is layered at several levels. Narrating the story of Praneshacharya, the devout and learned Brahmin and the epitome of everything "pure" in Brahmanical value system, *Samskara* takes the reader through the transformations within the Agraharaas also within Praneshacharya himself. The catalyst, both while alive and in death, is the "heretic" Brahmin Naranappa who drank liquor and ate meat—the man who has escaped and continues to override the regulatory discourses of Brahmanical way of life. Lakshmana, one of the Brahmins in the Agrahara, recounts:

> Naranappa abandoned his lawful wife after tying the wedding string round her neck … He went and got mixed up with some woman. …he didn't care to observe the death anniversaries of his own father and mother. …We tolerated things and sheltered him in our lap as long as we could. In return, what does he do? He comes to the river in full view of the Brahmins and takes the holy stone that we've worshipped for generations and throws it in the water and spits after it! Condone everything if you wish—but didn't he, willfully, before our eyes, bring Muslims over and eat and drink forbidden things in the wide open front yard? If any of us questioned him in good faith, he would turn on us, cover us with abuse from head to foot. As long as he lived, we just had to walk in fear of him" (Ananthamurthy, 1965, 7)

The immediacy in the plotline is driven by the dilemma regarding performing the last rites of a dead Naranappa that needs resolution, who, when alive, was not a "Brahmin" in the eyes of the Agrahara. Anyone who completes the cremation stands in fear of excommunication for cremating a man who was "un-Brahmin" all his life. This spirals into an annihilating "quasi-suicidal" (Derrida, 2003, 94) moment in the Agraharaas the plague takes over the space. Thus, both the dead body and the Agrahara are construed to be proliferating pathogens—a corporeal disease parallel to a cosmological corruption.

While Naranappa's rites as per prescribed Brahmin ways remain unperformed, the devout Praneshacharya is initiated into sexual desire, a rite of the flesh—a life he has never known in his voluntarily sacrificial celibate life. Eventually, a decadent

Brahmin village that fails to open itself to what it traditionally considers the Other (that is, everything that it deems un-Brahmanical) sees its own end. The novel closes on a note of uncertainty with Praneshacharya's dilemma now moving from how to cremate an "un-Brahmin," onto a plane that leads him to question his own identity as a Brahmin. I explore this uncertainty as a potentially liberating moment in the narrative construction of the events in the novel using the Derridean framework of "autoimmunity."

Locating Autoimmunity Vis-À-Vis Caste Spaces

My essay here imports the conceptual frames of "autoimmunity" vis-à-vis biopolitics as a mode to understand the caste-body as a diseased body, and also as a dead body—cast away. The frame of autoimmunity presents not just an incidental value to analyze disease alone, but it also functions as an analytic with considerable scope to negotiate systemic social orders that purport sterility and insularity as a mode of safeguarding the body's borders—both the biological, and the biopolitical kind. In fact, Mitchell (2007) delineates the interconnection between the biological and the socio-political inflections of "immunity" this way: "Whole theory of the immune system and the discipline of immunology is riddled with images drawn from the socio-political sphere—of invaders and defenders, hosts and parasites, natives and aliens, and of borders and identities that must be maintained" (Mitchell, 2007, 282). I extrapolate this framework to foreground the Agrahara as a space that legitimizes these insulations on the basis of caste and Brahmanical prescriptions of the inside/outside, us/them, Brahmin/un-Brahmin.

"Autoimmunity" is a biomedical term designated to the behavior of the immune system turning against its own tissues. This is often seen as an extremely terrifying and painful relation to one's own body. Philosophers including Jacques Derrida, Jean Baudrillard, among others, have theorized this notion by extrapolating the notion of "autoimmunity" to understand the social body politic. Baudrillard attributes the social body politic as being analogous to the biological one, in that, the "virality" in the community begins to attack its own structural social order (Baudrillard, 1993). This, Baudrillard observes, is because of a community's increased insularity that generates a sense of homogeneity. He writes: "At an early stage, the evil genie of otherness takes the form of accident, breakdown, failure. Only later does the viral, epidemic form make its appearance: a virulence ravages the entire system, and against which the system is defenseless precisely because its very integrity paradoxically engenders this alteration." (Baudrillard, 1993, 63). What Baudrillard cautions here is the potential annihilation of a system that practices sterile prophylaxis. In the elimination of any trace of otherness, Baudrillard warns, "total prophylaxis is evil" (1993, 64).

This line of conceptualization of autoimmunity is countered in Derrida's more open and potentially liberating notion of autoimmunity as both a necessity to maintain a vitality as well as possibly weakening the body politic. Positioned within the context of analyzing terrorism in general, and the 9/11 attacks in particular, Derrida builds an analogy between the body politic and the immune system, too (Derrida, 2003). He writes of autoimmunity in *Rogues* (2005) as the condition where a "living being can spontaneously destroy, in an autonomous fashion, the very thing within it that is supposed to protect it against the other, to immunize it against the aggressive intrusion of the other" (Derrida, 2005, 123). It is this kind of insularity that Ananthamurthy both cautions as an annihilation and also signals a liberatory potential in this destruction. Through Praneshacharya's dilemma, mirrored by all that he reproached in Naranappa, this insularity is dismantled through an uncertain, yet potential embrace of everything "un-Brahmin."[2]

I work with autoimmunity here at two levels: (1) at the level of the caste-body politic that regulates behavior, secures its ideology within space, and time, disallowing the Other to enter this spatially and ideologically guarded social order; and (2) the plague-pathogen that magnifies the effect of this insulation and is brought into the community by one of its own members whose disregard for the very body politic that seeks to regulate it also provides the body politic an opportune moment to turn inward (through Praneshacharya). It is in the interstices of these two vectors that I locate Naranappa's subjecthood, the first level augmenting the second one. When alive, Naranappa as the "heretic Brahmin" had turned into the community's autoimmunitary presence, *eatinginto* the Brahmin body politic from the insides of the community (consider how he is accused of having "initiated" many other Brahmin boys into "corrupt" and "un-Brahmin" ways (Ananthamurthy, 1965, 10)). In his death due to the plague, he brings the epidemic into the Agrahara, evidencing the Baudrillardian model of the community's annihilation. Ananthamurthy urges us to see the problems of the Brahmanical sterility and its spatial insulation through how the cast(e) body of Naranappa counters such spatial pathologies. In rendering his self as autoimmunitary both while alive and in death, Naranappa also sets the stage to reevaluate the Brahmanical biopolitics "analytically, systematically, and without moral tub-thumping, exactly as we would approach the diagnosis of a medical condition" (Mitchell, 2007, 283). Let us now explore this bipolar autoimmunitary existence enabled by Naranappa along specific thematic spatial frames. Here, I identify two paradigms to scaffold my argument. First is the spatiality of the Agrahara and everything it inheres and walls out. Second, the body's spatiality and the spatial pathologies that it charts in its being.

---

[2] In this light, Praneshacharya's initiation into the world of corporal desire is a transgression that still remains unknown to the community, and his turmoil is whether he must reveal his transgression or not.

Thematic 1: Physical Spaces and Caste Insularities—Agrahara and Its Biomedical and Cosmological Corruption

Spatiality immensely aids the comprehension of death including an attempt to grapple with its complexity. The place and space of death and dying frame our sense of death in different ways. "Deathscapes" (Maddrell & Sidaway, 2010) allude to the spatiality of death as bearing significance to our comprehension of death, and remembering the dead. Here, one can think about choices made of spaces for burial or about modes of memorializing the dead in materialities of public and private spaces (as photographs, or sculptures, or other art forms for instance), to fully understand the extent of significance that space plays in a culture's comprehension of death.

In *Samskara*, too, the spatiality of the Agrahara plays a crucial role in determining the community's negotiations after Naranappa's death. Here, caste inscriptions come to bear heavily on the physical space of the Agrahara, which in turn, also position Naranappa's dead body in a liminal location. To begin with, let us map the contours of the physical spatiality in the novel. "Agrahara" refers to a space of land traditionally granted to Brahmin communities by kings or the nobility in order to oversee temple administration in and around the area. This is always already a space regulated by the Brahmin body politic strictures including who can enter and exit the space, and how they do so. This public space is accorded centrality in the narrative.

Praneshacharya and the other caste-inhering members of the Agrahara live extremely well-mediated lives—a mediation enabled by scriptures. Naranappa's death in the "caste-pure" space of the Agrahara is central to the reading here. In fact, Naranappa is presented to the reader as the "hedonist" who took issue with the Brahminism followed in the Agrahara. Much of the Agrahara is characterized by greedy, lazy, and gluttonous Brahmins. The narrator observes:

> All twelve months of the year, they had vows to keep; they had calls for ritual meals occasioned by deaths, weddings, young boys' initiations. On big festival days, like the day of the annual temple celebrations or the death-anniversary of the Great Commemorator, there would be a feast in the monastery thirty miles away. The Brahmins' lives ran smoothly in this annual cycle of appointments. (Ananthamurthy, 1965, 16)

A space that ostensibly others any instance of transgression is constructed as an uncontaminated, pure space. The Agrahara represents a utopic spatiality of Brahmanical mores of purity and taboo, often concealing its own vices of human othering. Into this spatiality, Ananthamurthy introduces disease as a canvas onto which the community's caste anxieties are cast. Disease, then, is the discursive apparatus through which Ananthamurthy positions to invite moral and philosophical considerations of the Brahmin subject.

Naranappa, a resident of the Agrahara, enters it from outside (that is from Shivamogge) bringing along with him, the dreaded plague. Chandri informs Praneshacharya: "He came back from Shivamogge and took to bed in a fever. Four days of fever, that's all. He had a painful lump on his side, the kind they get with fever" (Ananthamurthy, 1965, 3). Naranappa's death due to the plague is inside his Brahmin home—a domestic, private space, situated in the public space of the Agrahara that does not come to be entered.[3] His "un-Brahmin" home is *unheimlich* (Freud, 2003) in the Freudian sense of liminality. Theodor Adorno's reflection on the Freudian uncanny is pertinent here: "The uncanny is uncanny only because it is secretly all too familiar, which is why it is repressed" (Adorno, 1997, 262). In this sense, Naranappa's unhomely private space is an uncanny reminder of the moral corruption of the entire Agrahara which it does not want to forthrightly acknowledge.

Death associated with disease, or an epidemic, is largely governed by medical discourses. In the Agrahara, the members in the community are ignorant of Naranappa's death as due to the plague. Hence, the death here is not rendered as scientifically knowable but is instead negotiated via the biopolitics of the community. Thus, Naranappa's mortal remains as a *heretic Brahmin* who has *died of the plague* draws onto itself the discursive terrain of religion and dogma on the one hand and rationality and biomedical discourse on the other. Even in his death, Naranappa brings the Agrahara to its knees. The narrator explains:

> The news of his death spread like a fire to the other ten houses of the Agrahara. Doors and windows were shut, with children inside. By god's grace, no Brahmin had yet eaten. Not a human soul there felt a pang at Naranappa's death, not even women and children. Still in everyone's heart, an obscure fear, an unclean anxiety. Alive, Naranappa was an enemy; dead, a preventer of meals; as a corpse, a problem, a nuisance" (Ananthamurthy, 1965, 3)

Through Naranappa, the community is increasingly worried about the risk of contamination due to Naranappa's "transgressions" that may now render an entire Agrahara defiled, if his last rites were to be performed. The defilement is also extended to the affective recognition of this anxiety as an "unclean anxiety," further consolidating the regulatory mores of caste control. The risk of potential self-contamination and caste-based ostracization that may follow is foregrounded every time one seeks answers to the question "how to continue to remain a Brahmin?" This is evidenced by the anxiety that pervades not only within the Agrahara but also in the neighboring Brahmin settlements. For instance, consider this: Parijatapura, an hour away from Durvasapura, housed the Smartha Brahmins who are considered lower than the Brahmins of Durvasapura. Naranappa often visited these households and "ate in their houses with no show of caste pride" (Ananthamurthy, 1965, 19). On learning about his death, and the dilemma of the Brahmins in Durvasapura, Shankarayya, the priest of Parijatapura is anxious, too. The narrator frames his exchange with the priests from Durvasapura thus:

> …Naranappa, who drank liquor and ate meat, who threw the holy stone into the river, is he a Brahmin or is he not? Tell me, which of us is willing to lose his Brahminhood here? Yet it is not all right, I agree, to keep a dead Brahmin's body waiting, uncremated.

[3] Although Brahmin by birth, Naranappa's home would be considered un-Brahmin by the Agrahara's standards. Apart from his sexual relations with a "low-caste" Chandri, the house was also the space of commune for other men and their "un-Brahmanical" indulgences including drinking liquor and consuming meat.

Shankarayya's heart panicked and missed a beat. His clan had already been classed low, and he did not like them to fall lower by doing something un-Brahmanical. So he said:

> 'If that's so, wait, we can't do anything rash. You, of course, have in Praneshacharya a man known all over the South. Let him look into it and tell us what's right in this crisis. (Ananthamurthy, 1965, 19)

The Agrahara continues to be ignorant about the possible contamination of their corporeal selves due to a potentially debilitating epidemic at the Agrahara's threshold, and the heightened sense of awareness about one's own mortality that death is likely to bring. Instead, they are inclined to think further around the immediate limits that dealing with this death can bring to their Brahminhood. Praneshacharya ruminates on the several encounters with Naranappa that had left him troubled. Even a few days before his death, a drunk Naranappa had challenged Praneshacharya thus:

> Let's see who wins in the end—you or me. I'll destroy Brahminism, I certainly will. My only sorrow is that there's no Brahminism really left to destroy in this place—except you. Garuda, Lakshmana, Durgabhatta—ahaha—what Brahmins! If I were still a Brahmin, that fellow Garudacharya would have washed me down with his *aposhana* water. Or that Lakshmana—he loves money so much he'll lick a copper coin off a heap of shit. He will tie another wilted sister-in-law round my neck, just to get my property. And I'd have to cut my hair to a tuft, smear charcoal on my face, sit on your verandah and listen to your holy-holy yarns. (24)

Naranappa presented for the community an increasingly harrowing anxiety of creating a breakdown in the social order. It is but evident that in an otherwise insulated social order, the autoimmune trigger within the body politic of the Brahmin Agrahara was Naranappa himself. In neither embracing Naranappa as a fellow-Brahmin, nor excommunicating him for his apparent "deviances," Naranappa takes on the autoimmune referent of the social body politic here. He triggers a nervousness by ushering the outside into the community, affecting its immunity. Here, one must note that not only does Naranappa bring non-Brahmanical acts into the community, but he also brings with him the disease vector, too.

Within the space of the Agrahara, Naranappa's death does not invoke the scientific imperative to understand the ramifications of his death to the community at the level of a public health discourse either. Naranappa leaves behind a plague in the Agrahara, and this is furthered by the Brahmins who go to neighboring towns only to either die *enroute* or bring the dreaded epidemic back into the Agrahara.[4] Naranappa has crossed the unthinkable in his transgression while alive and his death represents a finitude for the community that fears will be pushed into an abyss, dictated by caste-based mores. Naranappa while alive had made the Agrahara—the pure dwelling space of the Brahmins—unhomely. In his death, Naranappa is still designated by his life as an "un-Brahmin" rather than as a dead Brahmin. The members in the Agrahara then inhabit the ultimate liminality—living with death. This is not just the presence of Naranappa's body, but also the looming death in terms of the plague that has now entered the "morally" and "ritually pure" space of the Agrahara.

[4] In Naranappa's death, the "low-caste" Belli's household that loses its members to the plague does not invoke the same moral imperative, precisely because of its marginal position.

Thematic 2: Spatial Pathologies of the Caste-Body

The caste-body can be understood as the site that consolidates and carries with it the prescriptions of caste—high or low. Naranappa's caste-body is liminal caste-wise because he, a Brahmin by birth, has rendered it un-Brahmin with his sexual relationship with Chandri, whose caste-marked body is one of defilement owing to the lower hierarchy she occupies in the caste ladder. Naranappa has further indulged in actions that have defiled the purity of his caste identity by consuming meat and liquor. Within such complex renderings of caste-bodies, Naranappa's death complicates the spatiality his body comes to bear as a site of caste inscriptions.

Death in the novel is a spectacle that cannot be seen, its excesses not borne to witness in the corpse but allegorically made visual in all that is happening to the community. Its ostentatiousness is ironically rendered by ensuring the corpse is rendered invisible, since it represents a rogue-Brahmin in death, involved in surrogate pleasures unbecoming of a Brahmin. Yet, his death spontaneously enters public space through the Agrahara recalling his un-Brahmanical self. Thus, Naranappa as a corpse, slips in and out of space. His entry into the Agrahara's conversations after his death is as a Brahmin who has ceased to be one in the eyes of the Agrahara and has now literally ceased to be. He was also the anxiety-inducing Brahmin with the potential to render many others in the same way too.

In this context, Naranappa's not-yet-visible corpse becomes more than representational—it alludes to the death of the entire Agrahara which fails to cremate a fellow human. In his death, Naranappa foregrounds the complete absence of a democratic citizenship in the Agrahara and renders the space as devoid of a human bond constituted by human touch. Refusing to literally touch Naranappa's corpse fearing their own ostracization is the ultimate othering in death. Naranappa's body, therefore, contains and conceals: it contains the moral disease inscribed on him in an attempt to conceal the biopolitics of the community. However, it simultaneously reveals the moral epidemic of othering rampant in the community. There is a parallel invisibilization of the community's own moral corruption, that comes to the surface with the plague entering the Agrahara itself, and the moral dilemma that Praneshacharya finds himself in.

The novel introduces into its spatial negotiations two very different questions among the members of the community: "How does one remain a Brahmin?" and "to whose community does Naranappa's death belong?" In the face of death and disease, instead of asking how to remain human, the community instead searches answers to the question "how to remain a Brahmin?" Praneshacharya reflects:

> ... that Naranappa didn't behave as a well-born Brahmin, that he's a smear on the good name of the Agrahara, it's a deep question—I have no clear answer. For one thing, he may have rejected Brahminhood, but Brahminhood never left him. No one ever excommunicated him officially. He didn't die an outcaste; so he remains a Brahmin in his death. If we let someone else do it [the burial], we'd be sullying our Brahminhood. Yet I hesitate, I can't tell you dogmatically: go ahead with the rite. I hesitate because you've all seen the way he lived. What shall we do? What do the Law books really say, is there any real absolution for such violations? ... (Ananthamurthy, 1965, 9)

What is worthy of attention here is how Naranappa continues to be held as a Brahmin in death, and yet, his final rites cannot be performed. Occupying this liminal space, Naranappa's dead body is an ugly distortion of the mortal life he lived.

In its decomposed and degenerative state, the community is now concerned with the second question of spatiality and human death: To whom does this death belong? In his death, Naranappa is a non-human, nearly so while also alive, especially due to his indulgences. The Agrahara is now plagued by the crisis of positioning Naranappa within the community: Is he one of them, or not? The Brahmin body is construed to be the epitome of corporeal purity simply by being born a Brahmin. This is countered by the apocalyptic body of Naranappa. I call this apocalyptic not because Naranappa ended up being "un-Brahmanical," but because his recalcitrant remains continue to be unattended to, triggering further disease in the Agrahara, destabilizing the community's immunity from within. The moral sickness that Naranappa apparently carried with him when alive inscribes him in his death. In declining to cremate Naranappa fearing losing their own Brahminhood, Naranappa's uncremated corpse is also an allegorical representation of the sustenance of the plague epidemic, and the larger epidemic—of caste purity.

Often a dead body is comprehended by senses of smell, sight, and touch. In the case of Naranappa's death, it is not his body that is witnessed, but his death that is again only aurally announced by Chandri to Praneshacharya. Much later, the excesses of the dead body begin to seep into the Agrahara. Mary Douglas characterizes the dead body as a site of pollution, of total ruin and contagion. The dead body, theorizes Douglas, is an essence of impurity (Douglas, 1966). When the deceased is also diseased, it becomes a site of a disruption: Naranappa's flouting of the community's caste regulations results in a disruption in the Agrahara. Further, his unattended corpse in the Agrahara is a social and corporeal disruption, potentially becoming a site of grotesque decomposition. What Chandri encounters in Naranappa's house several hours after his death is this: "A horrid stench. Dead rotting rats. She was grief-stricken that she's left the body orphaned, unprotected, the body of the man who'd antagonized the whole Agrahara for her sake. …The dead body was reeking. The belly was swollen, the face of the dead man was grisly, disfigured" (69). Soon after, Chandri cremates his body with the help of Naranappa's Muslim friends in the dead of the night.

As Naranappa's "dead body" remains in its decomposed state in his house, private to Chandri—its "thereness" is only invoked by the kind of person Naranappa was and not as he lies in his house. The object-body of Naranappa becomes a subject-body in the discussions in the Agrahara. The disembodied, object-body is a corpse due to the plague. The subject-body continues to be embodied in the memory of a Brahmin man who escaped the biopolitical reduction of a Brahmin subject. This manner of invoking his presence even in his absence is layered further when the plague itself reaches the Agrahara in symbols of death, which the community still fails to grasp. The excesses of the biomedical epidemic (like dead rats, hovering vultures, and a corpse that can potentially generate further diseases) evince a sanction to generate a discourse around the cosmological corruption. The descent of vultures in huge numbers, seated atop the Brahmin houses, waiting to peck on dead rats flung from homes captures this sentiment sharply. The narrator recounts:

> The vultures … unfurled their wings and flew, with rats in their beaks … The Brahmins, tired out, uttering the holy name of Narayana, climbed their front yards, covered their noses with their upper cloths and wiped the sweat off their faces. Sitadevi and Anasuya went to their husbands and begged tearfully: 'Let the gold go to hell! Why do we need other people's property? Please take out the body and get to the rites. Naranappa's spirit is calling out these vultures.' There was not a breath of wind; the stench stood stagnant in each house, and like a bodyless ghost it troubled everyone—wretched already with heat, hunger and dread. The orthodox Brahmins were distraught, as if nothing in lives to come would wash the filth of their day. (61)

There is a significant reimagination of Naranappa's body that emphasizes the community's vulnerability not in terms of the spread of the disease, but in terms of a blemish on their Brahminhood. It is in this intersection of the Brahmin body and the diseased dead body that Ananthamurthy locates the ultimate dilemma for Praneshacharya. In Naranappa's death, what is laid bare is not the reality of the epidemic that is to soon grip Durvasapura, but the materiality and the sustenance of Brahmanical consciousness and the intersectional excesses of caste and disease.

Autoimmunity and Biopolitics of the Social Body

Death is apprehended only in the Other. The presence of a dead body is what provides a notional sense of death. In the context of an epidemic, the dead body in multitudes characterize human precarity. Therefore, epidemic is centered around disease, and around the possibility of end of life. In its in between lie the social, political, and cultural contours of epidemics and death. What strikes as warranting attention in the narrative is the complete eclipse of epidemics and scientific notions of death, by a delay in cremating the dead.

The novel deals with the power exerted by the community in determining who belongs within and who is outside the community. Drawing from Foucault's theorization of biopower (Foucault, 2003), the Brahmins in the Agrahara, led by the discourse of religion and caste hierarchy, had the power to excommunicate Naranappa for his transgressive actions but failed to do so.[5] However, in Naranappa's death, his diseased, dead, transgressive body literally controls the prescription of whether the members in the Agrahara will remain Brahmin or not. The more the community delays taking a decision about his cremation, the more their anxieties of remaining Brahmin are fueled. This, in turn, makes Naranappa's body all the more polluting, thereby foregrounding an extremely agential role that the diseased/deceased, cast-away body plays in determining the spatial politics of the Agrahara.

---

[5] I draw here from Foucault's theorization of biopower that deals with the regulation of life and mortality. See his *Society must be Defended* (2003) for more.

In this death, the community is not sequestered by a loss; it is, instead, sequestered by an intense othering process that encumbers a safe Brahmanical subject position for themselves. Disease and death both do not turn the space of the Agrahara into a humane one. It is continued to be ravaged by the disease of caste purity. In this instance, not only is the body of Naranappa cold and inanimate, the entire Agrahara, driven by the Brahmanical biopolitics is cold to a dead Naranappa. A dead body, without intervention, goes through significant degradation and changes. By having the space of the dead body unattended to, Ananthamurthy subtly shifts our attention to the unattended Brahmanical othering within the space of the Agrahara itself.

The outbreak of the epidemic is also attributed to cosmological reasons of moral corruption of the Brahmin in the Agrahara. This defilement is rendered not in the scientific manner that biomedical discourses do when an epidemic breaks out. Instead, its cosmological significance overwrites any possible scientific explanation to the event. In a similar vein, Priscilla Wald remarks in *Contagious Cultures* (2008): "Microbial invasions take a mythic turn when they are cast as the response of the Earth itself to human beings who have ventured into primordial places they should not disturb. Understood alternatively as defensive and vengeful, this primal action is a recurring feature of outbreak narratives ..." (Wald, 2008, 10). In the novel's context, this can be delineated in the manner in which the Agrahara responds to the dead rats, incoming vultures, and more Brahmin deaths in the Agrahara itself as a cosmological corruption of Naranappa's "sins" and not cremating a dead Brahmin in the Agrahara. Contrast this with Manjayya who begins to recognize, there is more to the successive death of Brahmins in the Agrahara. The narrator frames Manjayya's urgency as follows:

> Experienced in affairs as he was, in the exchanges, the markets, the law courts and offices of Shivamogge, he's just laughed at the other Brahmins' explanations. They all believed that these disasters were due to Naranappa's untimely death and the Brahmins' dereliction of duty in not performing his last rites. ... But when he hears that rats had been running out of the Agrahara and falling dead, and carrion birds had arrived to eat them, his suspicions became certainties. His guess was correct .... The Tayinadu newspaper that came yesterday, though a week old, had printed news in a corner: 'Plague in Shivamogge'. Naranappa did bring the plague into the Agrahara .... (103–104)

Ananthamurthy portrays a community that regimented and hierarchized notions of purity and defilement on the basis of simply being born a Brahmin. The iconoclast that Naranappa was, he sought to break these regimentations, which were, in turn, seen to be taboos and transgressions.[6] Ananthamurthy enables a fascinating intersection of the social experience of caste with the social experience of an epidemic. Naranappa becomes the unprecedented mobile carrier of the disease. He brings it from Shivamogge, having now passed it on to so many others in Durvasapura, many themselves unaware of the condition. Praneshacharya himself is forced to face this grim truth in the death of his wife, Bhagirathi. In such a scenario, the epidemic reconstitutes

[6] One cannot undermine the rather problematic position Naranappa wields when alive: an upper-caste patriarchal man whose transgressions are presented in the garb of "modernity." The focus here, however, is on the possibilities that Naranappa's death holds out to the community rather than his actions while alive.

the idea of a regimented space into a spatiality that makes human connections more amorphous in disease, simply by rendering everyone as victims of the epidemic.

Mitchell (2007) comments "The limits, borders, boundaries of the body (politic), its relations of inside/outside, friend/enemy, native/alien, literal/figurative are exactly what is in question in the metaphor of the immune system …" (2007, 281). Derrida foresees autoimmunity as presenting an opportunity to thrive with alterity. While both these conceptions promulgate the virus' capability to dismantle rigid structures, Baudrillard's model is reactionary to a complete sterility and homogeneity; Derrida's on the other hand, sees autoimmunity as a necessary condition to engage with the Other (Mcfadden, 2015). Only when a body politics' autoimmunity is triggered, Derrida urges, can there be an instance of potentially considering an alterity to thrive. In a sense, to prepare for a new future, Derrida's theorization of strengthening the body politics inside requires an internal attack that becomes an "outside" at the moment of autoimmunity. While Baudrillard's model invites contamination in walling in, Derrida's model demands contamination for the sustenance of the body politic.

Where, then, is the disease? Is it in the Brahmin subject or in its viral vector? Praneshacharya, the epitome of Brahmin subjecthood comes to question this himself, his potential trigger as an autoimmunitary presence in the community next. This is soon after his celibate subjecthood is dismantled through Chandri, the same outcaste female subject who was deemed responsible for Naranappa's lost Brahminhood, too.[7] Drawing from Mitchell's elucidation of Derrida's conceptual frames, we can observe that Ananthamurthy uses the plague as a heuristic device to suggest, that the caste community is anxiously wary of Brahmanical biopolitics that generates this kind of autoimmunitary, suicidal response.

Conclusion

Death is a point that brings about an experience of being together. However, in this context, Naranappa's death further alienates the Agrahara from his dead body. While the community feared ostracization in case Naranappa's rites were to be performed, a ravaging plague stood before them with almost the possibility of wiping out the entire Agrahara. Wald constitutes this "chaotic and recombinatory nature" of epidemics (Wald, 2008, 22) as the familiar being rendered strange, with the community's annihilation now imminent. In such an instance, Naranappa's caste-body and the dying Agrahara both become symbols of a community at the brink of its destruction. Ananthamurthy asks for not the sustenance of this community, but instead a need to take a sharp look at what really constitutes being human in such a community. Naranappa's

[7] The clear link between sexualizing desire and locating it in the outcaste female subject Chandri, and the diseaseing Brahmin body of Naranappa cannot be missed here.

death provides moments of potential alterity, an existence that celebrates remembering than concedes to an amnesia. A. K. Ramanujan, in his epilog to the translation signals this Derridean aporietic potential: "... the last phase of the Acharya's initiation is an anxious return, a waiting on the threshold; his questions seem to find no restful answers. What is suggested is a movement, not a closure. The novel ends, but does not conclude" (Ananthamurthy, 1965, 147). Naranappa is responsible for the seismic waves in the Agrahara, but he is also the seismologist tracing the disturbances.

References

Adorno, T. W. (1997). *Aesthetic theory*. The Athlone Press.
Ananthamurthy, U. R. (1965). *Samskara: A rite for a dead man*. Translated by A.K. Ramanujan.
Baudrillard, J. (1993). *The transparency of evil: Essays on extreme phenomena*. Verso.
Derrida, J. (2003). Autoimmunity: Real and symbolic suicide. In G. Borradori (Ed.), *Philosophy in a time of terror* (pp. 85–136). Stanford University Press.
Derrida, J. (2005). *Rogues: Two essays on reason*. Stanford University Press.
Douglas, M. (1966). *Purity and danger: An analysis of concepts of pollution and Taboo*. Penguin.
Foucault, M. (2003). *Society must be defended: Lectures at the College de France 1976–77*. In M. Bertani & A. Fontana (Ed.), (D Macey, Trans.) Picador.
Freud, S. (2003). *The uncanny*. Penguin.
Maddrell, A., & Sidaway, J. D. (Eds.). (2010). *Deathscapes: Spaces for death, dying, mourning and rememberance*. Ashgate.
Mcfadden, D. (2015). Viral possibilities: Media, the body, and the phenomenon of infection. https://ir.lib.uwo.ca/etd/2999/
Mitchell, W. J. T. (2007). Picturing terror: Derrida's autoimmunity. *Critical Inquiry, 33*(2), 277–290. https://doi.org/10.1086/511494
Wald, P. (2008). *Contagious: Cultures, carriers, and the outbreak narrative*. Duke University Press.

Part VI
Millennial Pandemics: AIDS, COVID-19 and Beyond

Chapter 10
"No Country for the Infirm": Reading *Angels in America* During COVID-19

Jessica C. Hume

Abstract In the 1990s, Tony Kushner's landmark two-part play, *Angels in America: A Gay Fantasia on National Themes*, burst through the forefront of national literary consciousness with its incisive exploration of a panoply of intersecting thematic threads of American culture: marginalization and isolation, illness, wellness, and queerness, politics and power, religious judgment and moral justice, and the complex nature of emotional and physical love. The play, which examined the lives of several couples during the AIDS epidemic, urged Americans to consider not only these concepts themselves, but also the ways in which they would shift, coalesce, or become even more compelling as Americans moved into the twenty-first century. Though it addresses a specific moment in American history, the play transcends its specificity by speaking to universal components of Americanism, components which have once again come into razor-sharp focus in the context of the COVID-19 pandemic; thus, contemporary readings of the play are germane and salient (even seemingly prophetical). Specifically, analysis of *Angels in America* which considers the parallels between the AIDS epidemic and the COVID-19 pandemic demonstrates that, in American cultural imagination and cultural discourse, marginalization is often constitutive of illness. Moreover, it indicates that illness and Americanism are, essentially, mutually exclusive.

Introduction

In her seminal work on illness, *Illness as Metaphor* and *AIDS and Its Metaphors*,[1] Susan Sontag notes, "Reality is bifurcated, into the real thing and the alternative

[1] Sontag (1989).

J. C. Hume (✉)
Department of Healthcare Administration and Public Health; Director, Health Humanities Program, Bellarmine University, Miles Hall, Rm. 317, 2001 Newburg Road, Louisville, KY 40205, USA
e-mail: jhume@bellarmine.edu

S. Venkatesan et al. (eds.), *Pandemics and Epidemics in Cultural Representation*,
https://doi.org/10.1007/978-981-19-1296-2_10

version of it, twice over."[2] She claims each apocalyptic reality looming over American culture (environmental disaster, war, pandemic) is "haunted" not only by the image of the event, which is projected, but also by the catastrophized predictions of the event's future outcomes; further, she claims that "future-mindedness" is a hallmark of twentieth-century American thinking: "Future-mindedness is…the distinct mental habit, and intellectual corruption of this century…"[3] The present analysis extrapolates from the first element of Sontag's premise that events in American culture are doubled, meaning that each event comprises the event itself, and the image of the event. Using *Angels in America* as a lens through which to examine the distinction between the "images" versus the "realities," this piece examines Americanism during the AIDS epidemic and the COVID-19 pandemic. In this discussion, *Angels* serves as a milestone to elucidate some ways in which the "future-mindedness" of the late twentieth century has manifested in the twenty-first century. In the spirit of Sontag's original work, the present examination maneuvers through multiple disciplines, weaving together literary analysis, health policy, health law, news media, and psychology.

The notion of "Americanism," represents a significant a concept which has bifurcated into an image and a reality in the way Sontag describes—there is what quintessential Americans are *perceived as* (or wish to be perceived as), versus what Americans *are.* In *Angels,* Kushner identifies many of these projected images, criticizing their limitations and contrasting them with reality. To appreciate this aspect of the play, readers must have a basic framework for a projected image of Americanism. A specific basic framework enables readers to extrapolate the mechanism of Sontag's bifurcation phenomenon and analyze the ways in which the bifurcation of Americanism it is evident the twenty-first century COVID-19 pandemic. L. Robert Kohls' 1984 piece for the Washington (now Meridian) International Center "The Values American's Live By"[4] offers a concise and synchronic instrumental framework with which to perform this analysis. Kohls and the Center, a non-profit specializing in public diplomacy and global leadership, produced the document from the amassed anecdotal experiences of the Center's work introducing visitors to America since the Cold-War Era 1960s.[5] The complete, original list is comprised of 13 essential values that Kohls claims describe "most, but not all" Americans: "…we can say that if the foreign visitor really understood how deeply ingrained these 13 values are in Americans, he or she would then be able to understand 95% of American actions."[6] Among these values, three are germane to this discussion: "time and its control; future orientation; action/work orientation; and practicality and efficiency."[7] Ironically, the very existence of Kohls' list of values proves Sontag's theory about bifurcation in American culture, as Kohls admits: "Even if Americans had considered [the] question [of

[2] Sontag, *Illness*, 177.
[3] See Footnote 2.
[4] Kohls (2021).
[5] Earnshaw (2020).
[6] Kohls, "The Values," 1.
[7] Kohls, "The Values," 7.

definitive American values], they would probably, in the end, decide not to answer in terms of a definitive list of values. The reason for this decision is itself one very American value—their belief that every individual is so unique that the same list of values could never be applied to all, or even most, of their fellow citizens."[8] Kohls argues that, though most Americans would identify their personal values as individualized and unique, that is not actually the case: most Americans, Kohls claims (citing an anthropological approach) demonstrate the collection of values he enumerates.[9] There is the *image* of Americans (individualistic and unique), versus the *reality* (that many Americans share similar values). This ironic truism substantiates Sontag's claim that American culture operates on a mechanism of bifurcation—projected image versus reality.

In *Angels*, these American values manifest themselves in two key themes threaded throughout the play: Christian religion (and its association with freedom) and thanatophobia. During the COVID-19 pandemic, these same themes appeared frequently and in rich layers of cultural and political discourse. Critical examination of these themes, comparing their manifestations in *Angels* to their appearances in contemporary cultural milieus during COVID-19, enables readers to distinguish images of Kohls' American values. Then, readers can identify the myriad cracks in the images and clearly view the reality of Americanism highlighted in the interstices, a reality in which any ill citizen is persona non grata.

Equality and Egalitarianism Through Religious Identity

Describing equality and egalitarianism as a key American value, Kohls notes that this value is closely tied to religious identity: "Equality is, for most Americans, their most cherished value…this concept is so important for Americans that they have given it a religious basis…Most Americans believe that God views all humans alike without regard to intelligence, physical condition or economic status. In secular terms, this belief is translated into the assertion that all people have an equal opportunity to succeed in life."[10] This reference to God connotes a monotheistic Christian God by virtue of the phrase, "most Americans." According to Gallup poll information from 1985, 85% of Americans identified themselves as Christian.[11] The narrative of America's founding is often proudly touted by Americans as a story of a struggle for Puritan religious freedom. Thus, in America, Christian religious identity is closely tied to freedom. Kohls' phrasing regarding the value of freedom implies an unspoken truth about American culture—the value of freedom of religion (and thus, equality

[8] See Footnote 6.

[9] See Footnote 6.

[10] Kohls, "The Values," 3.

[11] "Religion," Gallup.com (Gallup, July 20, 2021), https://news.gallup.com/poll/1690/religion.aspx.

and egalitarianism) extends specifically to those who identify with Christian beliefs, but beyond that demographic group, it may have limits.

In perception, America is emblematic of freedom, but the reality is that while Americans claim to value freedom, their behavior demonstrates that they truly value the freedom of Christians above that of non-Christians. Homosexuality is a classic example: generally, Americans claim that fellow citizens are endowed with freedom, which (presumably) extends to individual sex lives; but, according to many mainstream Christians, homosexuality violates the tenets of Christianity. The exact nature of this freedom as endowed by the due process clause in the Constitution has a long and complex history.[12] To avoid lengthy digression, this examination will briefly address particularly salient cases. One case contemporaneous with *Angels* and the AIDS pandemic, demonstrates this conflict: *Bowers v. Hardwick.*[13] This 1986 Supreme Court case established that "the Constitution does not confer a fundamental right upon homosexuals to engage in sodomy," and upheld the right of the state of Georgia to criminalize sodomy (with a penalty of up to 20 years).[14] Hardwick's suit argued that engaging in consensual sodomy was a fundamental right and asserted that these fundamental "liberties" were part of our nation's history.[15] After making its way to the Supreme Court of the United States (SCOTUS), this claim elicited a response from the court: "to claim that a right to engage in such conduct is 'deeply rooted in this Nation's history and tradition' or 'implicit in the concept of ordered liberty' is, at best, facetious."[16] Hardwick's suit argued that the law was arbitrary because it was based on the subjective moral view that homosexual sex is wrong. The court responded: "The law…is constantly based on notions of morality, and if all laws representing essentially moral choices are to be invalidated under the Due Process Clause, the courts will be very busy indeed."[17] However, the fundamental rights, liberties, and notions of morality to which it refers are significantly linked to, or based on, monotheistic Christian values related to nonprocreative sex.[18] The court euphemistically made clear that the American concept of "freedom" did not (at that time) extend to homosexuals. SCOTUS reversed this decision in *Lawrence v. Texas*, but not until 2003.[19]

Having an identity—religious or secular—inconsistent with the morality of mainstream Christianity places invisible limits on those living in America, as Kushner

[12] "Fundamental Rights (Noneconomic Substantive Due Process)," Legal Information Institute (Cornell Law School), accessed May 12, 2021, https://www.law.cornell.edu/constitution-conan/amendment-14/section-1/fundamental-rights-noneconomic-substantive-due-process#fn561amd14.

[13] "Bowers v. Hardwick, 478 U.S. 186 (1986)," Justia Law (Justia Law, 2021), https://supreme.justia.com/cases/federal/us/478/186/.

[14] "Bowers v. Hardwick, 1986," Justia Law.

[15] See Footnote 14.

[16] See Footnote 14.

[17] See Footnote 14.

[18] "Fundamental Rights (Noneconomic Substantive Due Process)," Legal Information Institute. (Cornell Law School).

[19] "Lawrence v. Texas," Legal Information Institute (Cornell Law School, June 26, 2003), https://www.law.cornell.edu/supct/html/02-102.ZS.html.

points out in *Angels*, when readers meet Roy and Joe in Roy's office. In real life, Roy Cohn exercised massive power as an attorney in conservative political circles since the 1950s, and by the 1980s, he mingled with the Reagans. Despite his conservative leanings, Cohn had sex with men and subsequently died of AIDS. In Kushner's text, readers witness a fictional version of Roy Cohn, but he is the only character with a non-fiction parallel. Joe works as chief clerk for the Federal Court of Appeals, Second Circuit. During the meeting, Joe meekly asks Roy not to take the Lord's name in vain. Roy responds… "principles count. I respect principles. I'm not religious but I like God and God likes me. Baptist? Catholic?"[20] Vitally, this interaction gives Roy the opportunity to establish that he identifies himself broadly and simply with mainstream Christianity (though he is ethnically Jewish). The fact that he follows up immediately by identifying two of the nation's predominant Christian religions positions him as having a stronger connection to the prevailing American culture. When Joe reveals that he is Mormon, Roy replies, "Mormon. Delectable. Absolutely. Only in America."[21] Significantly, the Mormon faith is the oldest and largest Christian faith engendered in the USA. So, while Roy's turn of an old phrase may seem to reference the uniqueness of American culture, its subtext demonstrates the strong association between cultural Americanism and Christianity.

This interaction between Roy and Joe is also notable in its coded revelations about Roy's sexual identity. Initially, readers operating from a heteronormative framework assume that Roy is a straight man, and he affirms this presentation by aligning himself with stereotypical gender norms while speaking to Judge Hollins and his wife on the phone, as Joe listens: "Sorry to keep you holding, Judge Hollins, I—Oh *Mrs.* Hollins, sorry dear, deep voice you got…(*Hand over mouthpiece again, to Joe*) She sounds like a truck driver and he sounds like Kate Smith, very confusing…"[22] While this remark solidifies the perception of Roy as a heteronormative straight man, his subsequent remarks to Mrs. Hollins, then Joe, about Broadway shows serve as a coded indicator of his actual identity: "No you wouldn't like *La Cage*, trust me, I know…(Button; to Joe) You see *La Cage*?…Fabulous. Best thing on Broadway. Maybe ever."[23] Here, Roy refers to *La Cage aux Folle, which* opened in New York in the early 1980s, and centered around a male gay couple who had to spend an evening with a conservative straight couple, with one of the men in drag.[24] (Readers may know *La Cage* as the basis for the film, *The Birdcage*). The show's popularity centered gay couples in the mainstream.[25] In *Angels*, Roy has the discernment to know that Judge Hollins and his wife do not make for an ideal audience, but then asks Joe if he has seen it. Here, he attempts to explore Joe's sexual identity, in much the same coded way that gay men used to ask others if they were "a friend of Dorothy." Roy is code-switching in order to maintain his power and status. Kushner's Roy Cohn is a manifestation

[20] Kushner (2013) 14–15.

[21] Kushner, *Angels*, 15.

[22] Kushner, *Angels*, 12.

[23] See Footnote 22.

[24] Staff (2020).

[25] Playbill Staff, "Look Back at La Cage," n.p.

of the bifurcation of American values: He presents as a straight Christian-affiliated man but is an ethnically Jewish gay man. Of all of Kohl's values, freedom is arguably situated on the highest pedestal, but even Roy Cohn is not truly free.

Cohn knows that openly embracing aspects of his identity that conflict with Christian-associated ideas of freedom would endanger his status and power, as it did for many other gay men. He makes evident the extent of his fear of status loss when he refuses to acknowledge his AIDS diagnosis, even to his doctor: "Like all labels they tell you one thing and one thing only: where does an individual so identified fit in the food chain, in the pecking order? Homosexuals are not men who sleep with other men. Homosexuals are men who in fifteen years of trying cannot pass a pissant anti-discrimination bill through City Council. Homosexuals are men who know nobody and nobody knows. Who have zero clout. Does this sound like me Henry?"[26] In his monologue, Roy refers to an actual proposed New York City Council bill that would have extended a discrimination ban to include discrimination on the basis of "sexual orientation."[27] Gay activists in New York City had attempted unsuccessfully to pass the bill since the 1970s.[28] His comment refers to New York City politics but echoes the sociopolitical upheaval occurring nationwide; the gay liberation movement of the 1970s had turned its energies toward politics, and the goal of expanding the rights and protections of gay Americans. To controvert the expansion of gay rights and diminish the power of the political platform of gay Americans, Jerry Falwell co-founded the Moral Majority, a political organization affiliated with the Christian right, which strongly advocated against gay rights and for an agenda the organization publicized as "pro-family."[29] The Moral Majority gained political power and aligned itself with Ronald Reagan, and while groups like the Gay liberation front made incremental progress in large, liberal cities, national politics remained an immovable monolith.[30] For Roy Cohn, national politics and the prevailing cultural zeitgeist affirmed his view of the powerlessness of the gay community.

The early and mid-1980s saw rising conflict around gay rights as a political issue, as advocates of the "pro-family" agenda attempted to erase the progress of the 1970s and further restrict the gay community. In the 1980 Democratic Convention, the Carter campaign had agreed to add a general platform opposing discrimination based on sexual orientation[31] but Carter lost to Ronald Reagan, a model of right-wing conservatism. Falwell and the Moral Majority claimed credit for Reagan's victory and subsequent success of the Christian Right throughout the early 1980s.[32] The Reagan era progressed alongside the spread of AIDS and, in order to slow it, officials attempted strategies that encroached on the freedom of gay men, such as the

---

[26] Kushner, *Angels,* 46.

[27] Carroll (1974).

[28] Carroll, "City Council's Bill," n.p.

[29] Shilts (2007), 44.

[30] Shilts, *And the Band Played On*, 44.

[31] Shilts, *And the Band Played On*, 30.

[32] See Footnote 30.

closing of bathhouses in major US cities like San Francisco and New York. Constitutionally, individual states have the power to impose mandates related to public health, even when those mandates may encroach on personal liberty.[33] The extent to which they may encroach on that liberty is fiercely debated, as was the case with bathhouse closures. While bathhouses and public gathering places for bathing have existed for many centuries,[34] by the mid-1980s, bathhouses in socially liberal cities had become a meeting place for gay men and had gained a reputation for frequent promiscuous activity.[35] As fear about AIDS grew, and cases increased, legislators, council members, and activists in large cities that included bathhouses considered the option of closing them in hopes of curbing the spread of AIDS. At the time, even gay activists were divided about the possible closures: some argued that the closures were unfortunate but necessary; others claimed that the closures were a political overreach that infringed on the rights of the gay men who frequented the bathhouses. Some feared the closures might reverse the progress made in the 1970s: "'We've found the cure for AIDS. Prohibit gay sex.' While you're at it, why not repeal the consenting adults legislation. Queer bashing and morals arrests will be making a comeback."[36] In New York State in late 1985, the State Public Health Council changed the sanitary code: "… no establishment shall make facilities available for the purpose of sexual activities where anal intercourse, vaginal intercourse, or fellatio take place. Such facilities will constitute a threat to the public health."[37] However, according to a Columbia University study, only 16 in 100 gay men visited bathhouses at all[38]; a fraction of those participated sexual activities in which body fluids were exchanged.[39] The bathhouse closures restricted all gay men from a part of their social community and limited or decimated the income of those who worked there (who were often gay). Closures made national news, and further aligned gay men with promiscuity and AIDS. This alignment was reinforced by President Ronald Reagan when he finally spoke at length publicly on AIDS in 1987: "But let's be honest with ourselves. AIDS information cannot be what some call 'value neutral.' After all, when it comes to preventing AIDS, don't medicine and morality teach the same lessons?"[40] The subtext of the remark is that medicine and morality both indicate that gay sex is wrong and dangerous, and that AIDS is the punishment for engaging in it. Here, the influence of Christian morality on health policy decision-making becomes evident. Reagan's pathologization of gay men fueled public outcry that resulted in government restrictions on the freedom of gay men, under the guise of controlling the

---

[33] Mariner et al. (2005).

[34] Rabin (1986).

[35] Rabin, "The AIDS Epidemic and Gay Bathhouses," 730.

[36] "The Bathhouse Wars," *The Washington Post*, March 4, 1984, sec. Lifestyle, https://www.washingtonpost.com/archive/lifestyle/1984/04/19/the-bathhouse-war/5c864455-2310-4f9e-ac70-e0f57083a511/).

[37] Prohibited Facilities, New York Codes and Regulations, 24–2.2 (1994).

[38] *New York Times*, September 22, 1985, p. 22, as cited in Rabin, "The AIDS Epidemic," 730.

[39] See Footnote 38.

[40] Curry (1987).

spread of AIDS. Importantly, scholars reflecting on this era feel that the bathhouse closures amounted to a significant overreach on behalf of the state.[41]

During the COVID-19 pandemic, the American public witnessed a phenomenon that similarly aligned Christianity with the value of freedom—the prominence of anti-mask activists. In the initial stages of the pandemic, governors across the nation instituted emergency mask mandates. The mandates varied from state to state, but generally required citizens to wear face coverings in public places. Though research published by the Centers for Disease Control demonstrated the effectiveness of mask-wearing to significantly reduce the spread of COVID-19,[42] anti-mask sentiment quickly bubbled up in public discourse and news media, and anti-maskers most frequently argued that forced mask mandates violated their freedom.

In the same way that Reagan's response to AIDS contributed to the pathologization of gay men, the sentiments of President Donald Trump fueled anti-maskers: "They [the CDC] suggested for a period of time, but this is voluntary. I don't think I'm going to be doing it."[43] Later, Trump associated anti-masking with a compelling image of Americanism, claiming that he did not see himself wearing a mask while sitting in the Oval Office, "behind that beautiful Resolute desk."[44] He continued to deny the severity of the pandemic and resist masking, while appealing to Christians via dog whistle politics: "It's going to disappear. One day, like a miracle, it's going to disappear."[45] Recent data confirms the effectiveness of this strategy. Christian nationalism—in part comprised of the belief that Americans are specially protected by God—was the leading indicator that people would be likely to participate in incautious behavior during the COVID-19 pandemic, and the second leading predictor of likelihood for disregard of preventive measures such as mask-wearing and hand washing.[46] Importantly, Christianity and Christian nationalism are not synonymous. Christian nationalists, a subset of Christians characterized by science denial, American nationalism, and belief in media conspiracy, make up about 20% of white Christians, approximately 30 million Americans.[47] However, they comprise a significant voting demographic, and influences of Christian nationalism colored sociopolitical discourse throughout the Trump administration. Some scholars claim that attitudes of Christian nationalism may have worsened the pandemic.[48] Notably, though anti-maskers claim the mandates encroach on their freedom, the general legal argument indicates they do not.[49]

---

[41] Rabin, "The AIDS Epidemic and Gay Bathhouses," 740.

[42] "Science Brief: Community Use of Cloth Masks to Control the Spread of Sars-Cov-2," Centers for Disease Control and Prevention (Centers for Disease Control and Prevention, June 29, 2021), https://www.cdc.gov/coronavirus/2019-ncov/science/science-briefs/masking-science-sars-cov2.html.

[43] Victor et al. (2020).

[44] Victor, Serviss, and Paybarah, "In His Own Words," n.p.

[45] Goldberg (2020).

[46] Perry, Whitehead and Grubbs (2020).

[47] Whitehead and Perry (2021).

[48] Perry, Whitehead, and Grubbs, "Culture Wars And COVID-19 Conduct," n.p.

[49] Kaganovich (2021).

While Trump's alignment with the Christian Right mimicked that of his 1980s predecessor, Ronald Reagan, a key contrast appeared between Reagan and Trump. Reagan pathologized and marginalized gay men, using them as a scapegoat. Trump's public stance on COVID-19 amounted to complete erasure of the susceptible, the sick, and the dying: "It affects elderly people, elderly people with heart problems and other problems…nobody young…But it affects virtually nobody."[50] At the time of his comment, nearly 200,000 Americans had died as a result of COVID-19, and roughly 6.8 million had been infected.[51] Contrasting the responses of Reagan versus Trump to these public health crises reveals that the bifurcation of American values changed significantly between the two administrations. During the Raegan administration, the projected value was freedom for Americans, but the reality was that that freedom only applied to heteronormative Americans. AIDS was used as a prop to support that dynamic in public life. During the Trump administration, freedom remained the projected American value, but, that freedom only applied to healthy Americans.

Thanatophobia as a Side Effect of Kohls' American Values

Thanatophobia ("death anxiety") presents another outstanding example of the bifurcation—image versus reality—Sontag notes. Among Kohls' values are control over time, action/work orientation (hereafter abbreviated as "work orientation"), and future orientation.[52] According to Kohls' the first two values distill down to a hyper-focus on productivity, getting as much done as possible in the most efficient way possible.[53] In describing future orientation, Kohls' notes that Americans value the future to the detriment of their attentiveness to other aspects of their lives: "Valuing the future and the improvements Americans are sure the future will bring means that they devalue the past and are…unconscious of the present…Americans have traditionally been hopeful that the future would bring even greater happiness."[54] Aging and illness negate the ability to achieve significant levels of productivity; thus, these phenomena conflict with Kohls' values, control over time, and work orientation. The concepts of "death" and "future" inherently oppose one another; thus, the notion of dying, or the acknowledgment of dying, conflicts with the American value of future orientation. Essentially, because of their conflict with the American values of productivity and future orientation, illness and dying are shoved to the margins of American life.[55] Kohls' American values exemplify a public that wishes to pretend that illness and dying do not exist when, in fact, they do.

---

[50] Thrush (2020).

[51] Thrush," It Effects Virtually Nobody," n.p.

[52] Kohls, "The Values," 9.

[53] Kohls, "The Values," 2 and 4.

[54] Kohls, "The Values," 4.

[55] Field and Cassel (1997).

In *Angels*, a distinct thread of thanatophobia highlights this paradoxical problem of the conflict between American values and the reality of illness and dying. The clearest example is Louis, who demonstrates his future orientation when he presents his anxieties about death to the rabbi: "Maybe because this person's sense of the world that it will change for the better with struggle, maybe a person who has this…sense of constant historical progress towards happiness or perfection…Maybe that person can't, um, incorporate sickness onto his sense of how things are supposed to go…he isn't so good with death."[56] Louis begins this conversation by indicating to the rabbi that "this person" might abandon someone they love who is in great need. It is clear that "this person" Louis identifies is himself, but the repeated use of the phrase "this person" indicates Louis represents an American Everyman. Louis demonstrates the future orientation of Americans and the overwhelming internal conflict presented by illness and dying. The struggle to reconcile future orientation and death is so great that the American psyche cannot handle the cognitive dissonance and dissociates itself from those who embody illness and dying.

Similarly, Roy exhibits typical American anxiety about death, but Roy's anxiety is less straightforward. Social-psychological theory about death and dying may serve useful in revealing the nature of Roy's death anxiety. The prevailing notion regarding the human attempt to cope with mortality is Becker's "Terror Management Theory (TMT)."[57] TMT asserts that humans suffer a constant and almost debilitating fear of dying (terror), and that they deal with this terror via two methods: cultural worldviews and self-esteem.[58] Cultural worldviews, such as the belief in an afterlife, provide a sense of "symbolic permanence."[59] Self-esteem is achieved through activities useful to those who share that cultural worldview, ensuring that a person will not be forgotten after death.[60] Cultural worldviews and achievements to bolster self-esteem both fall under the category of "immortality projects."[61] Roy's immorality project is politics: "…this is politics, Joe. The game of being alive. And you think you're…What? Above that? Above alive is what? Dead!"[62] On other occasions in the text, Roy claims he is unafraid of death: The ability to scoff at death is enabled by his immortality project, his political career. While Roy has established "clout," he is unafraid of dying. Once Joe destroys Roy's effort to install Joe as his personal advocate in the Justice Department, his immortality project is threatened and so he must deny his imminent death: "What the fuck are you talking about, Joe? I never said that. I'm in perfect health. There's not a goddamn thing wrong with me."[63] Because his immortality project is threatened, he must deny that death is near until he has reassured the security of his legacy. Roy demonstrates the truth of this theory just a few moments

[56] Kushner, *Angels*, 25.

[57] Menzies and Menzies (2020).

[58] Menzies and Menzies, "Death Anxiety," n.p.

[59] See Footnote 58.

[60] See Footnote 58.

[61] Coffey, Tufte and Nell (2019).

[62] Kushner, *Angels*, 117.

[63] Kushner, *Angels*, 114.

after his argument with Joe, when he shouts at the apparition of Ethel Rosenberg: "BETTER DEAD THAN RED!...From the throne of God in Heaven to the belly of Hell, you can all fuck yourselves and then go jump in the lake because I'M NOT AFRAID OF YOU OR DEATH OR HELL OR ANYTHING!"[64] As Roy writhes on the floor, seeing a hallucination of a woman he helped execute, he realizes death is imminent, Consequentially, he clings to his actions in the Rosenburg trial: Politically, he deems this a success because he combatted Communism, and that political legacy means more to Roy Cohn than anything else. Here, readers see the image versus the reality: Even as the end looms, Roy is work-oriented, clinging to his career to deny death.

Research shows that when people sense the threat of death, they feel more aggressive toward those with differing cultural backgrounds and those who criticize their nation.[65] Certainly, President Trump and Roy Cohn share this manifestation of death anxiety. Throughout the pandemic, Trump complemented his flair for Christian nationalism with xenophobia. He repeatedly referred to the illness as "the Chinese virus" and conveniently sidestepped a reporter's inquiry regarding a White House insider's calling it the "Kung Flu."[66] He frequently touted his closure of American borders to Chinese citizens[67] while continuing to deport immigrants seeking asylum in the USA, some of whom were infected with COVID-19.[68] This association of foreignness with illness harkens back to a McCarthy-era law passed during the "Lavender Scare," (a period of McCarthyism in which homosexuals were purged from government jobs and persecuted by Congress)[69]: The law claimed that foreign gay people could be prohibited travel to America on the premise that homosexuality was an illness.[70] The law remained in place during the AIDS epidemic in the 1980s.[71] As Sontag notes, the link between illness and foreignness has a long history, hundreds of years prior to McCarthy,[72] but she probes deeper into the concept: "...there is a link between imagining disease and imagining foreignness. It lies, perhaps, in the very concept of wrong, which is archaically identical with the non-us, the alien...a person judged to be wrong is regarded as a source of pollution."[73] Arguably, the most extreme example of the non-us to which Sontag refers is death, because death negates each of us. In the AIDS crisis, gay men were identified as a source of pollution and a vivid reminder of ever-present mortality and so, ostracized. During the

[64] See Footnote 62.

[65] See Footnote 58.

[66] Rogers, Jakes, and Swanson (2021).

[67] Katie Rogers, Lara Jakes, and Ana Swanson, "Trump Defends Using 'Chinese Virus' Label," n.p.

[68] Dickerson and Semple (2020).

[69] Adkins (2016).

[70] Shilts, *And the Band Played On*, 30.

[71] See Footnote 70.

[72] Sontag, *Illness*, 135.

[73] Sontag, *Illness*, 136.

COVID-19 pandemic, the Trump administration did not emphasize a specific American demographic to scapegoat; instead, Trump marginalized the very construct of illness. Thus, he effectively othered countries and citizens outside the US, historically pathologized groups within the US (such as the gay community), and the aging, sick, dying, and even those who acknowledge susceptibility to illness. By aligning sickness and death with foreignness, Trump emphasized that the constructs of sickness and death is un-American.

Perhaps the most significant example of death anxiety in Trump's long months of bellicosity during the pandemic is his reaction to his own diagnosis, publicly calling it "a blessing from God."[74] He did so even though he required oxygen during treatment, which indicates that he did not have a mild case.[75] Notably, in his comment he did not simply avoid negative or neutral descriptors; rather, he went so far as to associate his illness with Christianity. He did so because the preservation of his glory as President, as well as his Christian nationalist associations, serves as his own immortality project. His predecessor, Reagan, behaved similarly when asked about his cancer surgery, saying, "I didn't have cancer. I had something inside of me that had cancer, and it was removed."[76]Regan's response demonstrates his death anxiety, but Trump's comment about COVID-19 goes further: It simultaneously exhibits his own death denial and erases the deaths of hundreds of thousands of American citizens, as well as those whose experiences with COVID-19 were physically and mentally traumatic.

Conclusion

Kohlsclaims detrimental future orientation causes Americans to ignore the past.[77] Analysis of *Angels* in the context of the COVID-19 pandemic bears this out. Americans have focused so determinedly on the future that they may have missed the prologue to a significant cultural shift decades in the making. Donald Trump employed the real Roy Cohn as his attorney in the 1980s; Cohn circulated with the Reagans, politically and socially, and served as Joseph McCarthy's right hand during the Rosenberg trials. They share a sociopolitical lineage spanning the twentieth and twenty-first centuries, and certainly some figurative political DNA. From amidst the pandemic, looking backwards at history through the lens of *Angels in America,* readers can the powerful legacy of the immortality projects of these figures. Their work has compelled the bifurcation of American values such that the images and the realities of American values often oppose one another. Public discourse around illness in the twenty-first century makes this opposition evident. Americans have ostracized themselves from the construct off sickness, erasing the experiences and

---

[74] "Covid: Trump Describes His Illness 'a Blessing from God'," BBC News (BBC, October 8, 2020), https://www.bbc.com/news/world-us-canada-54455040.

[75] Schwartz (2020).

[76] Sontag, *Illness*, 154.

[77] See Footnote 54.

lives of their ill and dying fellow citizens, and diminishing their Americanness. Or, as Roy observes while he suffers from the final stages of AIDS: "Americans have no use for the sick…It's just no country for the infirm."[78]

References

Adkins, J. (2016). These people are frightened to death: Congressional investigations and the lavender scare. In *Prologue Magazine*. National Archives. https://www.archives.gov/publications/prologue/2016/summer/lavender.html

"Bowers v. Hardwick, 478 U.S. 186 (1986)." Justia Law. Justia Law, 2021. https://supreme.justia.com/cases/federal/us/478/186/

Carroll, M. (1974). City Council's Bill on Homosexual Rights. *New York Times*. May 5, 1974. www.newyorktimes.com

Coffey, M., Tufte, J., & Nell, T. (2019). *Ernest Becker Foundation, 2019*. http://www.ernestbecker.org/

"Covid: Trump Describes His Illness 'a Blessing from God'. (2020). *BBC News*. BBC, October 8, 2020. https://www.bbc.com/news/world-us-canada-54455040

Curry, G. (1987). Reagan says AIDS solution rests with morals. *The Chicago Tribune*. April 2, 1987. https://www.chicagotribune.com/news/ct-xpm-1987-04-02-8701250240-story.html

Dickerson, C., & Semple, K. (2020). U.S. Deported thousands amid Covid 19 outbreak: Some proved to be sick. *The New York Times*. April 18, 2020. https://www.nytimes.com/2020/04/18/us/deportations-coronavirus-guatemala.html

Earnshaw, S. (2020). Exclusion and erasure in 'the values majority culture Americans live by'. *Migrant Knowledge*, September 3, 2020. https://migrantknowledge.org/2020/09/03/exclusion-and-erasure/

Field, M. J., & Cassel, C. K. (1997). A profile of death and dying in America" essay. In *Approaching death: Improving care at the end of life*. Washington, D.C.: National Academy Press

"Fundamental Rights (Noneconomic Substantive Due Process)." Legal Information Institute. Cornell Law School. Accessed May 12, 2021. https://www.law.cornell.edu/constitution-conan/amendment-14/section-1/fundamental-rights-noneconomic-substantive-due-process#fn561amd14

Gallup. "Religion." Gallup.com. Gallup, July 20, 2021. https://news.gallup.com/poll/1690/religion.aspx

Goldberg, D. (2020). It's going to disappear': Trump's changing tone on Coronavirus. *Politico*, March 19, 2020. https://www.politico.com/news/2020/03/17/how-trump-shifted-his-tone-on-coronavirus-134246

Kaganovich, K. (2021). Mask mandates are constitutional. Here is why. *Columbia Political Review*, March 3, 2021. http://www.cpreview.org/blog/2021/1/mask-mandates-are-constitutional-here-is-why

Kohls, L. R. (2021). Fordham.edu. Fordham University. Accessed March 1, 2021. https://www.fordham.edu/download/downloads/id/3193/values_americans_live_by.pdf

Kushner, T. (2013). *Angels in America: a gay fantasia on national themes*. Theatre Communications Group.

"Lawrence v. Texas." Legal Information Institute, June 26, 2003. https://www.law.cornell.edu/supct/html/02-102.ZS.html

Mariner, W. K., Annas, G. J., & Glantz, L. H. (2005). "Jacobson v. Massachusetts": It's not your great-great-grandfather's public health law. *American Journal of Public Health 2005*, April 2005. https://www.ncbi.nlm.nih.gov/pmc/articles/PMC1449224/

---

[78] Kushner, *Angels*, 151.

Menzies, R. E., & Menzies, R. G. (2020). Death anxiety in the time of Covid-19: Theoretical explanations and clinical implications. *The Cognitive Behaviour Therapist, 13*(E19). https://doi.org/10.1017/s1754470x20000215

Perry, S. L., Whitehead, A. L., & Grubbs, J. B. (2020). Culture wars and Covid-19 conduct: Christian nationalism, religiosity, and Americans' behavior during the Coronavirus Pandemic. *Journal for the Scientific Study of Religion, 59*(3), 405–416. https://doi.org/10.1111/jssr.12677

Rabin, J. (1986). The AIDS epidemic and Gay Bathouses. *Journal of Health Politics, Policy. and Law, Winter, 10*(4), 729–743. http://williamapercy.com/wiki/images/The_AIDS_Epidemic.pdf

Rogers, K., Jakes, L., & Swanson, A. (2021). Trump defends using 'Chinese virus' label, ignoring growing criticism. *The New York Times*. March 18, 2021. https://www.nytimes.com/2020/03/18/us/politics/china-virus.html

Schwartz, M. (2020). Live updates: Trump tests positive for Coronavirus. *NPR*. October 11, 2020. https://www.npr.org/sections/latest-updates-trump-covid-19-results

"Science Brief: Community Use of Cloth Masks to Control the Spread of Sars-Cov-2." (2021). *Centers for disease control and prevention*, June 29, 2021. https://www.cdc.gov/coronavirus/2019-ncov/science/science-briefs/masking-science-sars-cov2.html

Shilts, R. (2007). *And the band played on*. St. Martin's Griffin.

Sontag, S. (1989). *Illness as metaphor; and aids and its metaphors*. Picador.

Staff, P. (2020). *Look back at the original broadway production of La Cage Aux Folles*. PLAYBILL INC., August 21, 2020. https://www.playbill.com/article/look-back-at-the-original-broadway-production-of-la-cage-aux-folles

"The Bathouse Wars." (1984). *The Washington Post*. March 4, 1984, sec. Lifestyle. https://www.washingtonpost.com/archive/lifestyle/1984/04/19/the-bathhouse-war/5c864455-2310-4f9e-ac70-e0f57083a511/)

Thrush, G. (2020). It affects virtually nobody' Trump says, minimizing the effect of Coronavirus on young people as U.S. death toll hits 200,000. *The New York Times*. September 22, 2020. https://www.nytimes.com/2020/09/22/us/politics/trump-coronavirus-virtually-nobody.html

Victor, D., Serviss, L., & Paybarah, A. (2020). In his own words, what Trump Has said about wearing a mask. *The New York Times*. October 2, 2020. https://www.nytimes.com/2020/10/02/us/politics/donald-trump-masks.html

Whitehead, A., & Perry, S. (2021). The growing anti-democratic threat of Christian nationalism in the U.S. *Time Magazine*. Time USA, May 27, 2021. https://time.com/6052051/anti-democratic-threat-christian-nationalism/

Chapter 11
"Bleeding" into Reality: Popular Representations of Ebola and the 2013–2016 West African Ebola Epidemic

Rebecca Henderson

Abstract To help contain Ebola and to save lives, during the 2013–2016 epidemic aid agencies and healthcare workers traveled to affected countries to help in the response. As they arrived in West Africa, however, they brought Ebola with them—in their imaginations. In this chapter, I will explore the ways that these fictional representations of disease may "bleed" into reality. I will engage with popular representations of Ebola circulating prior to the epidemic—specifically the Robin Cook (Outbreak. Penguin, 1987) thriller *Outbreak* (and the subsequent 1995 film) and Richard Preston (The Hot Zone. Random House, 1994) "The Hot Zone." I will focus on the ways Ebola was represented in these works of fiction and will compare these features to descriptions of the disease in contemporary scientific articles, and protocols used for the diagnosis of Ebola. Finally, I will link these ideas to the experiences of Ebola responders through interviews. I will argue that the centrality of bleeding to popular representations of Ebola may have influenced the ways that critical diagnostic decisions were made in the absence of quick diagnostic testing. Using this example, I will explore the ways that fictional representations of disease may have crucial impacts in the low-certainty, high-risk setting of epidemics.

Introduction

In each case, the opening scene is Africa: a lush, tropical, sometimes conflict-torn jungle. Monkeys leap through the branches, their screams a portent of things to come. The narrative immediately focuses on the disease's first victim, inevitably a white man in a place where he does not belong. Robin Cook's *Outbreak* (1987) invents a Frenchman on safari, Charles Monet, who unwisely delves too deeply into a contaminated cave and develops a mysterious illness. Richard Preston's "terrifying true story" *The Hot Zone* (1994) opens onto a Yale student backpacking through Zaire, who suddenly begins to feel unwell. In the film *Outbreak* (1995), directed by

R. Henderson (✉)
Department of Anthropology, University of Florida, Gainesville, FL, USA
e-mail: rrhenderson@ufl.edu

S. Venkatesan et al. (eds.), *Pandemics and Epidemics in Cultural Representation*,
https://doi.org/10.1007/978-981-19-1296-2_11

Wolfgang Peterson and loosely based on Preston's book, the first victims are white soldiers fighting an unexplained conflict in the Democratic Republic of Congo. One of them, in a horrifying infirmary full of the victims of this mysterious new disease, reaches toward a doctor: "You're Americans!" He cries. "Please get me out of this shithole!". In this case, however, his privilege as a white American male is not enough to save him; seeing the danger inherent in the virus, the military decides that his life is expendable, ultimately firebombing the hospital to contain it.

Ebola: a terrifying, mysterious new Africa disease which arose in the popular media of the late 1980s and early 1990s. Audiences were terrified by the idea of it invading their homes even as the first few hundred cases were seen in real patients in Africa. In retrospect, it is easy to forget that in the context of the mid-1990s, Ebola was a very rare disease, and a very unlikely threat; in fact, 1990s audiences' fascination and horror at Ebola seems out of proportion to the limited nature of its early spread. Cook's, 1987 *Outbreak* was published only a decade after the first cases of Ebola were described in the medical literature, when only two relatively small outbreaks had been reported (Heymann et al., 1980; WHO, 1976). In 1995, as audiences in the global north sat enthralled in theaters watching a fictionalized "Ebola-like" disease decimate a small California town in *Outbreak*, the real disease was claiming its first several hundred victims in Zaire (Muyembe-Tamfum et al., 1999). Ebola is a disease that managed to capture the attention of the American public despite being extremely rare and affecting almost exclusively Africans. At the time of these popular depictions, and the fear they generated, the real disease remained confined to just two countries. Despite this fact, Ebola had already spawned two widely read "medical thrillers" and an action movie starring the likes of Dustin Hoffman, Rene Russo, and Morgan Freeman, as well as a made-for-TV movie and several documentary news broadcasts (Semmler, 1998).

Thus, even when there was very little actual Ebola in the world, our fascination with Ebola proved contagious. Clare Herrick (2019) describes Ebola's "unique charisma," igniting "fear, anxiety, and disgust at a scale greater than the lives it claims" (p. 1488). Media representations were already widely spreading the ideas which would come to define popular conceptions of the disease. In one of the first scenes of *Outbreak* (1995), a helicopter discharges men wearing perhaps the most recognizable uniform in all of medicine: the yellow Personal Protective Equipment (PPE) that would become inextricable from Ebola. As is suggested by the titles of these books and films, Ebola terror was defined by its contagion, by the frightening ease with which it seemed to spread, and by fears it would cross oceans and borders. Monkeys, as a vector, proliferate through this media. And, Ebola was set apart from other diseases by its terrifying disease course, in which the body seemed to become unglued, to liquify in a horrific explosion of fluids. Ebola, as one of a rare group of "viral hemorrhagic fevers," was set apart by a cardinal symptom: uncontrollable bleeding. Ebola victims, in popular media depictions, gush, drip, ooze, and flow—in some cases they "liquify" with contagious bloody secretions.

In 2014, the world watched in horror as Ebola began to spread throughout West Africa at unprecedented levels as a part of the largest, longest, and most widespread outbreak on record. In August, after 33 weeks of Ebola spread, the World Health

Organization (WHO) declared the epidemic to be a Public Health Emergency of International Concern, based on continuing high rates of transmission in Guinea, Libera, and Sierra Leone (Briand et al., 2014). To help contain the epidemic and to save vulnerable lives, foreign healthcare workers traveled to affected countries. As they arrived, however, they brought Ebola with them—in their imaginations. Few, if any, had ever seen a real case of Ebola before, and thus, their ideas of what Ebola, as a disease, looked like were highly shaped by the fictionalized representations they had encountered.

In this chapter, I will suggest that this "imaginary Ebola" had important effects on the experiences of international medical aid workers, and may have even impacted the treatment of Ebola patients. Specifically, I will suggest that fictional descriptions of Ebola patients, which spectacularized hemorrhage as a cardinal symptom, may have influenced attempts to triage suspected Ebola patients in the 2014 epidemic.

In the absence of timely diagnostic testing in overtaxed and under-resourced health systems in West Africa, international health workers had to decide case by case which patients they suspected of having Ebola amid a host of other endemic health problems. The consequences of a wrong decision could be catastrophic; admitting an infected patient to the hospital would risk their lives, those of their colleagues, and those of other admitted patients, and would threaten the safety of the barely functioning health system. On the other hand, inappropriately quarantining a patient consigned them to wait for lifesaving treatment, without which they could die. To make these difficult decisions without laboratory diagnosis, international aid workers constructed diagnostic algorithms that relied upon descriptions of Ebola in order to properly diagnose cases. While Ebola had, by 2014, become known popularly as a disease of uncontrollable bleeding, a "hemorrhagic fever," scientific evidence available before and during the epidemic established that bleeding is actually a relatively rare and late symptom of Ebola—a poor criteria to use in triaging early suspected cases. Despite this established scientific fact, hemorrhage as a cardinal symptom was incorporated into triage protocols which attempted to separate suspect Ebola cases. In this chapter, I suggest that the seemingly inextricable association of Ebola with bleeding as a cardinal symptom may be in part due to the centrality of bleeding in sensationalized depictions of the disease, which emerged in the 1990s as a part of larger circulating fears about blood, contagion, Africa, and AIDS.

In examining the ways that fictional representations of disease may "bleed" into reality, I will engage with popular representations of Ebola circulating prior to the epidemic—specifically Cook's, (1987) thriller *Outbreak*, Preston's, (1994) *The Hot Zone*, and the subsequent 1995 film *Outbreak*. I focus on the ways the disease itself was represented in these works of fiction, and compare these features to descriptions of the disease in medical textbooks, contemporary scientific articles, and protocols used for the treatment of Ebola. I also follow earlier scholars (Semmler, 1998) in linking Ebola fears with terror and fetishism of blood f stemming from the epidemic of HIV/AIDS, and contrast the bloody Ebola of fiction with the more amorphous disease of scientific fact.

Finally, I will briefly link these ideas to the experiences of Ebola responders through 30 interviews with those on the front lines during the 2014 epidemic.

Conducted by phone between 2017 and 2019, these interviews were collected by the author as a part of the larger Ebola 100 project (Post, 2019), and specifically targeted medical professionals concerned with treating patients who were pregnant during the epidemic. Interviews were conducted by phone, transcribed, and coded thematically within QSR NVivo. While these interviews did not specifically target the impact of popular media depictions of Ebola, the topic of bleeding as a symptom spontaneously emerged as a theme from these interviews and was linked to media representations by interlocutors. Using the interview data and the example of bleeding, I will explore the ways that fictional representations of disease may have crucial impacts in the low-certainty, high-risk setting of epidemics.

Pools of Blood: Ebola in Popular Media

Writing from the late 1990s, before the 2014 realization of Ebola as pandemic threat, Alexandra Semmler noted that Ebola had "taken the place of the nuclear bomb as a threat of destruction" (1998, p. 161). However, in these media depictions, it was not just the virus that was the villain, but conspiratorial elements within American culture. Pop-fiction writer Robin Cook took full advantage of this "virus-as-weapon" trope in *Outbreak* (1987), in which a CDC scientist tracks unexplained outbreaks of Ebola through several American cities, only to link the outbreaks to a group of disgruntled conservative doctors seeking to discredit MMOs and foreign-born physicians. Likewise, the film *Outbreak* (1995) features an "Ebola-like virus" in a costal California town; although initially spread by an infected monkey, the virus is revealed to be a carefully guarded secret biological weapon. Protagonists must race not only to cure the virus, but also to stop the government from destroying the town as a part of a cover-up. Preston's work, *The Hot Zone*, by contrast, features the "true story" of Ebola accidentally introduced to a Virginia primate laboratory, only to be successfully contained through the heroic efforts of government agencies.

As a "nuclear bomb" level threat, Ebola was represented as not just a terrifying disease, but *the most* terrifying and dangerous disease that could be imagined. As *Outbreak* (1995) opens, the audience is taken deeper and deeper into a biocontainment center, progressing through different levels of security and protection. As the locks on the doors get bigger and the precautions move from gloves to full PPE, subtitles flash across the screen, listing the diseases that are stored on each level-moving from common entities the audience recognizes as benign (level 1—pneumococcus, salmonella), to disease they know are terrifying (level 3—anthrax, typhus, HIV). Ebola is on level 4, which is, we are led to believe, the final level. The lethality of Ebola and the havoc it could wreck if introduced to an unprotected population are repeatedly referenced, and Ebola is framed as the ultimate disease.

In each representation of Ebola, the crux of the horror is the virus itself, and the terrible destruction it wreaks within the body of its victims. Depictions of those infected with Ebola are the climactic moments of terror and disgust. In *Outbreak* (1995), for example, a suspenseful early scene depicts the protagonists arriving at an

African clinic overrun by Ebola. In full yellow PPE, the investigators walk through the eerily still clinic; bloody bodies are strewn everywhere, as if there has been a massacre. Following the sound of a child crying, they throw aside a curtain to reveal a young boy, covered in sores, sobbing next to the bloody bodies of his parents. At the horrific sight, a protagonist vomits, panics, and rips off his protective equipment. In another example, Preston graphically describes the death of his first Ebola victim:

> Suddenly he goes into the last phase—the human virus bomb explodes...He leans over, head on his knees, and brings up an incredible quantity of blood from his stomach and spills it on the floor with a gasping groan. He loses consciousness and pitches forward on the floor. The only sound is the choking in his throat as he continues to vomit blood and black matter while unconscious. Then comes a sound like a bedsheet being torn in half, which is the sound of his bowels opening and venting blood from the anus. The blood is mixed with intestinal lining. He has sloughed his gut. The lining of his intestines have come off and are being expelled along with huge amounts of blood.... Pools of blood spread out around him, enlarging rapidly... everything had gone wrong inside this man, absolutely everything, any one of which could have been fatal: the clotting, the massive hemorrhages, the liver turned to pudding, the intestines full of blood (The Hot Zone, p. 28).

Belling (2016) describes the horror of an Ebola-stricken body as an "opaque receptacle subject to rupture," overflowing its boundaries and uncontainable (p. 43). The bloodiness of Ebola, and fears of contagious, uncontainable, and especially African blood, has been linked to underlying cultural fears of the ongoing epidemic of HIV/AIDS. Semmler (1998) ties Ebola media to "the new blood culture" identified by Rich (1992), and also perhaps also reflected in the rash of 1990s vampire media, including *Bram Stoker's Dracula* (1992), *Buffy the Vampire Slayer* (1992), *Interview with a Vampire* (1994), and *Blade* (1998). In Cook's book *Outbreak* (1987), protagonists self-consciously compare Ebola and AIDS, concluding: "Given the fact that hemorrhagic fever could spread between strangers who shared a towel, or by the most casual close touch, Ebola made the AIDS scare seem like a tempest in a teapot" (p. 336). Ebola, then, was more frightening than AIDS—a deadly, incurable virus which could not be contained within international borders, within "at-risk" populations, or within bleeding bodies.

In addition to gratuitous descriptions of the bloody deaths of Ebola patients, bleeding was repeatedly singled out as the most pertinent diagnostic feature of the virus. For example, Robin Cook's *Outbreak* features a scene in which his protagonist, a young CDC investigator, describes the new outbreak to an expert: "She sensed that [he] was particularly interested in the fact that two of the patients had vomited blood, that another had passed bloody diarrhea, and that three had conjunctival hemorrhages in their eyes... [he] exclaimed, "My God, do you know what you're describing?... Viral hemorrhagic fever!" (p. 48). While a prodrome of clinical symptoms is depicted, it is the bleeding which is most salient. Bleeding is the feature which sets Ebola apart and defines it as an illness.

As Semmler (1998) notes, although popular depictions of Ebola were obviously not intended as medical fact, "most of us feel we have acquired some expertise as a result of the medical information we are exposed to through the media we watch and the books we read" (p. 158). In fact, efforts were made in every case to ensure that the fictional depictions would seem, at least at first glance, to be as biomedically accurate as possible. Richard Preston's *The Hot Zone,* based off an earlier article for *The New Yorker*, was described as non-fictional and written to be biomedically accurate popular science. Likewise, despite its use of a scientifically unsound deus-ex-machina plot device which appeared, at the last moment, to save the day, the film *Outbreak* employed a costly top-rate medical advisor to ensure that medical details of the scenes, down to biomedical equipment in laboratories and hospitals, was as accurate as possible. Robin Cook, a trained physician, likewise makes his works more compelling by the inclusion of realistic medical details (Semmler, 1998). Fictional depictions of Ebola, therefore, were often made to seem as realistic as possible, even as they spectacularized the horrors of a hemorrhagic fever. American medical practitioners, like the rest of the public, encountered Ebola almost entirely through such depictions.

A Mysterious Disease: Diagnosing Ebola

In March 2014, in Guinea, practitioners reported encounters with "a mysterious disease characterized by fever, severe diarrhea, vomiting, and an apparent high fatality rate" (Baize et al., 2014, p. 1418). The disease was eventually determined to be Ebola, and it rapidly spread through Guinea to neighboring Liberia and Sierra Leone. The Ebola pandemic threat had been realized, but not in the form depicted in the movies; in an account of these early cases published several months later in the New England Journal of Medicine, the authors note that "hemorrhage was not documented for most of the patients with confirmed disease." They go so far as to suggest a name change for the virus, recommending the use of "the term Ebola virus disease (rather than the earlier term Ebola hemorrhagic fever) [which] takes into account that hemorrhage is not seen in all patients and may help clinicians and public health officials in the early recognition of the disease" (Blaize et al., 2014, p. 1423). The fact that Ebola was a hemorrhagic fever that could occur without bleeding had, in fact, already been established; a 2011 review of available evidence from all previous epidemics noted that early cases of Ebola manifest with non-specific symptoms, developing only minor hemorrhagic manifestations until late in the infection (Kortepeter et al., 2011). Yet in mid-2014, as the epidemic raged, *The New Yorker* re-published an excerpt from Richard Preston's 1992 article "Crisis in the Hot Zone," the precursor to his book, in which he describes Ebola. In this article, Preston writes:

> a sort of melting occurs…. connective tissue, skin, and organs, already peppered with dead areas and heated with fever, begin to liquefy, and the slimes and uncoagulated blood that run from the cadaver are saturated with Ebola-virus particles…. (Preston, 2014)

Meanwhile, scientific descriptions of the virus from the unfolding 2014 epidemic reveal a disease that, in its early stages, was exceedingly non-specific, varied considerably in its presentation, and was a far cry from the easy diagnosis of a hemorrhagic fever from massive hemorrhage depicted in *Outbreak*. Especially in the context of multiple endemic diseases, including malaria, HIV, typhoid, and others, real Ebola was exceedingly difficult to pin down. For example, according to one attempt to characterize the presentation of Ebola:

> Symptoms may begin abruptly and progress from non-specific (fever, fatigue, headache, myalgias, and malaise) to include abdominal pain, nausea, high-volume vomiting, and diarrhea. Hiccups are uncommon but well described in the present and previous outbreaks. On examination, a characteristic finding in Ebola infection is elevated temperature, but this can oscillate markedly... A maculopapular rash (that may subsequently desquamate) has also been described but is sometimes hard to identify in dark-skinned patients. In the current outbreak, rash has not been a prominent feature. Some patients have mucosal hemorrhage and oozing from puncture sites; gastrointestinal hemorrhage also occurs and can be massive. However, hemorrhage has been reported in fewer than 50% of cases previously. In the current outbreak, only 19% of patients have symptoms of unexplained bleeding, and overt bleeding at presentation seems to be much less frequent.... (West & von Saint André-von Arnim, 2014, p. 1342)

Furthermore, the epidemic was occurring in settings a far cry from the high-tech CDC and military laboratories of pandemic fiction. Instead, it occurred in low-resource hospitals with limited diagnostic technology, where existing diagnostic resources were quickly overwhelmed. Certainty in Ebola diagnosis required diagnostic testing, and, in the case of a negative result, confirmatory retesting. Yet test results were commonly delayed more than 24 h, with some interviews describing delays of up to two days, leaving physicians in the untenable position of not knowing if their patient had Ebola and was contagious, or was simply critically ill with something else. One clinician interviewed described the uncertainty surrounding the virus as "*the most challenging thing about Ebola... I felt like it was the most difficult disease to predict of any I'd ever seen. I think it's because I had so little data of what was going on. No lab tests really to speak of and all that. That was very challenging.*" Adding to this uncertainty was the low level of medical evidence available about Ebola, and the fact that few, if any, clinicians had treated Ebola cases before. In this setting of uncertainty, clinicians drew on their ideas of what they thought Ebola should look like.

On the one hand, physicians risked admitting an Ebola positive patient to the hospital, a mistake that would endanger the lives of patients and healthcare providers, including themselves and their colleagues. Given the high levels of mortality in the healthcare workforce during the epidemic, this risk was extremely serious and real, and threatened not only the lives of workers but the ability of the healthcare system to continue to function (Evans et al., 2015). On the other hand, however, if patients were deemed to be a risk for Ebola, they were held in quarantine in an Ebola holding unit (EHU) until test results arrived. During this period, patients risked exposure to Ebola while in the EHU, and treatments for other conditions were often unavailable while awaiting test results.

It was thus critically important to determine, in the moment, who needed to be quarantined and who could be admitted to the hospital. Clinicians at work in containing Ebola thus developed and used a "case definition" which formed the basis for a triage protocol used on the front line to screen patients, and to determine which patients warranted quarantine. In one example of such a guideline, patients must be quarantined if they have a fever after contact with a known case, three or more symptoms along with their fever, or any kind of unexplained bleeding/miscarriage—regardless of contact or fever (WHO, 2014). Thus, bleeding is treated as an exceptional symptom—no other symptoms or contact with known cases are required.

The case definition retained bleeding as a signal characteristic of Ebola throughout the epidemic, despite preexisting and mounting evidence which suggested it was not a common early symptom, and thus not potentially suitable for inclusion in triage protocols. In interviews, experts involved in the use and creation of triage documents reported that while diagnostic guidelines were adapted and refined over the course of the epidemic, due to low levels of evidence, they were often poorly standardized. As an expert responsible for making and using diagnostic guidelines acknowledged:

> Ebola carries the name hemorrhagic fever around with it and there is a conception that there's more bleeding than there actually is- and media accounts and the movies don't help- and so the absence of bleeding makes people wonder a moment, do they really have Ebola? While for 80% people, that's the hardest part, is that you don't have bleeding.

Yet, despite this expert knowledge, physicians on the front line of the epidemic continued to treat bleeding as a signal symptom for triaging Ebola patients. The persistence of unexplained bleeding in triage protocols throughout the epidemic, and the continued vigilance about blood in diagnosis may have been due to practitioner's exposure to popular media depictions of Ebola, an idea that some health workers endorsed. As one physician on the front lines in 2014 acknowledged:

> Before seeing a patient, I thought that patients would die of bleeding, while this is not the case… patients die mostly because of kidney insufficiency, not because of bleeding. So this is true… I don't think we'll ever get the word 'hemorrhage' out of it.

Given the complex prodrome of Ebola reported in prior epidemics and seen in 2014, it is possible to argue that bleeding, as a relatively uncommon and late symptom, should not have been incorporated into decisions to triage patients whose status was uncertain. However, by 2014, the idea of Ebola had become inextricable from hemorrhage.

This inextricable association of Ebola with bleeding had particularly dire consequences for a vulnerable population of patients: women who were experiencing complicated pregnancies during the epidemic. Sierra Leone, for example, had one of the highest rates of miscarriage and pregnancy complications in the world (WHO, 2011). Since hemorrhage remained a key a diagnostic criterion, pregnant women who presented with "unexplained" vaginal bleeding due to complications during pregnancy were triaged to EHUs throughout the epidemic, delaying care with sometimes devastating consequences. Interviewees who cared for pregnant women returned again and again to the difficulty of having to quarantine women who were likely miscarrying. As one obstetric provider put it:

> That was morally, ethically, really difficult knowing that you're taking in this patient, in this setting where the care they receive will be worse than-- I don't know if you'd call it

> sentencing-- but by saying that they meet case definition, they're going to be just not receiving the care that they need. It's often pretty time critical in obstetrics. Three days in a holding unit often makes them miss that time window with bad results both for mother and child, sometimes. That was difficult.

Throughout the epidemic, women who were pregnant or postpartum were stigmatized in healthcare environments and in their communities as the result of fears of contamination due to uncontrollable bodily fluids (Strong & Schwartz, 2019). Pregnant bodies were seen as potential threats, overflowing with contagious blood. One interviewee described hearing them referred to in the hospital as "*Ebola bombs,*" ready to release a flood of virus whenever they went into labor. Strong and Schwartz (2019) link this stigma to cultural taboos surrounding pregnancy, women's bodies, and contamination, as well as fears of HIV/AIDS. Pregnant bodies became even more of a threat when it was discovered that Ebola could be retained in the placenta even after a patient had cleared and recovered from the virus; this discovery meant that pregnant women could test negative for Ebola and the products of conception would still carry infectious Ebola (Baggi et al., 2014). The link between pregnancy and blood, and between blood and Ebola contagion, caused clinicians to view pregnant women as an almost unmanageable patient population. As one expert put it: "*Before I took on the job of [caring for pregnant women with Ebola] I was told by many, many, many people not to do it because I would die.*"

Since the 2014 epidemic, more recent papers have attempted to refine clinical heuristics by, for example, allowing the calculation of a score for use in risk stratification; notably absent from this calculation is any mention of bleeding, including vaginal bleeding or miscarriage, as useful for diagnosis (Levine et al., 2015). A New England Journal of Medicine paper, published by the WHO Ebola Response team in October of 2014, found that only 3.2% of the 4507 total Ebola cases reported up to that point had exhibited bleeding from the vagina as a symptom (WHO ERT, 2014).

Evidence about the epidemic prior to 2014 was limited, and especially in the early epidemic it was difficult to be scientifically certain of anything about the constantly shifting new disease. However, I argue that this void of evidence provided an opening for popular depictions, in which uncontrollable bleeding was the inevitable hallmark of Ebola, a disease worse than AIDS. Clinicians who had never seen Ebola outside of the movies carried images of *Outbreak* and *The Hot Zone* which tied the disease to blood. The fears of a clinician working with pregnant women seem to echo the fate of the anonymous soldier at the beginning of the film *Outbreak* (1995): "*We thought we will all get Ebola and we will die in Sierra Leone. They will not even be allowed to retrieve the bodies.*"

Conclusion: Pandemic Fiction, Pandemic Reality

The 2019 National Geographic production of The Hot Zone suggests that not much has changed in depictions of Ebola since the 2014 epidemic. The series opens with Richard Preston's fictional white man in Africa, vomiting blood into an airplane doggy bag, bleeding from open sores all over his body, and eventually erupting blood into the startled face of his now-doomed African doctor. Again, the series takes us to a biocontainment unit and shows us the men working in PPE in a room with "Level 4" written on the door. From the series' opening scenes, parallels are again drawn to HIV/AIDS; AIDS, unlike Ebola is "actually hard to contract," says the protagonist to her son. "You've seen something worse [than AIDS]?" He asks her, terrified. Again, Ebola is depicted as a nightmare version of AIDS, a virus that makes blood uncontainable and thus leaps through the boundaries of bodies and borders.

In this chapter, I suggest that popular depictions of Ebola as a disease of bleeding may have influenced the way clinicians encountered the disease during the 2014–2015 epidemic. The 2014 Ebola epidemic has been described as a setting in which "cultural beliefs" played a role in spreading harmful misconceptions about the virus (Manguvo & Mafuvadze, 2015); these cultural beliefs are frequently attributed solely to Africans, who refused to comply with public health measures due to cultural misunderstandings. Yet, biomedical healthcare providers are far from immune to the power of cultural beliefs, in this case ideas carried in fictional or media depictions of disease and illness. They may be particularly susceptible in cases, like Ebola, of frighting, rare, or new diseases which they may not have ever personally encountered before. They may also be susceptible in cases which are dramatically, if inaccurately, depicted in ways that link to underlying anxieties—such as those surrounding blood, contagion, and HIV/AIDS. These cultural representations may have consequences. In her analysis of non-fictional accounts of Ebola as "urban legend," Rebecca Weldon warns that perpetuating the legend of the predatory virus as the "true villain" of Ebola stories hides the ways that inequity is the real cause of pandemic tragedy (2001). Similarly, the linkage of bleeding with Ebola may have exacerbated the vulnerability of already at-risk groups by associating Ebola with common causes of bleeding, including miscarriage or complications of pregnancy.

In suggesting that anxieties about bleeding shaped ideas of Ebola in 2014, I recognize the limited nature of biomedical evidence during this period, and the high level of uncertainty which permeated almost every aspect of Ebola care in early 2014. During this time, clinicians made the best decisions they could quickly, with high potential consequences and limited evidence; this often involved best guesses or heuristics. However, pandemic settings are often characterized by urgency, criticality, and limited or uncertain evidence, thus leaving more room for unconscious associations built from popular representations to impact ideas of disease. Media representations of pandemics may therefore have particularly dramatic effects on care, filling a void left by scientific certainty during a time when anxieties are heightened, and fast action is crucial. COVID-19 has once again demonstrated the power of popular fears—such as vaccine hesitancy—on shaping beliefs about a new disease and its treatment. In the face of pandemic threat, biomedical practitioners should not just look outwards to identify the impact of culture or media representations on ideas of disease, but should also reflect on the potential popular media origins of their own ideas and preconceptions.

References

Baggi, F. M., Taybi, A., Kurth, A., Herp, V., et al. (2014). Management of pregnant women infected with Ebola virus in a treatment centre in Guinea, June 2014. *Eurosurveillance, 19*(49), 20983.

Baize, S., Pannetier, D., Oestereich, L., et al. (2014). Emergence of Zaire Ebola virus disease in Guinea. *New England Journal of Medicine, 371*(15), 1418–1425.

Belling, C. (2016). Dark zones: The Ebola body as a configuration of horror. In *Endemic*, pp. 43–66. Palgrave Macmillan.

Briand, S., Bertherat, E., Cox, P., Formenty, P., Kieny, M.P., Myhre, J.K., Roth, C., Shindo, N., & Dye, C. (2014). The international Ebola emergency. *New England Journal of Medicine, 371*(13), 1180–1183.

Cook, R. (1987). *Outbreak*. Penguin.

Evans, D. K., Goldstein, M., & Popova, A. (2015). Health-care worker mortality and the legacy of the Ebola epidemic the lancet. *Global Health, 3*(8), e439–e440.

Herrick, C. (2019). Geographic charisma and the potential energy of Ebola. *Sociology of Health & Illness, 41*(8), 1488–1502.

Heymann, D. L., Weisfeld, J. S., Webb, P. A., Johnson, K. M., Cairns, T., & Berquist, H. (1980). Ebola hemorrhagic fever: Tandala, Zaire, 1977–1978. *Journal of Infectious Diseases, 142*(3), 372–376.

Jones, S. A., Gopalakrishnan, S., Ameh, C. A., White, S., & van den Broek, N. R. (2016). ‘Women and babies are dying but not of Ebola’: The effect of the Ebola virus epidemic on the availability, uptake and outcomes of maternal and newborn health services in Sierra Leone. *BMJ Global Health, 1*(3), e000065.

Kortepeter, M. G., Bausch, D. A., & Bray, M. Basic clinical and laboratory features of filoviral hemorrhagic fever. *Journal of Infectious Diseases, 204*(3), S810–S816.

Levine, A. C., Shetty, P. P., Burbach, R., et al. (2015). Derivation and internal validation of the Ebola prediction score for risk stratification of patients with suspected Ebola virus disease. *Annals of Emergency Medicine, 66*(3), 285–293.

Manguvo, A., & Mafuvadze, B. (2015). The impact of traditional and religious practices on the spread of Ebola in West Africa: Time for a strategic shift. *The Pan African Medical Journal, 22*(Suppl), 1.

Muyembe-Tamfum, J. J., Kipasa, M., Kiyungu, C., & Colebunders, R. (1999). Ebola outbreak in Kikwit, Democratic Republic of the Congo: Discovery and control measures. *The Journal of Infectious Diseases, 179*(1), S259–S262.

Post, K. (2019). The Ebola 100 project: A flexible approach to collaborative research. *Practicing Anthropology, 41*(1), 12–14.

Preston, R. (1994). *The hot zone*. Random House.

Preston, R. (2014). Takes: Richard Preston on Ebola. *The New Yorker.* July 14.

Semmler, I. A. (1998). Ebola goes pop: The filovirus from literature into film. *Literature and Medicine, 17*(1), 149–174.

Strong, A. E., & Schwartz, D. A. (2019). Effects of the West African Ebola epidemic on health care of pregnant women: Stigmatization with and without infection. *Pregnant in the Time of Ebola* (pp. 11–30). Springer.

West, T. E., & von Saint André-von Arnim, A. (2014). Clinical presentation and management of severe Ebola virus disease. *Annals of the American Thoracic Society, 11*(9), 1341–1350.

Weldon, R. (2001). An “urban legend" of global proportion: An analysis of nonfiction accounts of the Ebola virus. *Journal of Health Communication, 6*(3), 281–294.

WHO/International Study Team. (1978). Ebola haemorrhagic fever in Sudan, 1976. *Bulletin of the World Health Organization, 56*(2), 247–270.

World Health Organization. (2011). *Atlas of African Health Statistics 2016*. https://apps.who.int/iris/bitstream/handle/10665/206547/9789290232919.pdf?sequence=1&isAllowed=y

WHO, Ert. (2014). Ebola virus disease in West Africa—The first 9 months of the epidemic and forward projections. *New England Journal of Medicine, 371*(16), 1481.

WHO, "Clinical Management of Patients with Viral Haemorrhagic Fever". (2014). *World Health Organization*. May 31. http://apps.who.int/iris/bitstream/10665/205570/1/9789241549608_eng.pdf?ua=1

Chapter 12
Wearing Masks: Living and Coping "WITH CORONA" in Japan Under the Pandemic

Akinori Hamada

Abstract By examining two cultural representations of the pandemic from anthropological perspectives, this paper aims to reveal the characteristics of the "With Corona" policy and the COVID-19 experience in Japan from January 2020 to March 2021. The lively creative artwork series of "WITH CORONA," continuously created by one of the top Japanese contemporary artists Yokoo Tadanori, represents some aspects of daily experiences under the pandemic. These works seem to echo the Japanese "With Corona" policy and also shed light on the vulnerability and potentials of bodily becomings. On the other hand, the seemingly mechanical and dull graphs of the number of new cases have allowed people to reflect on their situation and constantly adjust their daily actions. This widespread self-restraint has been met with numerous criticisms, from defenses of freedom to respect for economic activity. However, paying attention to the vulnerability and potentials that Yokoo points out, it will become clear that participating in collective care practices is not the same as being controlled. Through focusing on the people's subjectivity and becomings, this paper attempts to describe some aspects of the experience of the early phase of the COVID-19 pandemic in Japan.

Introduction

On March 2, 2020, Reuters reported 19 new cases of COVID-19 in Japan, along with one photograph captioned, "Crowds wearing protective masks, following an outbreak of the coronavirus, are seen at the Shinagawa station in Tokyo, Japan."[1] Reuters also selected it as a "Top Photo of the Day." This story caused some controversy in Japan. At that time, WHO had not yet declared a pandemic, and the outbreak of the new

[1] "Japan confirms at least 19 new coronavirus infections on Monday: Kyodo," *Reuters*, March 2, 2020, web edition, https://jp.reuters.com/article/us-health-coronavirus-japan-infections/japan-confirms-at-least-19-new-coronavirus-infections-on-monday-kyodo-idUKKBN20P1MR.

A. Hamada (✉)
University of Tokyo, Tokyo, Japan
e-mail: hamada@anthro.c.u-tokyo.ac.jp

S. Venkatesan et al. (eds.), *Pandemics and Epidemics in Cultural Representation*,
https://doi.org/10.1007/978-981-19-1296-2_12

infectious disease was considered an event on the other side of Europe. The European press took a somewhat discriminatory view of crowds of people going to work while wearing surgical masks whose scientific efficacy was yet to be determined. It is not certain that this gaze was discriminatory, but there is no doubt that this photograph captured both the biological and the social aspects of the outbreak. It is impressive that for us who now know what happened in the past year worldwide, the news that there were only 19 cases in an island nation in the Far East attracted so much attention. However, this image, highlighting the presence of a small number of infected people and the practice of wearing masks, surely prefigures the COVID-19 experience in Japan through March 2021.

From March 2020 through March 2021, Japan is said to have experienced four epidemic waves. The government and mass media monitored the severity of the outbreaks using the number of cases, not deaths, as the primary indicator. The number of COVID-19 deaths in Japan was not very high, with 9,176 confirmed deaths by March 31, 2021. The National Institute of Infectious Diseases reported that Japan was experiencing fewer, not excess, deaths this year, based on analysis through the end of March 2021.[2] Due to the small number of deaths, which is also a lagging index, the number of cases has been used exclusively to capture the infection trends sensitively by the government and mass media. At the same time, people living in Japan also understand their situation through a particular representation of the outbreak, the trend graphs of cases, and adjust their actions accordingly.

Meanwhile, wearing masks emerged as a consistent focal point of public discussion on the Japanese response to COVID-19. In the spring of 2020, surgical masks disappeared from store shelves, and their escalating prices became a social problem. Shinzo Abe, the prime minister at the time, spent 26 billion Japanese yen (approximately 24 million USD) to implement the "Abeno Mask" ("Abe's Mask") policy of distributing two cloth masks to every household. People frowned upon this policy because the masks were distributed too slowly and were of poor quality. After the second half of 2020, some politicians recommended a strange procedure called "masked dinners" to balance infection control and economic measures. Instead of closing restaurants, people were encouraged to wear masks when holding drinking sessions, except at the brief moment of bringing food or drink to their mouths. This recommendation seems to have significantly contributed to the re-emergence of the outbreak. These examples clearly show that wearing masks became a core practice of the "With Corona" policy, imposing a new way of life on people under the pandemic, in which they were to continue daily life and economic activities as usual while also attempting to control the spread of infectious diseases.

Tadanori Yokoo was sensitive to this situation and created a new artwork series using masks as a motif. Born in 1936 in Hyogo Prefecture, Japan, Yokoo emerged as a graphic designer in his early twenties and gained an international reputation with a

---

[2] "Wagakuni niokeru Subeteno Shiinn wo Fukumu Cyoukashibousha-suu oyobi Kashoushiobusha-suu (2021 nenn 3 gatsu made no data bunseki) (Number of excess and fewer deaths including all causes of death in Japan [data analysis up to March 2021)]," National Institute of Infectious Disease, Japan, accessed July 8, 2021, https://www.niid.go.jp/niid/ja/from-idsc/493-guidelines/10488-excess-mortality-210630.html.

solo exhibition at MoMA in New York in 1972. After seeing Picasso's paintings in 1981, he declared his intention to become a painter in 1982 and has since become one of the top Japanese contemporary artists. Even in his old age, Yokoo has continued to make artworks vigorously. After turning 64 in 2000, he began creating the "Y-Jiro" ("Y-junction") series, which bears the artist's initials. The composition of the picture plane in this series changes little by little under the influence of surrealism, based on his photographs in various cities in Japan. On the occasion of the COVID-19 pandemic, this 84-year-old painter began to create a new series of works called "WITH CORONA." Back in his twenties, Yokoo had gained experience making cloth masks on which he printed a large mouth with the tongue sticking out. The artist then pulled this mask out of storage and used it in digital collages to cover the people's mouths depicted in his past works. His production speed was overwhelming. He presented the first work of this series on his Twitter account (@tadanoriyokoo) on May 22, 2020, and by March 31, 2021, the number of pieces visible on his Twitter account alone was well over 600. The relationship between the policy of "With Corona" as imposing a new way of life on people and Yokoo's work series "WITH CORONA" is not limited to the similarity of their names. As a prominent cultural representation of the pandemic in Japan, "WITH CORONA" not only brings joy to the viewer but also provides us with important clues for understanding the "With Corona" policy.

Therefore, by examining from anthropological perspectives two cultural representations of the pandemic, namely, the lively creative artwork series of "WITH CORONA" and the seemingly mechanical and dull graphs of the number of cases, this paper aims to reveal the characteristics of the "With Corona" approach and the COVID-19 experience in Japan.

The Hyogo Prefectural Yokoo Emergency Hospital Exhibition and "WITH CORONA"

On January 31, 2020, the opening event of a unique exhibition called the "Hyogo Prefectural Yokoo Emergency Hospital Exhibition" was held. This exhibition coincided with Yokoo's new book "The Benefits of Illness" on his injury and illness experiences (Yokoo, 2020). The collection included about 80 artworks, mostly related to the episodes introduced in the book, as well as pieces from the picture diary of the artist's stay in the hospital, which were divided into sections such as pediatrics, surgery, eye clinic, ENT doctor, inpatient wards, geriatrics, and the sports outpatient department.

The exhibition required that masks be worn inside the facility even before the new coronavirus infection spread fully in Japan. A photograph of the opening event posted on Yokoo's official website shows all the guests sitting with masks.[3] Yokoo

[3] Yokoo Tadanori, "YOKOO'S VISION," February 4, 2020, http://www.tadanoriyokoo.com/vision/2020/02.html.

said that wearing masks was mandatory because "I thought masks symbolize the difference between the inside and outside of the hospital."[4] The subsequent outbreak of COVID-19 forced various changes to the exhibition. Not only were the scheduled lectures and curator's talks canceled, but the museum itself was temporarily closed for most of the period from March 4 to May 31. Since June 1 fell on a regular closing day, the exhibition did not reopen until June 2, and at that time, the museum changed the exhibition's content.

The most significant change was the installation of "WITH CORONA," in which a new mask was added to the original work, around it, as a contrast. This deployment was an exhibition technique that highlighted the differences between pre-pandemic and pandemic lives. The proliferation of "WITH CORONA" and curious masks seemed to represent the uniqueness of the pandemic experience, particularly a universal mask policy. The fact that "WITH CORONA" was not framed in any special way but was simply printed on a paper in color and pasted directly on the wall also worked to catch the viewer's attention. This rough presentation seemed to express his attitude well that "the goal of painting is not completed product, but the process."[5]

Yokoo is a painter who has positioned the experience of illness as a crucial opportunity for his transformation. In the book that inspired this exhibition, he writes, "I have been involved in a traffic accident or suffered a serious injury every ten years, but each time I have seen a significant change in my life and work. It's hard when I'm on a sickbed, but if I think that without that period of pain, I would not have grown. I have no choice but to accept pain and danger as part of being human" (Yokoo, 2020, 73–74). If this is the case, what kind of experience triggered his transformation in the "WITH CORONA" series? Furthermore, what can we learn from his process of creating such a large number of artworks?

Although the latter question requires extensive engagement in the remainder of this paper, the former question is relatively easy to answer. The pandemic of the new coronavirus led to the birth of "WITH CORONA." However, Yokoo did not say that he personally contracted COVID-19. In this respect, the experience of the new coronavirus infection is different from his previous experiences of injury and illness in a critical manner. The first lesson from the painter is that COVID-19 is a disease that affects people even if they are not infected.

The experience of a pandemic has different characteristics from the experience of a sustained regional epidemic or a non-communicable disease. The primary reason why a pandemic is considered a pandemic is that it causes simultaneous outbreaks in different parts of the world. Nevertheless, having experienced the pandemic of a new coronavirus infection, we are compelled to find more meaning in *pandemic* than simply an infection that spreads around the world. When an infection spreads fast

[4] This and all other Japanese quotations are translated by the author. Tanake Erena, "Korona Ka De Mugen Ni Zousyoku Suru Masuku Art: Yokoo Tadanori-san (Mask Art Multiplies Infinitely with Corona Disasters: Tadanori Yokoo)," *The Asahi Shinbun*, July 1, 2020, https://digital.asahi.com/articles/ASN6Y4TN0N6QPTFC01B.html.

[5] Yokoo Tadanori, "YOKOO'S VISION," January 28, 2019, http://www.tadanoriyokoo.com/vision/2019/01.html.

enough to cause a pandemic, those considered healthy in ordinary times also must change their behavior to prevent disease to at least some extent, similar to the ways many others live with illness or disability daily.

"With Corona" as a Method of Mitigation

There is no doubt that the title "WITH CORONA" has something to do with the "With Corona" policy aimed at slowing down the spread of infectious diseases while going about daily life in Japan. It is difficult to determine who started using the term "With Corona" and with what original connotation in Japan; the phrase, which appeared on the web around March 2020, first was printed in major Japanese newspapers in late May. After the "With Corona Declaration" by Tokyo Governor Yuriko Koike on May 29, 2020, the term became more commonly used. It is a bit lengthy, but the following quote from Koike's statement introduces the critical points:

> We must not forget that there are no effective treatments or vaccines available. Unless this situation is improved, the only way to prevent the spread of this virus is through our own strong will and action. The fight against the new coronavirus might last for a long time, so we must recognize that we will have to live with the new coronavirus. Therefore, I would like to issue the "With Corona Declaration" in the sense of "living with corona." By living with corona and firmly creating a "new normal" together, we will continue to live in this era along while appropriately modulating our fear of it. We need to have a common understanding of the concept of "With Corona" as an inevitable reality, as the basis for various things, such as promoting scientific research, including the development of vaccines and pharmaceuticals, and the establishment of tests and medical care systems.[6]

This declaration asserts two things: (1) that elimination or eradication of this infection will be difficult, at least in the short term, because of the lack of established pharmaceutical interventions, and (2) that this will require a solid commitment to a "new normal" way of life that includes non-pharmaceutical interventions. These two intertwining understandings, which form the basis of "With Corona" here, are not so far-fetched. Japanese infectious disease experts were aware early that it would be tough to eliminate new coronavirus infections from the country and that achieving temporary elimination was not as important as keeping the number of new infections at a low level. For example, in an interview published in *The Asahi Shimbun* on March 11, 2020, Taro Yamamoto of Nagasaki University, while stressing the importance of saving the patients in front of him, said, "I believe that it is more desirable to achieve 'coexistence' with infectious diseases than eradication."[7] To achieve this, he

---

[6] "GOVERNER'S OFFICE," Toukyouto, May 29, 2020, https://www.metro.tokyo.lg.jp/tosei/governor/governor/kishakaiken/2020/05/29.html.

[7] Ota Hioroyuki, "Kansensyo to Shakai Mezasubeli ha "Kyouzon": Yamamoto Tarou Nagasaki Nettai Igaku Kenkyujyo Kyoujyu ni Kiku (Infectious Diseases and Society: What Should We Aim to Is "Coexistence" Interview with Taro Yamamoto, Professor, Institute of Tropical Medicine, Nagasaki University)," *The Asahi Shinbun*, March 11, 2020, https://digital.asahi.com/articles/DA3S14397923.html.

emphasized the need to slow down the infection rate and promote the weakening of the virus through mutation by implementing specific infection prevention measures.[8] Shigeru Omi, who has consistently served as primary medical advisor to governments during this pandemic, expressed the same view. Omi, a former Director-General of the World Health Organization's Western Pacific Regional Office, stated at a press conference held on March 9, 2020, "If all goes well, the outbreak will be brought under control rapidly to some extent. I am hoping so. However, it will not be zero. It will never be zero. It is impossible to eliminate all infections in Japan in many ways, and there will be infections from abroad."[9]

At the same time, infectious disease experts analyzed the locations where super-spreading events had occurred and discovered certain shared features of the environments of such events. They summarized the main risk factors as the 3Cs (Closed spaces, Crowded places, and Close-contact settings), and the avoidance of these 3C conditions became a core component of the "With Corona" policy. The Japanese government asked each industry association to prepare guidelines outlining measures to avoid 3Cs conditions, and up to 271 associations implemented such measures by the end of March 2021. This industry-association-based reaction has mainly depended on rearranging customers' and staff's environments to reduce the quantity of human-to-human contacts and to change the quality of contacts by removing existing elements and adding new objects in particular areas, such as reducing the number of seats and desks in restaurants and universities or calling for the wearing of masks in various stores and public facilities. In light of Michel Foucault's discussions on the milieu and dispositive (Bussolini, 2010; Foucault, 2004), this type of approach could be interpreted as attempting to restyle the particular physical space in which the body moves around and to thereby guide human actions in specific directions, as has been done in treating other infectious diseases such as tuberculosis. In fact, a focus on deploying concrete objects and influencing the bodies that live among them has been a central subject of anthropological studies on infectious diseases since before the pandemic (Hamada, 2017; Nishi, 2017; Prince, 2014).

It is thus easy to see this approach as a type of biopower and criticize its will to control people under the name of necessary public health measures. However, what moves people to do something is not necessarily limited to state power or medical sciences, represented respectively by politicians and infectious disease specialists. To simplify disciplinary power in this way would lead to depicting the lives of people living in a pandemic as severely flattened. Instead, it should be acknowledged that people can also change their environment to make their daily lives better through the same procedure of deploying concrete objects. In fact, Japanese infectious disease experts have also been wary of asserting total control over people's daily lives through

---

[8] Yamamoto's statement may seem a bit optimistic to those who already know that the mutation caused increased infectivity and increased virulence at the same time. However, it would have been quite difficult to foresee this possibility when it was not clear that the severity of the new coronavirus infection was due to the associated cytokine storm rather than the virus infection itself.

[9] FNN Prime Online, "Shingata Korona Uirus Kansen Kakudai: Senmonnka Kaigi Kaiken (The Spread of the New Corona Virus: Press Conference of Expert Committee," YouTube, March 9, 2020, https://www.youtube.com/watch?v=6nMypdClk7A.

infectious disease control measures, fearing that too much restraint would lead to a lack of cooperation or to economic collapse.[10] For example, these experts allowed for travel as long as travelers followed specific guidelines, although from the perspective of infection control, people should generally abstain from traveling. This concern for people's freedom or trust in subjective decision-making without strict regulations, perhaps enabled by biological factors such as acquired immunity and genetic traits, is a critical feature of the infectious disease control approach in Japan under the pandemic.

This "With Corona" policy can therefore be regarded as a combination of containment and mitigation methods. In contrast to the containment method, which aims to eliminate the virus by finding infected people through testing and tracing their contacts, the mitigation method seeks to slow down the spread of infection by reducing human-to-human contacts (OECD, 2020). "With Corona" is characterized by the fact that it does not simply rely on the closure of specific facilities but is implemented by changing how people live their daily lives. It is worth noting that politicians and infectious disease experts repeatedly emphasized the importance of thinking and responding voluntarily with a "strong will." In other words, this policy encourages people to transform their daily lives side by side with containment measures, which are carried out only by medical personnel and mainly at public health centers. The "With Corona" policy might have played some role in controlling the spread of new coronavirus infections in Japan. Indeed, although climatic factors such as temperature and humidity likely played a significant role, the number of new cases decreased in August and September 2020 without vigorous enforcement, often only requesting that restaurants limit their business hours.

Experiencing the Number of Cases

Nevertheless, it is not accurate to attribute this partial success to the achievements of only politicians and infectious disease experts. There were various measures taken from the bottom up, close to the bodies living in particular environments. The importance of these bottom-up responses will be evident when noting that the number of cases often began to decrease before the state of emergency was declared. What lies behind this phenomenon is similar to the reflexivity that led Yokoo to sublimate his perception of the situation. People grasped their situation through another cultural representation, the graphs showing the trends in the numbers of cases, and changed their actions based on this perception.

As a matter of course, we cannot see the virus with our own eyes. We cannot even tell who is infected with the virus in many cases. Nevertheless, the presence of the

[10] The Page, "Seifu no Senmonka Kaigi Kansensha Gen Demo "Gaishutsu Jishuku Nado no Keizoku" Teigen (Government Expert Committee Recommends "Continuing to Request Self-restrain from Going out, etc." Even if the Number of Infected People Declines)," YouTube, May 1, 2020, https://youtu.be/pnh1Kw4ozhM?t=4086.

virus is associated with visual images. SARS-CoV-2, which causes COVID-19, is a type of coronavirus, and the name "corona" comes from the Greek (or Latin) word for it, as the shape of the virus resembles a crown. Similarly, the Ebola virus that causes Ebola fever is a type of filovirus, and the name "filo" is also derived from the Latin word for thread, referring to its shape. It is likely that many people saw the shape of the virus during the outbreak in photographs taken with an electron microscope. The website of the Japanese public broadcaster NHK (Japan Broadcasting Corporation) repeatedly illustrated the shape of the virus as the icon for COVID-19 at this time.

On the contrary, in depicting the Zika epidemic in Brazil, feminist anthropologist Debora Diniz employed the strategy of portraying the virus not through the virus itself but through the suffering of the patients infected with the virus. At the end of the short film *Zika*, which is available on YouTube, nine women with their babies give impressive testimonies of how they contracted Zika fever during their pregnancies and how they are now the mothers of children with microcephaly.[11] This type of representation focusing on people's experiences of suffering has also been prevalent during the pandemic in Japan.

However, the most important images of the virus to understand people's bottom-up response to COVID-19 in Japan are not the representations captured by the electron microscope or the video camera but the graph that visualizes the infection status by the increasing or decreasing number of new cases. Of course, people did not only look at the numbers during the COVID-19 pandemic. Instead, the public has been exposed to a large amount of information about particular people who survived COVID-19, people who lost their lives, and people who were severely affected by various measures to prevent the spread of infection. More to the point, by its very nature, contact tracing treats an infected person not as one among millions but as an entity with individuality, moving from place to place at a particular time. Nevertheless, the characteristics and trends of COVID-19 have been monitored primarily by changes in the number of new cases.

The differences in how these viruses have been represented may owe a great deal to the cultural perceptions of the people trying to represent them. Still, they also depend on the characteristics of the relationship between viruses and people, how each virus affects the human body. There are appropriate ways to portray the Zika virus, which does not cause severe damage to the infected person's body but significantly impacts the fetus. There are also fitting ways to show the SARS-CoV-2 virus, which does not have a notably high fatality rate among infected people like the Ebola virus, but can infect a large number of people in a short period, causing a systemic breakdown of medical care. It is therefore not surprising that there are differences. These differences reflect not just cultural perceptions but also what people have learned from the different viruses.

Since the 2000s, medical anthropology has emphasized the importance of biosociality when examining infectious diseases and people's responses (Farmer et al. 2013). This idea suggests that we should not only focus on the cultural diversity of perceptions of the same entity but also on the interactive relationships in which

[11] Debora Diniz, "Zika," 2016, video, https://filmfreeway.com/Zika.

the existence of pathogens affects human activities and human activities affect the prevalence of infectious diseases. The biological characteristics of the virus, as identified by the changes in cases, already include the features of this or that human society. Human societies, in turn, have already been reconstructed in preparation for the emergence of possible pathogens. The biological and the social are intertwined and have unfolded in multiple ways.

The COVID-19 pandemic has again brought to light the interdependent relationship between the number of people infected and their individual experiences. COVID-19, which causes an exponential explosion of infections, produces more critically ill patients than the medical system can handle, and infected people die without proper medical care. Therefore, the very numbers themselves have a significant impact on the prognosis and experience of the infected. The status of the numbers provides information for determining whether there has been adequate preparation for dealing with infectious diseases. The graph of the quantitative trends of cases is a vital element of this recursive cycle.[12]

What is characteristic of COVID-19 when portrayed as a change in the number of infected people is that it has the potential to significantly impact those not yet infected. In Japan, the government regularly monitors the number of people infected with influenza and has often issued epidemic alerts and warnings. In the past, these alerts and notifications may have prompted some people to be more vigilant in preventing infection, but the COVID-19 pandemic has required more people to be more sensitively self-reflective in response to the number of cases. The experience of COVID-19 is not only one of fever, suffocation, and isolation. It is also an experience of being forced to constantly adjust one's actions, paying attention to and responding to the increase or decrease in the number of cases. In this sense, as Yokoo's "WITH CORONA" implies, we are all already patients of COVID-19. Based on this experience, people have been coping with the pandemic by subjectively self-restraining their movement.

[12] COVID-19 is not the first time that the number of infected people has had a significant impact on the disease experience of individual infected people. It is not hard to imagine that the development of prevention and treatment methods for diseases with a small number of infected people tends to take a back seat. The first confirmed case of Ebola fever was in 1976. However, no preventive or curative measures were developed for a long time, as until the outbreak in West Africa in 2014, there were 2410 confirmed cases of Ebola fever worldwide and 1,594 deaths. Ebola, which kills 50–70% of infected people in a short time, has never had a significant outbreak because it kills off infected people before they can spread the disease. Developing treatments and prevention methods require a lot of money and effort. No matter how dangerous the disease is, if the total number of patients is in the thousands, the benefits will not be worth the cost. Although the calculations are cruel to the infected, it is easy to see that estimates based on numbers can significantly impact the perceived danger of infectious diseases and the experience of being infected.

Body and Vulnerability

This widespread self-restraint has been met with numerous criticisms, from defenses of freedom to respect for economic activity. The "With Corona" policy calls for everyone to act accordingly, and it has a distinctly normative coloration. Although it seems to slow the spread of infection, some scholars have continuously criticized this approach as involving unjustified and excessive interventions in people's daily lives by politicians and infectious disease experts (e.g., Mima 2020). This criticism is similar to Georges Agamben's early objections to restrictions on freedom of movement when the lockdown was introduced in Italy (Agamben, 2020). There is a certain persuasiveness to these arguments, especially when considering specific worrying cases, such as the Singapore government's announcement that it will use an app developed to track the contacts of infected people for criminal investigations. In light of the repression of citizens by state agencies in Hong Kong, Thailand, Nigeria, and Myanmar during this pandemic, the fear that something similar will happen in Japan is not entirely unfounded.

Considering this point, we may learn something from the painter who turned even the daily challenges under the pandemic into a source of inspiration for his creativity. It seems that Yokoo suspended the evaluation of the experience of wearing masks in his "WITH CORONA." Even though it depicted the experience of universal masking, this series was not an advertising poster recommending that everyone should wear a mask. Instead, his practice of adding the curious cloth mask to people's faces seems to be an ironic commentary on the fact that all people are now wearing masks. Furthermore, the museum posted the precautions not for the visitors of the Yokoo Tadanori Museum of Contemporary Art but of the non-existent Hyogo Prefectural Yokoo Emergency Hospital. Here, we can see the attitude of making "With Corona" an object of parody.

On the other hand, Yokoo was critical of young novelists expressing excitement over the uncertainty caused by the pandemic, stating that he personally was not excited but afraid. "The expression 'excited about the unknown' is an idealistic view of Corona. Corona is not a metaphysical problem but a straightforward physical one. As a painter, I can't see it as an emotion like excitement. Corona is a matter of the body. This position is where I differ from the conceptual literary man."[13] Here, he establishes the dichotomies of excitement-fear, novelist-painter, and idealistic-physical. Yokoo continued that a novelist cannot acquire a body as long as he remains a novelist: "So the novelist and the painter are originally different beings. That's fine. Some novelists try to acquire a body, but it is only a body as an idea. If you want to acquire a body like a painter, you have to turn to painting or destroy your own body and declare the end of your career as a novelist. Or you can die."

Yokoo would probably say that even the author of this paper, as an anthropologist, cannot acquire a body. Nevertheless, suppose it is permissible to examine the significance of the body, which Yokoo emphasizes to such an extent, with the help of his

---

[13] Yokoo Tadanori, "YOKOO'S VISION," March 31, 2020, http://www.tadanoriyokoo.com/vision/2020/03.html.

description of his experience of illness (Yokoo, 2020), along with existing discussions of the body in anthropology. In that case, we can point out the importance of embodiment as (1) the vulnerability of the body (e.g., Mol, 2008; Robbins, 2013), (2) being a bodily presence that occupies a specific space–time (e.g., Hall, 1966), and (3) being able to leave traces of physical movements, for example, as paintings (Ingold, 2013).

These three features of human bodily existence are, in fact, also the ones focused on in "With Corona" as a method: the body that (1) is affected by COVID-19, (2) occupies a specific space and time, which may place it in a 3Cs situation with a high risk of infection, and (3) could become the object of contact tracing. Nevertheless, when "With Corona" is criticized for restricting freedom, large parts of the body are rendered invisible. Instead, what emerges is a politics of "who" without pointing fingers about precisely "who" is trying to limit our freedom. It is worth remembering that Annemarie Mol has highlighted the impasse in the politics of "who" and raised the need to move instead to the politics of "what." Mol argues that the point of caring for the vulnerable body is not about who is controlling whom but about what should be done: the question is not "who is in charge, but whether or not the various activities involved are well attune to each other" (Mol, 2008, 56). Participating in collective care practices is not the same as being controlled.

The End of "With Corona"

This paper has shown that the Japanese "With Corona" policy is not simply seen as an expansion of biopower but as a way for people to participate in collective care. The graph showing the number of infected people plays an essential role in this process. However, it is also true that this policy has its limitations. As time goes on, the baseline number of cases when a state of emergency is declared or lifted has been consistently increasing. Rather than the number of infected people remaining low, severe outbreaks in some areas have exceeded medical capacity. People seem to have grown tired of responding to inconsistent political decisions rather than to have grown used to a new way of life. Yokoo's "WITH CORONA" also foreshadows the end of "With Corona."

After the "Hyogo Prefectural Yokoo Emergency Hospital Exhibition" ended on August 30, 2020, "Tadanori Yokoo's Declaration of Emergency" was held at the same museum from September 19 to December 20. Here, too, "WITH CORONA" was exhibited along with the works the artist had painted thus far. However, the museum changed the style of the exhibition from the previous one. It is unclear whether this change in the exhibition was the artist's intention or the curator's idea. In any case, "WITH CORONA" gave a completely different impression to the viewers.

In the "Declaration of Emergency," representations of the crisis were categorized based on the type of crisis. It was not immediately apparent what kinds of trouble each work represents. Therefore, viewers were encouraged to think about the meaning seriously. Unlike in the previous exhibition, "WITH CORONA" was not displayed in its raw state as a color print on paper but in a more stylized form. It is worth noting that there were no direct connections between the large earlier painting and "WITH CORONA" placed around it. Due to this exhibition method, viewers experienced "WITH CORONA" receding into the background when they thought about the original works. When they focused on "WITH CORONA," they stopped considering the crisis. Two types of artworks led to two different series of the flow of thought and were arranged like ground and figure, which are reversed depending on the viewer's perspective. At this point, if we are paying attention to the representation of the crisis, "WITH CORONA," as the experience of the pandemic and the universal mask becomes something that we cannot see, even though it is undoubtedly there.

This experience of habituation of the universal mask can be traced not only by how it is displayed but also by how the painter has made it. Initially, by collaging his cloth mask with a red mouth and tongue on a white background, Yokoo placed it as something that dissimilates everyday lives, making the viewer feel uncomfortable. However, the painter gradually started to distribute color to the mask from September 2020. This trend became more evident in November 2020, as people got used to the presence of surgical masks. By adding color, the presence of the cloth mask no longer caused discomfort but harmonized with the original works. Even though the mask was still there, the sense of strangeness faded.

Despite most citizens' opposition, the Japanese government and the Tokyo Metropolitan Government decided to host the Olympic Games this year, in 2021. Behind this decision is the observation that holding the Olympics will provide an advantage to the ruling party in the election of the House of Representatives this fall. The fifth wave has arrived, and the Olympic Games will be held under a declared state of emergency. These situations have raised a public discussion on the relationship between politics and public health measures and have weakened people's willingness to exercise self-restraint. There is no doubt that the decisions of politicians will continue to affect the situation of infectious diseases. We should, however, learn from the 84-year-old painter still creating "WITH CORONA" because our physical actions can also influence the spread of COVID-19.

References

Agamben, G. (2020). *A che punto siamo? L'epidemia come politica.* Macerata: Quodibet. Japanese edition: Agamben, G. (2021). *Watashi Tachi ha Doko ni Irunoka?* Trans. Takakuwa Kazumi. Tokyo: Seidosha.

Bussolini, J. (2010). What is a dispositive? *Foucault Studies, 10*, 85–107.

Famer, P., Kleinman, A., Kim, J., & Basilico, M. (Eds.). (2013). *Reimagining global health: An introduction.* University of California Press.

Foucault, M. (2004). *Security, territory, population: Lectures at the college de France 1977–1978.* Trans. Graham Burcjell. New York: Palgrave Macmillan.

Hall, E. T. (1966). *The hidden dimension.* Anchor Books.

Hamada, A. (2017). Restyling the Milieu: On Milieu Making practices around tuberculosis treatment projects in Southern Ghana. In A. Hamada & M. Toda (Eds.), *How do biomedicines shape*

people's lives, socialities and landscapes? Senri ethnological report (Vol. 143, pp. 141–161). Osaka: National Museum of Ethnology.
Ingold, T. (2013). *Making: Anthropology, archeology, art and architecture*. Routledge.
Mima, T. (2020). *Kansensho Shakai: Afutaa Korona no Seiseiji* (*Infectious Disease Society: Biopolitics after the Pandemic*). Kyoto: Jinbun Shoin.
Mol, A. (2008). *The logic of care: Health and the problem of patient choice*. Routledge.
Nishi, M. (2017). Reconsidering therapeutic citizenship in the era of universal treatment: (Dis)connectedness and (non)transformation among HIV positive people in ethiopia. In A. Hamada & M. Toda (Eds.), *How do biomedicines shape people's lives, socialities and landscapes? Senri ethnological report* (Vol. 143, pp. 3–21). Osaka: National Museum of Ethnology.
OECD. (2020). *Flattening the COVID-19 peak: Containment and mitigation policies*. Accessed July 8, 2020. https://www.oecd.org/coronavirus/policy-responses/flattening-the-covid-19-peak-containment-and-mitigation-policies-e96a4226/
Prince, R. (2014). Navigating global health in an East African City. In R. Prince & R. Marsland (Eds.), *Making and unmaking public health in Africa: Ethnographic and historical perspectives* (pp. 208–230). Ohio University Press.
Robbins, J. (2013). Beyond the suffering subject: Toward an anthropology of the good. *Journal of the Royal Anthropological Institute, 19*(3), 447–462.
Yokoo, T. (2020). *Byouki no Goriyaku* (*The Benefits of Illness*). Tokyo: Popurasha.

Chapter 13
<3 I Am Tracy: Meme Culture, Coping, and Community During the COVID-19 Pandemic

Julie M. Powell

Abstract <3 I AM TRACY examines the social, cultural, and psychological impact of the COVID-19 pandemic through Internet memes. It considers how meme culture has been utilized to weather the COVID-19 emergency and how this practice fits within a larger history of mobilizing humor to cope with crisis. By analyzing coronavirus memes shared via Instagram in the spring of 2020, I shed light on the ways in which they contributed to the resilience of those who created and circulated them. I argue that meme culture helped participants to cope with the psychological stress of the crisis—providing a means to defuse anxiety, reassert control, establish a sense of continuity between past, present, and future, and develop and maintain bonds within communities. I establish, moreover, that, despite the novel form, the features and functions of meme culture during the pandemic share important characteristics with the coping strategies—prominently, dark humor—of Lost Generation writers in Britain and the USA. In doing so, the chapter contributes to existing work in the fields of communication studies, literature, anthropology, popular culture and folklore studies, psychology, science and technology studies, and history.

Introduction

On April 14, 2020, Instagram meme account @beigecardigan reposted a video from TikTok user @gracecuratolo. The nineteen-second clip featured an exasperated, middle-aged woman standing on her front porch, calling out breathlessly through a megaphone, "I just want you to know that I miss you and I love you. I can't wait for this shitshow to be over so that I can touch people, drink with people, and have the best life ever." Grace, giggling from her vantage point in the street, yells back:

J. M. Powell (✉)
School of History, IRC Postdoctoral Research Fellow, University College Dublin, Dublin, Ireland
e-mail: julie.powell@ucd.ie

S. Venkatesan et al. (eds.), *Pandemics and Epidemics in Cultural Representation*,
https://doi.org/10.1007/978-981-19-1296-2_13

"Miss you too, Tracy!"[1] The captured conversation, shared with @beigecardigan's 3.8 million followers, made light of the COVID-19 pandemic and the trying social distancing measures that it imposed. The clip is funny; the viewer laughs along with Grace at the woman's woeful lament—a message comically at odds with its method of delivery. But it is also sad and relatable. The post garnered over 900 comments, many of which linked friends to the video and noted "I'm having a bottle of wine tonight and swinging by all your houses and yelling this," or, more succinctly, "<3 I AM TRACY". By the time the video was shared, the topic of the pandemic was dominating the feed—or supply of posted content—of @beigecardigan and similar Instagram meme accounts. Seventy-eight percent of content posted by @mytherapistsays from March 1 to May 31 served as direct commentary on the COVID-19 pandemic and its impact, particularly on the experience of confinement and social distancing. The same is true for 58% of content posted by @daddyissues_, 55% of content from @insta.single, and 53% from @whitepeoplehumor. Like the "<3 I AM TRACY" clip, the posts can be darkly comedic, but can also be sentimental and hopeful, projecting into a future in which COVID-19 is but a memory and things are as they should be, or better.

Despite the ostensible frivolity of the form, Instagram meme accounts like @beigecardigan raise important questions about humor, technology, and resilience that I aim to investigate. The concept of a "meme" has been with us since 1976, when ethologist Richard Dawkins coined the term to describe a unit of cultural information spread by imitation. In the last two decades, "meme" has increasingly come to refer to the proliferation and adaptation of online multimedia artifacts that combine text, image, audio, and/or video elements derived from popular culture to express public commentary.[2] Memes are both humorous and thoroughly social. Transmitted from person to person, their intelligibility relies on "shared cultural experiences and practices" that likewise "define certain affinity spaces."[3] Indeed, decontextualizing memes, divorcing them from their affinity spaces, renders them meaningless. In 2011, Giselinde Kuipers removed 9/11 "Internet jokes" from their online fora and presented them to people in interviews for scrutiny. The memes proved unintelligible and were misinterpreted by those outside of the community from which they had originated.[4] The point is perhaps made most appropriately, though, by an @insta.single post from April 10, 2021. Bold text reads: "Me: Mom, look at this meme; Mom:". Below the text is "mom," a picture of a befuddled Winnie the Pooh staring at a sheet of paper.

---

[1] @beigecardigan. April 14, 2020. Miss you too, Tracy!. Digital video clip.Instagram.com. Accessed April 6, 2021.

[2] Milner, Ryan M. 2013. Pop Polyvocality: Internet Memes, Public Participation, and the Occupy Wall Street Movement. *International Journal of Communication* 7: 2357–2390.

[3] Lankshear, Colin and Michele Knobel 2007. Online memes, affinities, and cultural production. In *A New Literacies Sampler*, ed. Colin Lankshear and Michele Knobel, 199–228. New York: Peter Lang Publishing, 217.

[4] Kuipers, Giselinde. 2011. "Where Was King Kong When We Needed Him?": Public Discourse, Digital Disaster Jokes, and the Functions of Laughter after 9/11. In *A Decade of Dark Humor: How Comedy, Irony, and Satire Shaped Post-9/11 America*, eds. TedGournelos and Viveca Greene, 20–46. Jackson, MS: University Press of Mississippi.

Questions swirl around him, including: "do you know this person?", "who's this?", "did you make this?", and, of course, "what does this mean?".[5]

Memes have become a way for Millennials and Gen Z to communicate from a place of shared experience and, in doing so, to build and reinforce communities both real and virtual. Within the Instagram platform, users repost or like memes that resonate with them as a way to signify identification with, or endorsement of, an expression and send them to friends as a form of care, communicating "I'm thinking about you." It is these aspects of meme culture, as well as its ability to respond rapidly and in real time to events, that make it a fruitful site for exploring the social, cultural, and psychological impact of the COVID-19 pandemic. "<3 I AM TRACY" asks: How was meme culture—the creation and circulation of, and engagement with, memes—used to weather the COVID-19 emergency? And, how does this practice fit within a larger history of mobilizing humor to cope with crisis?

By analyzing coronavirus content shared by Instagram meme accounts in the spring of 2020, I aim to shed light on the ways in which such content contributed to the resilience of those who created and circulated it. I argue that Millennial meme culture in the USA helped participants to cope with the psychological stress of the first months of the COVID-19 crisis by providing them a means to defuse anxiety, reassert control over their circumstances, develop and maintain bonds within communities both real and imagined, and establish a sense of continuity between past, present, and future. What is more, I establish that, despite the novel form, meme culture during the coronavirus pandemic shared important characteristics with the coping strategies of Interwar and Depression-era writers in Britain and the USA, American prisoners of war in Vietnam, and frontline soldiers of the First World War. In all cases, a culture of dark humor served important social and psychological functions. According to literary scholar Lisa Colletta, "dark humour is not the humour of change; it is the humour of survival, the humour of getting through another day with as little trauma as possible."[6] It is this self-preservative humor that pervaded much of Millennial Instagram meme culture even before the COVID-19 outbreak, providing participants with a familiar language and an intelligible framework for confronting and managing the crisis—a psychological as well as an epidemiological one, according to the American Psychological Association, with potentially profound impacts on mental health and well-being.[7] Fortunately, as behaviorist Linda Henman writes, "resilience is contagious, as humour seems to be."[8] Memes, defined by their virality, most certainly are.

---

[5] @insta.single. April 10, 2021. Mom, look at this meme.Digital image.Instagram.com. Accessed April 6, 2021.

[6] Colletta, Lisa.2003. *Dark Humor and Social Satire in the Modern British Novel.* New York: Palgrave Macmillan, 114.

[7] American Psychological Association. October 2020. Stress in America™ 2020: A National Mental Health Crisis. Press Release.https://www.apa.org/news/press/releases/stress/2020/report-octoberAccessed April 14, 2021.

[8] Henman, Linda D. June 2001. Humor as a Coping Mechanism: Lessons from POWs. *Humor* 14: 83.

"<3 I AM TRACY" is an interdisciplinary study that engages with existing work in the fields of communication studies, literature, anthropology, popular culture and folklore studies, psychology, science and technology studies, and history. Together, these disciplines have generated two rich, multi-disciplinary literatures that explore, independently, the nature and utility of humor and of Internet memes.[9] This chapter brings the two literatures into dialogue and, in so doing, contributes something new to each. Moreover, by evaluating meme culture within the context of the Instagram platform, rather than through YouTube, Web sites such as Reddit, Tumblr and 4chan, or through e-mail, online discussion forums, and chat spaces, the chapter allows us to broaden our understandings of what memes are and what communicative functions they serve.[10]

Meming the Pandemic

On January 21, 2020, the USA confirmed its first case of COVID-19. On the same day, Chinese authorities reported two deaths from the virus, bringing their total of casualties to six. New cases were announced in Beijing, Shenzhen, and Shanghai; Hong Kong and Taiwan confirmed their first cases. One day later, North Korea became the first country to seal its borders, and over the next five days, cases were reported in Malaysia, France, Vietnam, Nepal, Canada, Australia, and Germany. In the last days of January, initial cases of the novel coronavirus, which had originated in Wuhan, China, and belonged to the same family as the SARS coronavirus, appeared throughout the world and travel bans spread. Within a month, there were over 75,000 cases of COVID-19 confirmed in twenty-seven countries and 2000 deaths. Yet, only eleven of those deaths were reported outside of China, the epicenter of the pandemic, and many beyond the region believed that the disease was little more than a flu. Nevertheless, February saw San Francisco and Washington State declare states of emergency; California and New York followed in the first days of March, and by March 10, states of emergency had been declared in twenty-three states. As schools began to close, public gatherings were cancelled, and borders across the globe were sealed, it became clear that the virus was a much larger threat than anyone had

[9] Useful works on the utility of humour, not directly cited in the chapter, include: Gehring, Wes. 1996. *American Dark Comedy: Beyond Satire*. Westport, CT: Greenwood Press; Martin, Rod A. and Thomas Ford. 2010. *The Psychology of Humor: An Integrative Approach*. Cambridge, MA: Academic Press; McGraw, Peter and Joel Warner. 2015. *The Humor Code: A Global Search for What Makes Things Funny*. New York: Simon & Schuster; Pratt, Alan R., ed. 1993. *Black Humor: Critical Essays*. New York: Garland Publishing, Inc.; and Rourke, Constance and Greil Marcus, eds. 2004. *American Humor: A Study of the National Character*. New York: New York Review Books.

[10] For YouTube, see Gal, Noam, Limor Shifman, and Zohar Kampf. "It Gets Better": Internet Memes and the Construction of Collective Identity. 2016. *New Media & Society* 18, 8: 1698–1714; for Tumblr and 4chan, see Milner, Pop polyvocality; and for the evaluation of memes in e-mail, online discussion forums, and chat spaces, see Lankshear and Knobel.

supposed and that business as usual was no longer a viable option. The first US stay-at-home order was issued by California on March 19. By the end of the month, full or partial stay-at-home orders were implemented in 78% of states plus the District of Columbia and nearly one-third of the world's population was in some form of lockdown.[11]

Memers began engaging almost instantaneously with what would become a singular and defining global event. From February 26, 2020, to March 10, the topic of the coronavirus pandemic debuted in the feeds of popular Instagram meme accounts, eight of which are closely analyzed here. Accounts were selected for examination according to their influence, (all had between four million and fifteen million followers at the time of investigation), their location (all were based in the USA, with the exception of @mytherapistsays, whose administrators travel between the USA and Canada and cite 70% of their engagement as coming from American accounts), and age (administrators of all accounts either identify as Millennials or were between twenty-five and thirty-eight years of age in the spring of 2020). @sarcasm_only (14.9 M followers), @lmao (9 M), @insta.single (6.3 M), @memequeen (6 M), @shitheadsteve (5.8 M), @mytherapistsays (5.2 M), @whitepeoplehumor (4.4 M) and @daddyissues_ (4.2 M), together boast nearly 56 million followers. While, inevitably, there is some overlap in the subscriber bases, the collective visibility and popularity of such accounts is undeniable. They were, and remain, entertaining and engaging for tens of millions of followers, who connect with the content and regularly reaffirm their identification with the material by continuing to subscribe to the feed. Compared with digital newspaper or magazine subscriptions, the number of followers on an Instagram account is a high-fidelity reflection of interest—if a follower no longer enjoys the content, they unsubscribe instantaneously, with the touch of a button. In seeking to understand the reaction of Millennial meme culture to the pandemic, such accounts are useful because they are popular, but also because they are prolific; accounts with large followings but relatively infrequent posts were excluded from the study.

From the beginning of March to the end of May 2020 (a period that covers the initial and heightened awareness of the pandemic, the introduction of social distancing and shelter-in-place policies, the routinization of confinement, and the relaxing of social distancing measures), the accounts collectively added 4327 posts to their feeds. Of these, 45% or 1,952 posts were related to the pandemic. In these three months, five of the seven accounts dedicated approximately half or more (47–78%) of their content to aspects of the COVID-19 crisis. @sarcasm_only, @lmao, and @memequeen dedicated 30%, 34%, and 38%, respectively. This is, in part, a reflection of the total number of posts contributed during the period by these accounts; the top four posters, including those above, contributed nearly 3000 posts while the bottom four contributed less than half that, 1,351. More productive accounts padded

[11] For an exhaustive timeline of the pandemic, see Kantis, Caroline, Samantha Kiernan, and Jason Socrates Bardi. 26 March 2021. UPDATED: Timeline of the Coronavirus. Thinkglobalhealth.org. https://www.thinkglobalhealth.org/article/updated-timeline-coronavirus. Accessed April 6, 2021.

their feeds with diverse, and generally recycled, content. Nevertheless, the accounts posted, on average, over 240 coronavirus memes each for the period.

The majority (74%) of pandemic-related content, 1,448 posts in total among the eight accounts, was added between March 11 and April 25 (see Figs. 13.1 and 13.2). More than half of those posts, 843, were added in the twenty days between March 16 and April 5, 2020, when 60–74% of the collective content of the eight

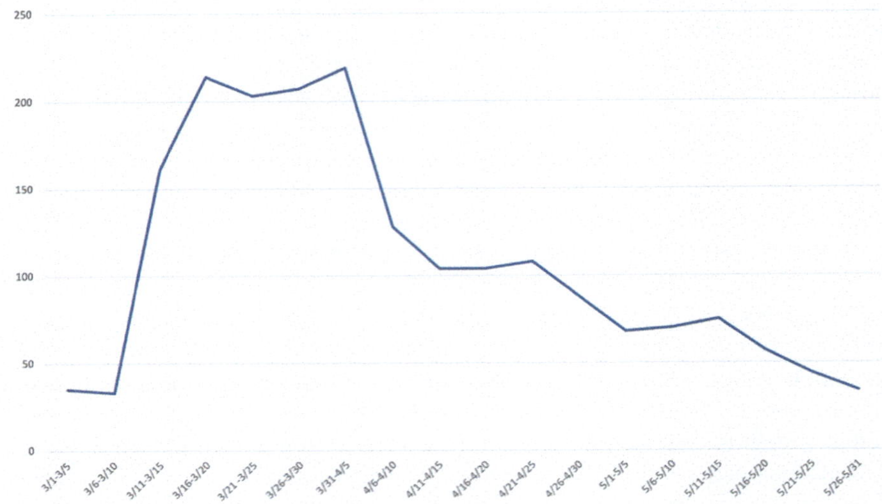

Fig. 13.1 Total coronavirus posts for all accounts by 5-day intervals (non-cumulative)

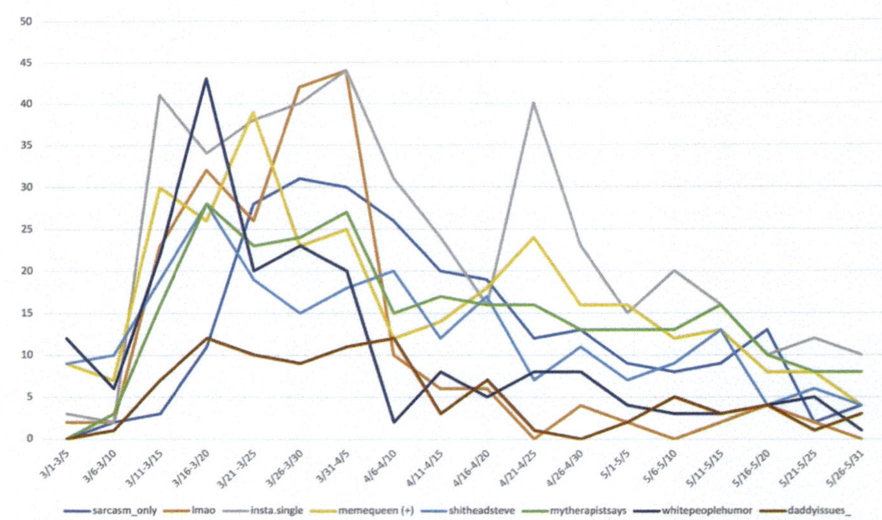

Fig. 13.2 Total coronavirus posts for individual accounts by 5-day intervals (non-cumulative)

accounts focused exclusively on the pandemic. Five accounts dedicated in excess of 80% of their feeds to pandemic-related content for portions of this 15-day window (see Figs. 13.3 and 13.4). Not incidentally, stay-at-home orders were implemented throughout the USA in this period, the earliest beginning March 19 and the latest on

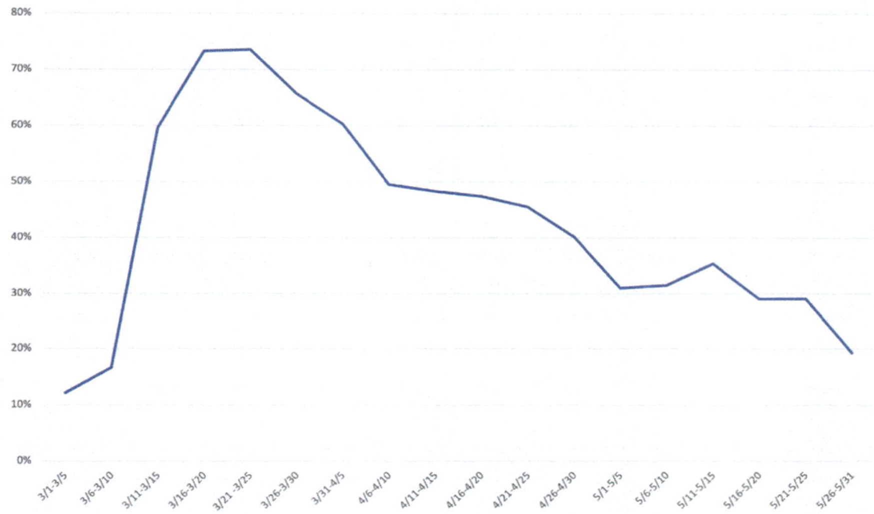

Fig. 13.3 Coronavirus posts for all accounts as a percentage of total posts by 5-day intervals (non-cumulative)

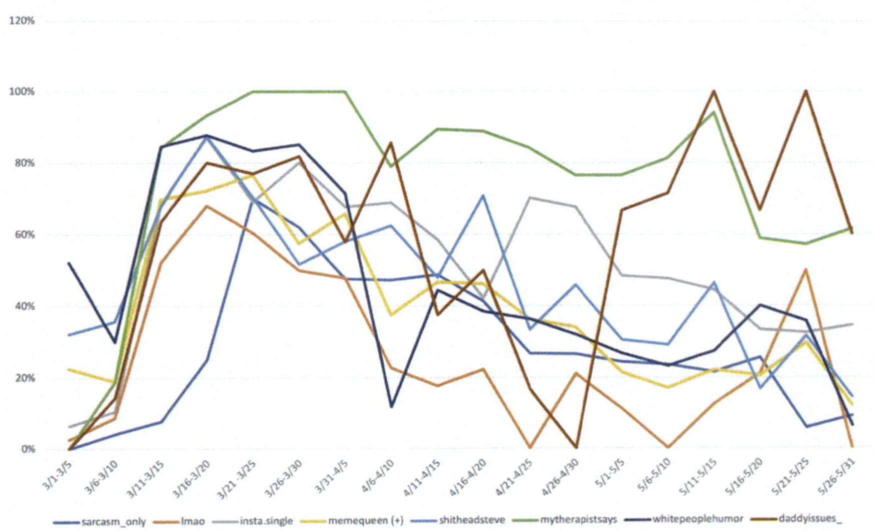

Fig. 13.4 Coronavirus posts for individual accounts as a percentage of their total posts by 5-day intervals (non-cumulative)

April 7 with an average implementation of March 28. Production and circulation of pandemic-related content (which would continue into 2021) slowed from April 26 to May 31 as state lockdowns approached their—arguably premature—conclusion. As of June 2, 2020, the majority of states had reopened; seven did so partially. Only New Jersey and DC remained under stay-at-home orders. There is no doubt that Instagram meme accounts engaged with the pandemic, particularly during the initial phases of lockdowns, in quantitatively significant fashion. Qualitative analysis of the content and a thorough understanding of the Instagram platform, however, are required to fully explore the ways in which meme culture contributed to Millennial resilience during the COVID-19 crisis.

The popularity and functionality of the Instagram platform have changed the landscape of meme culture. While, for instance, approximately 800,000 new posts are made on the 4chan website each day, 95 million photographs and videos are shared daily on Instagram. Active monthly users for the platform total 1 billion, as compared with 27 million for 4chan. Maintaining and growing audiences for meme accounts—which can be rewarded with lucrative licensing or advertising deals—means that new content must be delivered at a near constant rate. Thus, the majority of successful meme accounts repost content from other social media platforms, including TikTok, Reddit, and, most frequently, Twitter to augment their feeds.

While previous scholarly definitions of Internet memes would exclude a screen-captured Tweet, I contend that, taken in the aggregate, such posts must be considered mimetic. Limor Shifmanhas defined an Internet meme as a "group of digital items sharing common characteristics of content, form and/or stance that were created with awareness of each other and were circulated, imitated, and/or transformed via the Internet by many users."[12] The form of such posts, a close-cropped screen grab revealing only the original Twitter user's name and the text of the Tweet has become ubiquitous in the feeds of Instagram meme accounts. At times, the images are manipulated or annotated for added comedic effect. Uniformity in composition pairs with uniformity in "keying," one aspect of stance identified by Shifman that comprises the tone and style of the communication.[13] The messages captured from Twitter and posted to Instagram closely resemble one another in structure, grammar, punctuation and tone. They are witty, pithy, easily manufactured, easily shared, and, not only intelligible, but resonant and relevant for a growing number of participants in meme culture, who ensure their proliferation by engaging with the material.

The transition of meme-sharing to the Instagram platform has shaped not only its form but, in many cases, its function. Participation in the original forums for meme development and sharing—'exclusionary enclaves' like 4chan, Reddit, etc.,—was predicated on anonymity and technological savvy.[14] Instagram, however, is an accessible platform for general users, a mechanism for self-branding, where real-life social networks and virtual ones regularly intersect—aspects of the application

[12] Shifman, Limor. 2013. *Memes in Digital Culture*. Boston: MIT Press, 6–7.

[13] Ibid., 40.

[14] Milner, Ryan M. 2012. The World Made Meme: Discourse and Identity in Participatory Media. PhD Dissertation. University of Kansas, 306.

that have important implications for meme culture. Scholars recognize with near unanimity that memes are not simply artifacts but modes of communication that play a role in constituting social relations through shared values, symbols, and cultural sensibilities that define community boundaries in both the real and virtual spaces.[15] Precisely what is being communicated, however, depends on the "community" and remains a field wide open for inquiry. Scholars evaluating memes within the context of their original, anonymous online fora pinpointed the utility of the medium for expressing political commentary, but this is not the primary communicative function of Instagram memes which, for the largest accounts, remain openly apolitical in an effort to avoid alienating followers.[16] Shifman rightly notes that Internet users share memes for both "social and self-presentation purposes" and has identified humor as a critical component of a meme's virality. Humor, either "quirky and situational" or as "biting social commentary," he argues, elicits positive emotions and Internet users "prefer spreading content that makes others feel good and at the same time reflects on themselves as upbeat and entertaining."[17] Shifman's analysis raises questions, however, about the role of dark humor, which characterized much of the content of the largest Instagram meme accounts both before and during the COVID-19 crisis. Sharing memes about isolation, depression, and existential ennui hardly reflects on the sharer, or the receiver, as "upbeat," but it does create and sustain bonds based on shared experience and allows all parties that are in on the joke, that are part of the community, to laugh at their circumstances.

Catharsis and Control

Scholarship on humor provides useful insight into the therapeutic and communicative work that dark humor does in times of crisis. Koenraad Du Pont argues that humor in trench journals of the First World War provided "a psychological escape valve" that allowed groups to bond and cope with "negative events and loss of control." Soldiers at the front were able to acknowledge the grimmer realities of their situation but treat them "in a whimsical manner that elided or vicariously ameliorated their implications."[18] Henman found that for American POWs in Vietnam, humor was a means of "taking control of one's reaction when he could not have control of the situation" and remaining connected to others who found themselves in the same situation.[19] It

---

[15] See Gal, Noam, Limor Shifman, and Zohar Kampf; Wiggins, Bradley E. 2019. *The Discursive Power of Memes in Digital Culture: Ideology, Semiotics, and Intertextuality.* New York: Routledge.

[16] See, for example, Milner, Pop Polyvocality and Rintel, Sean. 2013. Crisis Memes: The Importance of Templatability to Internet Culture and Freedom of Expression. *Australasian Journal of Popular Culture* 2, 2: 253–271.

[17] Shifman, 66; 78; 96.

[18] Du Pont, Koenraad. 2015. Nature and Functions of Humor in Trench Newspapers (1914–1918). In *Humor, Entertainment, and Popular Culture during World War I*, ed. Clementine Tholas-Disset, and Karen A. Ritzenhoff, 107–122. Basingstoke: Palgrave, 111.

[19] Henman, 88.

was the "laughter of despair," notes Alan R. Gates, that rang through Depression-era literature, a temporary cure for the "creeping paralysis of will" experienced by many Americans who related to the loneliness and sense of futility highlighted by contemporary humourists.[20] Similarly, British novelists between the World Wars were, according to Colletta, "haunted by a sense of anxiety and powerlessness" born of the uncertainty and trauma of being socially redundant that they explored through a form of satire whose purpose had "more to do with the coping devices of gallows humour" than anything else.[21]

Dark humor is the humor of "running to remain in place"; it is about preservation, the casting off of pain and suffering and the refusal to grant them the power to overwhelm and destroy.[22] In particular, the work of Lost Generation writers that Colletta explores abandoned "any hope of understanding the world" and reflected the traumatic state of those individuals who found themselves superfluous within postwar society.[23] Millennials have become the new "lost generation," indeed, the "unluckiest generation in US history," according to studies by *The Washington Post* and *The Wall Street Journal*.[24] Wage stagnation over the last thirty years, combined with a dramatically increasing cost of living, job market instability, and political institutions that have proven both unable and unwilling to provide relief have shaped the Millennial experience into one of precarity. Hobbled by the experience of entering the workforce following the 2008 financial crisis, Millennials—'indebted and a step behind on the career ladder'—have taken a "second pummeling'" due to the pandemic, further hamstringing their ability to achieve financial independence, accrue wealth, and reach what have historically been considered important life milestones.[25] Millennials have dealt with their lot through a "weird" brand of surrealist humor, according to Elizabeth Bruenig, that is "tangibly dark." "Leaning in to feelings of worry, failure, and dread," Millennial humor, she writes, has created a "world of the surreal and bizarre, [where] horror mingles with humour, and young people have space to play with emotions [of 'emptiness and despair'] that proceed from ordinary life…a doubting [of] the essential meaningfulness of the world."[26]

---

[20] Gates, Robert Allan. 1999. *American Literary Humor during the Great Depression.* Westport, CT:Greenwood Press, 2.

[21] Colletta, 1; 6.

[22] Ibid., 7; 13.

[23] Ibid., 6; 32.

[24] Adamy, Janet. August 9, 2020. Millennials Slammed by Second Financial Crisis Fall Even Further Behind. *The Wall Street Journal.* https://www.wsj.com/articles/millennials-covid-financial-crisis-fall-behind-jobless-11596811470. Accessed March 30, 2020; Van Dam, Andrew. June 5, 2020. The Unluckiest Generation in U.S. History. *The Washington Post.* https://www.washingtonpost.com/business/2020/05/27/millennial-recession-covid/. Accessed March 30, 2021.

[25] Adamy, np.

[26] Bruenig, Elizabeth. August 11, 2017. Why is Millennial Humor so Weird? *The Washington Post.* https://www.washingtonpost.com/outlook/why-is-millennial-humor-so-weird/2017/08/11/64af9cae-7dd5-11e7-83c7-5bd5460f0d7e_story.html. Accessed March 30, 2021.

Arguably, Millennial meme culture—making light of one's own insecurity, powerlessness, isolation, and depression—was primed to articulate the psychological stresses imposed by social distancing, economic downturn, and stay-at-home mandates that, in some ways, constituted a lifestyle difference of degree rather than of kind. Indeed, before the spread of coronavirus occupied our collective consciousness, the @beigecardigan feed featured posts that now seem either eerily prescient or indiscernible from the confinement-related content yet to come. Witty posts took reflective aim at emotional eating, isolation and alcohol consumption, temporal disorientation, and existential crisis. On January 10, 2020, the account posted a screen capture of a Tweet from @robots_feel that read: "[guy who's about to invent parties] *drinking alone* i wish this was worse."[27] The post garnered over 57,000 likes. A December 13, 2019 post, endorsed by followers to the tune of 148,055 likes, featured a screen-captured tweet from @IsabelSteckel proclaiming: "i don't eat at midnight because i'm hungry, i eat because i'm empty inside, now would you like to have nachos with me or not rebecca."[28] Another screen-captured tweet from @Jordan_Coombe posted on December 29, 2019, and receiving 63,363 likes, noted: "It's the period between Christmas and New Year. No one knows what day it is. Time doesn't really exist. Can we start drinking at 10 am? Why tf not. Existence is a confusion."[29] For many, such coping behaviors and sensations are recognizable features of living through the first months of the COVID-19 pandemic, during which, as *Forbes* contributor Jack Kelly found, Americans "just trying to hang on and make it through the day" largely did so by "excessively eating, drinking, smoking pot, playing video games, and watching porn."[30]

Perhaps unsurprisingly, the same themes of the pre-pandemic period ran through the feeds of Millennial Instagram meme accounts from March to May 2020. On May 27, @mytherapistsays posted a screen-captured Tweet from @psybermonkey: "Friend: I'm worried about you; Me: *dunks oreo in whiskey* why".[31] The post received 88,570 likes, a fraction of those (183,189) received when @insta.single posted a Tweet on April 4 from @hannahmlplanet declaring: "Quarantine has shown me that you really don't need fun to have alcohol," a darkly comedic reversal of the old adage "you don't need alcohol to have fun."[32] On April 29, @mytherapistsays

[27] @beigecardigan. January 10, 2020. i wish this was worse. Digital image. Instagram.com. Accessed April 6, 2021.

[28] @beigecardigan. December 13, 2019. i eat because i'm empty inside. Digital image. Instagram.com. Accessed April 6, 2021.

[29] @beigecardigan. December 29, 2019. Existence is a confusion. Digital image. Instagram.com. Accessed April 6, 2021.

[30] Kelly, Jack. April 6, 2020. Americans Are Excessively Eating, Drinking, Smoking Pot, Playing Video Games and Watching Porn While Quarantined. *Forbes*. https://www.forbes.com/sites/jackkelly/2020/04/06/americans-are-excessively-eating-drinking-smoking-pot-playing-video-games-and-watching-porn-while-quarantined/?sh=2297757a404e. Accessed 8 April 2021.

[31] @mytherapistsays. May 27, 2020. I'm worried about you. Digital image. Instagram.com. Accessed April 6, 2021.

[32] @insta.single. April 4, 2020. you really don't need fun to have alcohol. Digital image. Instagram.com. Accessed April 6, 2021.

posted a Tweet from @jillyhendrix that received 103,778 likes, asking: "Trying to decide if I should buy my feelings today or eat them?"[33] On May 3, the account posted, and received 153,576 likes for, a screen capture from the show "Who Wants to Be a Millionaire." In the still image, the contestant ponders the question "What day is it?"; options, displayed in the image, include: "A. Monday, B. Tuesday, C. 1982, and D. Saturday." The text above the image states pithily: "All of us right now."[34]

Other posts considered the pandemic within the larger context of Millennial experiences of precarity. On April 4, @whitepeoplehumor posted a screen capture from the Netflix documentary "Tiger King." Subtitles for the film were visible in the image and read: "I am never going to financially recover from this." Above the image was the text, endorsed with 133,693 likes: "Like 98% of Americans when they don't have a pay check for literally one week."[35] Twelve days later, @insta.single posted a video clip, viewed over 2.5 million times, of actor Paul Rudd interviewing himself. Cracking open a beer, one Rudd says to the other: "hey, look at us." "Look at us," the other responds; "who would've thought?" "Not me," the first Rudd replies. The text imposed above the image reads: "millennials living through their 2nd 'once in a generation' economic collapse."[36] A Tweet by @kimberlycongdon, reposted by @whitepeoplehumor on March 22 made light of the criticism that many Millennials have faced for delaying parenthood. As parents around the world struggled to educate and manage their children during lockdowns, Congdon chimed in: "Not to brag but what a time to be childless."[37] The repost earned 198,585 likes.

Millennial precarity has made the generation, along with Generation Zers, "more anxious than previous generations" and the mental health struggles that Millennials have faced have frequently and unapologetically found their way into meme culture.[38] In a 2019 article for CNet, Abrar Al-Heeti discovered that "memes tackling existential crises and mental health have become an outlet for people…[and] topics like depression, debt, and failed relationships are regularly explored" using dark humor.[39] Millennial meme accounts on Instagram, true to form, responded to the COVID-19 crisis candidly. On April 21, @mytherapistsays posted a screen capture of a Tweet from @farringt0n that received 102,509 likes and asked: "gonna hit rock bottom does

---

33 @mytherapistsays. April 29, 2020. Trying to decide if I should buy my feelings today or eat them. Digital image. Instagram.com. Accessed April 6, 2021.

34 @mytherapistsays. May 3, 2020. All of us right now. Digital image. Instagram.com. Accessed April 6, 2021.

35 @whitepeoplehumor. April 2020. I am never going to financially recover from this. Digital image. Instagram.com. Accessed April 6, 2021.

36 @insta.single. April 16, 2020. Look at us. Digital video clip. Instagram.com. Accessed April 6, 2021.

37 @whitepeoplehumor. March 22, 2020. What a time to be childless. Digital image. Instagram.com. Accessed April 6, 2021.

38 McMaster, Geoff. January 28, 2020. Millennials and Gen Z are More Anxious than Previous Generations: Here'sWhy. *Folio: University of Alberta.* https://www.ualberta.ca/folio/2020/01/millennials-and-gen-z-are-more-anxious-than-previous-generations-heres-why.html. Accessed April 8, 2021.

39 Al-Heeti (2019).

anybody want anything," as though plummeting into the depths of despair was as *de rigueur* as a run to the corner shop.[40] A post by the account on May 16 featured a medieval painting of a saint menaced by demons with the caption: "when everyone is lonely from isolation but you have your demons to keep you company." Text superimposed on the image, understood as dialogue from the saint, read: "haha u guys are the best."[41] The post received 143,908 likes and one from @whitepeoplehumor posted on March 31, which similarly "boasted" that existing difficulties provided some sort of inoculation against the worst stresses of the pandemic, received 95,389. The post included a photo with three collaged images of Tom Cruise laughing heartily. The text read: "When people with physical health issues are vulnerable to coronavirus but you only have mental health issues."[42] If Millennial meme culture, shot through with a brand of humor in which "lol, nothing matters," provided participants with tools to weather the COVID-19 crisis, so too, it seems, did their dedication to cultivating rich online lives.[43]

Maintaining and Building Bonds

When stay-at-home orders went into effect in March 2020, meeting with friends and family became strictly forbidden in all but digital spaces and the time that Millennials had spent online began to pay dividends. Nicole Argiris, one of the creators behind @mytherapistsays, argues that Millennials and GenZers were advantaged going into quarantine, having solidified their online presences and developed "powerful online communit[ies] where it's normalized and encouraged to seek help." The initial shock of isolation was less pronounced, she maintains, for those whose "social lives already were [intertwined] with social media."[44] The nature of the Instagram platform—which largely eschews anonymity—meant that if offered an important lifeline, not just to virtual social networks, but to the real-life ones that the pandemic rendered otherwise inaccessible. Communication with friends remained, in some sense, uninterrupted. Millennial and Gen Z participants in meme culture have long made a practice of connecting with, and caring for, real-life friends via tagging—when an Instagram user comments on a particular post with a friend's username. As

[40] @mytherapistsays. April 21, 2020. gonna hit rock bottom. Digital image. Instagram.com. Accessed April 6, 2021.

[41] @mytherapistsays. May 16, 2020. haha u guys are the best. Digital image. Instagram.com. Accessed April 6, 2021.

[42] @whitepeoplehumor. March 31, 2020. When people with physical health issues are vulnerable. Digital image. Instagram.com. Accessed April 6, 2021.

[43] Bruenig, np.

[44] Tempesta, Erica. December 30, 2020. Creators of one of Instagram's most popular meme accounts say Gen Zers and millennials are "better equipped" to deal with the stress of the pandemic because they're glued to their phones. *Daily Mail*. https://www.dailymail.co.uk/femail/article-9064365/My-Therapist-Says-creators-say-younger-generations-better-equipped-deal-pandemic.html. Accessed March 30, 2021.

with the "<3 I AM TRACY" declaration, tagging publicly links the user, the friend(s) that they tag, and the post. Jaja Liao notes that tagging friends in meme posts constitutes a "passive and semi-public way of telling our friends that we are thinking about them or that we miss them." "We want other people to see that we share these certain feelings," he continues, "it's a way of showing affection."[45] A post by Instagram account @memezar on April 30, 2020, reproduced an anonymous quote: "Being on FaceTime, sending memes to each other on [Instagram] like 'look what I just sent you'." "THIS is my love language", @memezar responded in supplementary text.[46] To the tune of 265,058 likes and 2,772 comments (almost exclusively tags), many agreed. On October 10, 2020, the @pubity account posted a Tweet from @xforcadess that received 463,687 likes and nearly 4,000 comments and read: "honestly being tagged in things makes me so happy because it's like ??? u think of me? u remember me? wtf i don't even remember me sometimes ily [i love you]".[47] Such posts are far from scarce and engagement with them tends to be high.

In the spring of 2020, dark humor memes got a lot of attention in the form of likes, but the posts from meme accounts that garnered the most comments—almost exclusively tags—were those that positively connected friends. During the three-month window, account administrator Violet Benson added 157 posts to the @daddyissues_ feed. Of those, twelve earned in excess of 8,000 comments. The post with the most comments (12,251) was from March 21 and featured a video clip of two small children running toward one another and embracing. The accompanying text read: "Me and my best friend reuniting once coronavirus has disappeared."[48] The same post shared on the same day by @mytherapistsays received an astounding 44,718 comments. Just seven of the 341 posts shared by the account in the spring of 2020 received in excess of 20,000 comments. Another post that broke the 20,000-threshold, with 22,676 comments, was from April 1 and included a video that cut between two cats in t-shirts furiously typing on laptop computers. The caption was: "Me and my bestie ferociously tagging each other in quarantine memes."[49]

Many of these posts, and others besides, look forward, allowing users to imagine a happy ending to present hardships. Video clips posted by @daddyissues_ on April 9 and 10 showed 1) two women dancing in the dark with glow sticks affixed to their clothes, and 2) a woman holding the base of a wine glass between her lips and tipping

[45] Liao, Jaja. December 17, 2016. How Memes Conquered Our Hearts and News Feeds. *Venture Beat*. https://venturebeat.com/2016/12/17/how-memes-conquered-our-hearts-and-news-feeds/. Accessed March 30, 2021.

[46] @memezar. April 30, 2020. THIS is my love language. Digital image. Instagram.com. Accessed April 6, 2021.

[47] @pubity. October 10, 2020. U think of me? Digital image. Instagram.com. Accessed April 6, 2021.

[48] @daddyissues_. March 21, 2020. Me and my best friend reuniting. Digital video clip. Instagram.com. Accessed April 6, 2021.

[49] @mytherapistsays. April 1, 2020. Quarantine memes. Digital image. Instagram.com. Accessed April 6, 2021.

the contents into the mouth of her partner seated behind her. The captions, respectively, were: "when quarantine ends I am doing this with my best friends [laugh-crying emoji; sobbing emoji]" and "if this isn't me and my future wife in 20 years, I don't want it [laugh-crying emoji; wine glass emoji]."[50] Both posts received over 8,000 comments from Instagram users, as did a March 16 post from @mytherapistsays that reminded followers, "let's not forget that Rapunzel was quarantined and met her future husband so let's think positively here."[51] Posts by @mytherapistsays that blended dark humor and hopefulness were likewise top comment-getters. A April 5 post (9,983 comments) included a video clip from the film Forrest Gump in which Tom Hanks reflexively jumps off of a moving shrimping boat as he spots his friend, Lieutenant Dan, on the dock. The text reads: "When we're finally off lockdown in 2025 and I see another human being."[52] Another from May 12 (15,011 comments) paired the text "When my friends and I are finally allowed to hang out after spending 44 years in quarantine" with a video clip of elderly Eastern European women singing Britney Spears's "…Baby One More Time" to accordion accompaniment.[53]

Even virtual-only networks, rooted in Instagram meme culture, provide a sense of community, the utility of which became only too clear during the pandemic. Sherry Turkle has been critical of the capacity of social media to provide important, "real" connection, arguing that Millennials "flight from conversation" is also a "flight from empathy," cultivating, not community but a simulative "sense of community."[54] Yet such theories fail to consider the indispensability of social media platforms when isolation is involuntary or face-to-face interaction is otherwise prohibited. As @dominickevans tweeted on January 23, 2021, "a lot of disabled people who are isolated…rely on social media for any type of human connection."[55] A reply from @MrHotWheelz confirmed, "People don't understand this. They talk about being isolated since March but I've been at home for far longer than that. I don't know how I'd handle it without the people I've met online."[56]

---

[50] @daddyissues_. April 9, 2020. when quarantine ends. Digital video clip. Instagram.com. Accessed April 6, 2021; @daddyissues_. April 10, 2020. If this isn't me and my future wife. Digital video clip. Instagram.com. Accessed April 6, 2021.

[51] @mytherapistsays. March 16, 2020. let's think positively here. Digital image. Instagram.com. Accessed April 6, 2021.

[52] @mytherapistsays. April 5, 2020. When we're finally off lockdown. Digital video clip. Instagram.com. Accessed 6 April 2021.

[53] @mytherapistsays. May 12, 2020. When my friends and I are finally allowed to hang out. Digital video clip. Instagram.com. Accessed April 6, 2021.

[54] Turkle, Sherry. 2015. *Reclaiming Conversation: The Power of Talk in a Digital Age*. New York: Penguin, 11; 173. See also Turkle, Sherry. 2011. *Alone Together: Why We Expect More from Technology and Less from Each Other*. New York: Basic Books.

[55] @dominickevans. January 23, 2021. People often say, just delete social media…But you see, the thing is, a lot of disabled people who are isolated, sometimes trapped at home, sometimes trapped in bed rely on social media for any type of human connection. Twitter.com. Accessed January 25, 2021.

[56] @MrHotWheelz. People don't understand this. They talk about being isolated since March but I've been at home for far longer than that. I don't know how I'd handle it without the people I've met online. Twitter.com. Accessed January 25, 2021.

The truth of the relationship between social media and the mental well-being of users is that it is not all bad nor all good.[57] In many cases though, the good can outweigh the bad and administrators of @mytherapistsays have "seen friendships form in the comments section of posts as people offer words of support to one another."[58] What is more, according to @dankrecovery account administrator Timothy Kavanagh, who found "my people" in the recovery meme community, memes "convey a shareable feeling or emotion through comedy without the need for face-to-face interaction."[59] The distance is, in some ways, critical. As a Tweet from @Satansplusone shared by @insta.single on February 28 2021 expresses, memes might be the best, or only, way to express and share dark emotions: "Sometimes I forget sad jokes are only funny on the internet. I told this guy I wished the earth was flat so I could yeet [throw] myself off the edge and he just got really concerned."[60]

Conclusion

Scholars have always been, and should continue to be, skeptical of the role that new technologies play in our lives. In *The Shallows*, Nicholas Carr argues that the Internet has changed the way that we think and read; he laments that the mental functions required to "traverse a lengthy narrative" are becoming lost and that, as both cause and effect, written works are undergoing fragmentation.[61] That may be, and Millennial memes may appear a poor substitution for Lost Generation literature, but they are not necessarily "shallow." Indeed, if you are part of the community and know the language, memes can, and do, speak volumes. Memes constitute a new, more responsive and transmissible format for old ideas, communicating through the same dark humor that the previous "lost generation" used a century ago—a humor that, in Freud's analysis, protects the individual by making "comic sense out of overwhelming non-sense," making a joke out of powerlessness, loneliness, and chaos and, in so doing, "allowing them to be mastered for a moment."[62] And, at least where the pandemic is concerned, dark humor and meme culture seem to have fulfilled their remit. On December 21, 2020, Instagram meme account @Iamthirtyaf posted that "A bunch of our followers have kindly told us that we've gotten them through

---

[57] McMaster discusses some of the negative impacts of social media in "Millennials and Gen Z are More Anxious than Previous Generations: Here'sWhy."

[58] Al-Heeti, n.p.

[59] Knight, Ivy. September 25, 2020. How Memes and a Good Laugh Can Help People in Recovery. *The New York Times*. https://www.nytimes.com/2020/09/23/style/recovery-memes.html. Accessed March 30, 2021.

[60] @insta.single. February 28, 2021. Sometimes I forget sad jokes are only funny on the internet. Digital image. Instagram.com. Accessed March 1, 2021.

[61] Carr, Nicholas. 2010. *The Shallows: How the Internet is Changing the Way We Think, Read and Remember*. New York: W. W. Norton and Company, 142; 91.

[62] Cited in Colletta, 7.

this disaster of a year."[63] On the 27th, @insta.single reposted a follower's comment, which declared that, "being…stuck at home your page is getting me through [hugging face emoji]."[64]

I have endeavored here to understand the communicative and therapeutic functions of COVID-19 memes, to begin to uncover the ways in which they embody the various coping strategies that psychologists and scholars have suggested are inherent to dark humor, and to shed light on their capacity to build and maintain important social support networks. This is a beginning not an end, and additional studies with more expansive temporal and geographic scopes promise to add to our understanding of the utility of memes in times of crisis, particularly during the coronavirus pandemic of 2020–2021. Nevertheless, as this chapter demonstrates, Millennial meme culture in the USA helped participants to weather the psychological stress of the early months of the COVID-19 crisis by providing them a means to defuse anxiety, reassert control over their circumstances, develop and maintain bonds within communities both real and imagined, and, in the final analysis, to establish a sense of continuity between past, present, and future. As a coping strategy, writes Judit Ujlaky, jokes work by "guarantee[ing] the feeling of continuity" following crucial events that mark discontinuities in the life of groups or communities.[65] Pandemic memes both projected optimistically into a future beyond the crisis and, by mobilizing a dark humor that was close at hand, created a continuity with the past that situated the COVID-19 crisis as essentially more of the same, another "pummeling" for Millennials who had necessarily become accustomed to weathering the storm. Millennial meme culture reminded contributors at once that they were resilient, as ever, and that it would get better, if they stuck together. Everyone, Tracy promised, would soon gather again and could still have "the best life ever."

References

Adamy, J. (2020, August 9). Millennials slammed by second financial crisis fall even further behind. *The Wall Street Journal*. https://www.wsj.com/articles/millennials-covid-financial-crisis-fall-behind-jobless-11596811470. Accessed March 30, 2020.

Al-Heeti, A. (2019, August 23). Memes on instagram, reddit bring comfort to people struggling with depression. CNet. https://www.cnet.com/news/memes-on-instagram-/reddit-bring-comfort-to-people-struggling-with-depression/. Accessed March 30, 2021.

[63] @iamthirty. December 21, 2020. '2020'. Digital 'story'. Instagram.com. Accessed December 21, 2020.

[64] @insta.single. December 27, 2020. Your page is getting me through. Digital 'story'. Instagram.com. Accessed
December 27, 2020.

[65] Ujlaky, Judit. 2007.Evaluation of the Historical Recent Past: Humour as a Possible Collective Coping Strategy. In *Suffering the Slings and Arrows of Outrageous Fortune: International Perspectives on Stress, Laughter and Depression*, ed. Bernie Warren, 31–42.Amsterdam: Editions Rodopi, 35.

American Psychological Association. (2020, October). *Stress in America™ 2020: A National Mental Health Crisis*. Press Release. https://www.apa.org/news/press/releases/stress/2020/report-october. Accessed 14 April 2021.

Bruenig, E. (2017, August 11). Why is millennial humor so weird? *The Washington Post*. https://www.washingtonpost.com/outlook/why-is-millennial-humor-so-weird/2017/08/11/64af9cae-7dd5-11e7-83c7-5bd5460f0d7e_story.html. Accessed March 30, 2021.

Carr, N. G. (2011). *The shallows: How the internet is changing the way we think, read and remember.* Ocala, FL: Atlantic.

Colletta, L. (2003). *Dark Humor and Social Satire in the Modern British Novel*. Palgrave Macmillan.

Du Pont, K. (2015). Nature and Functions of Humor in Trench Newspapers (1914–1918). In T.-D. Clementine, & K. A. Ritzenhoff (Eds.) *Humor, entertainment, and popular culture during world war I* (pp. 107–122). Basingstoke: Palgrave.

Gal, N., & LimorShifman, and Zohar Kampf. (2016). "It gets better": Internet memes and the construction of collective identity. *New Media & Society, 18*(8), 1698–1714.

Gates, R. A. (1999). *American literary humor during the great depression*. Greenwood Press.

Gehring, W. (1996). *American dark comedy: Beyond satire*. Greenwood Press.

Henman, L. D. (2001). Humor as a coping mechanism: Lessons from POWs'. *Humor, 14*(1), 83–94.

Kantis, C., Kiernan, S., & Bardi, J. S. (2021, March 26). UPDATED: Timeline of the coronavirus. Thinkglobalhealth.org. https://www.thinkglobalhealth.org/article/updated-timeline-coronavirus. Accessed April 6, 2021.

Kelly, J. (2020, April 6). Americans are excessively eating, drinking, smoking pot, playing video games and watching porn while quarantined. *Forbes*. https://www.forbes.com/sites/jackkelly/2020/04/06/americans-are-excessively-eating-drinking-smoking-pot-playing-video-games-and-watching-porn-while-quarantined/?sh=2297757a404e. Accessed April 8, 2021.

Knight, I. (2020 September 23). How memes and a good laugh can help people in recovery. *The New York Times*. https://www.nytimes.com/2020/09/23/style/recovery-memes.html. Accessed March 30, 2021.

Kuipers, G. (2011). Where was King Kong when we needed him? Public discourse, digital disaster jokes, and the functions of laughter after 9/11. In T. Gournelos & V. Greene (Eds.), *A decade of dark humor: How comedy, irony, and satire shaped post-9/11 America* (pp. 20–46). Jackson, MS: University Press of Mississippi.

Lankshear, C., & Knobel, M. (2007). Online memes, affinities, and culturalproduction. In C. Lankshear & M. Knobel (Eds.), *A New Literacies Sampler*, pp. 199–228. New York: Peter Lang Publishing.

Liao, J. (2016, December 17). How memes conquered our hearts and news feeds. *Venture Beat*. https://venturebeat.com/2016/12/17/how-memes-conquered-our-hearts-and-news-feeds/. Accessed March 30, 2021.

Martin, R. A., & Ford, T. (2010). *The psychology of humor: An integrative approach*. Academic Press.

McGraw, P., & Warner, J. (2015). *The humor code: A global search for what makes things funny*. Simon & Schuster.

McMaster, G. (2020 January 28). Millennials and gen Z are more anxious than previous generations: Here's why. *Folio: University of Alberta*. https://www.ualberta.ca/folio/2020/01/millennials-and-gen-z-are-more-anxious-than-previous-generations-heres-why.html. Accessed April 8, 2021.

Milner, R. M. (2013). Pop polyvocality: Internet memes, public participation, and the occupy wall street movement. *International Journal of Communication, 7*, 2357–2390.

Milner, R. M. (2012). *The world made meme: Discourse and identity in participatory media*. PhD Dissertation, University of Kansas.

Pratt, A. R. (Ed.). (1993). *Black Humor: Critical Essays*. Garland Publishing Inc.

Rintel, S. (2013). Crisis memes: The importance of templatability to internet culture and freedom of expression. *Australasian Journal of Popular Culture, 2*(2), 253–271.

Rourke, C., & Marcus, G. (Eds.). (2004). *American humor: A study of the national character*. New York: New York Review Books.

Shifman, L. (2013). *Memes in digital culture.* MIT Press.

Tempesta, E. (2020, December 30). Creators of one of instagram's most popular meme accounts say Gen Zers and Millennials are 'better equipped' to deal with the stress of the pandemic because they're glued to their phones. *Daily Mail.* https://www.dailymail.co.uk/femail/article-9064365/My-Therapist-Says-creators-say-younger-generations-better-equipped-deal-pandemic.html. Accessed March 30, 2021.

Turkle, S. (2011). *Alone together: Why we expect more from technology and less from each other.* Basic Books.

Turkle, S. (2015). *Reclaiming conversation: The power of talk in a digital age.* Penguin.

Ujlaky, J. (2007). Evaluation of the historical recent past: Humour as a possible collective coping strategy. In B. Warren (Ed.), *Suffering the slings and arrows of outrageous fortune: International perspectives on stress, laughter and depression* (pp. 31–42). Amsterdam: Editions Rodopi.

Van Dam, A. (2020, June 5). The unluckiest generation in U.S. history. *The Washington Post.* https://www.washingtonpost.com/business/2020/05/27/millennial-recession-covid/. Accessed March 30, 2021.

Wiggins, B. E. (2019). *The discursive power of memes in digital culture: Ideology, semiotics, and intertextuality.* Routledge.

Chapter 14
Comics, Cartoons, and Vignettes: The Graphic Narratives of the COVID-19 Pandemic

A Personal Reflection

Mónica Lalanda

Abstract Comics and cartoons have been showing a great capacity to describe illness in a far richer way than plain texts do. When COVID-19 pandemic was declared in Spain back in March, I started collecting these cartoons expecting to end up with a large collection of Graphic Medicine material. I created a folder called COVID graphic medicine and started to save any cartoon I came across in the social media that contemplated the situation from any of its many angles: the virus, the biology, the disease, the human and professional suffering and the way people and the health system were confronting such a new and complicated situation. The expectation was to find rich and complex descriptions of the disease but surprisingly once I started analyzing in depth the creation of hundreds of authors (mainly Spanish), I was surprised to see that there was little Graphic Medicine in them. I took up the challenge of doing a revision, analysis, and classification of the cartoons for various talks. I found out that here were several themes that came up in a recurrent and persistent ways: (a) anthropomorphism of the virus, (b) the suffering of the health care workers, their hero-like attitude, and their physical and emotional struggle, (c) death portrayed as a classic skeleton dressed in a cape and cloak hood but little about the process of dying, (d) the "flatten the curve" as an inspirational idea for many cartoons, (e) the confinement and life in the balconies (Spanish people live mainly in apartments rather than houses), (f) facial masks in various and fascinating ways. Surprisingly, in spite of the suffering of so many, there has been very little on the disease itself, the admission of COVID-19-afflicted persons to a hospital, the ICU patients or the deceased. Reflections and thoughts about a collection of nearly 500 cartoons are presented here.

M. Lalanda (✉)
Doctor and Illustrator, Valladolid, Spain
e-mail: monica.lalanda@yahoo.com

S. Venkatesan et al. (eds.), *Pandemics and Epidemics in Cultural Representation*,
https://doi.org/10.1007/978-981-19-1296-2_14

Introduction

The world, as we knew it, came to a halt in March 2020. A deadly minuscule virus turned our lives upside down from one day to the other, changing our routines and altering our peace and tranquillity. An invisible and fast spreading coronavirus entered our communities at a time when most adults lack a common memory of any large-scale life-threatening event.

As we quickly entered a general panic, our governments reacted in different ways trying to ease the fateful situation that would eventually cost millions of lives. Here in Spain, we were made to stay at home for months and became used to daily information about impersonal numbers telling a horrible story of death, illness, and collapse of our health care system. Our diminutive enemy took over our mostly comfortable lives.

For months, the knowledge about the virus was limited, even the way you could get infected was not clear, so personal contact was strictly limited. People with symptoms were strongly advised to lock themselves in their rooms at home, and relatives were forbidden to visit their loved ones in hospitals. Many people died but furthermore, they died alone. The situation in most nursing homes was dreadful. With the healthcare system about to succumb, many elderlies were denied admission to hospitals but the situation in their own residencies was horrible, with many of them very ill and dying, and their carers leaving off sick; thousands of them died lonely deaths, locked in their rooms, receiving minimum care.

March and April 2020 were months that most people will never forget. Illness, fear, loneliness, restrictions to personal freedom, confusion, disbelief, chaos, and death. After that, there have been many ups and downs and several waves of morbidity and mortality, many things have happened, changes in restrictions, face masks, vaccines, increase knowledge about the virus and how it spreads and works. As I write this in the summer of 2021, the pandemic is not over but hopefully we've seen the worst of it.

Most of us have luckily not suffered a civil or world war, so these months may have been the nearest we've lived to a global disaster. Even if most people have remained healthy throughout the pandemic, the common sense of anxiety and impending doom have left a mark on us.

Analysis of Graphic Materials

Any event affecting people's health can be translated into the creation of illness narratives in various forms. I'm particularly interested in Graphic Medicine, the use of comics, illustrations, and infographics to communicate issues of health and illness. So, nowadays the amount of graphic material regarding common and uncommon diseases is enormous. The value of a comic to tell a personal story of an illness is increasingly appreciated by both the public and the medical community. Learning from real patients about the experience of being ill provides a more humanistic, practical, and authentic perspective, allowing us to move from the illness to the ill. These are what we call Graphic Pathographies created by patients or their closed ones. There is also another kind of comics, the educational ones, aimed at transmitting information to patients or to the general society.

The academic world of Graphic Medicine has been undergoing an exponential growth in the last decade. Telling stories from a different point of view shift the cultural perception from the passivity of being a sufferer, a victim, to the activity of being the center of the story. Graphic Medicine is a term coined by British doctor and comic artist Dr Ian Williams who soon managed to attract a small team of professionals with similar interests. In a few years, a webpage, a manifesto, (1) and international conferences happened, and this little world continues to evolve and become larger and more international. There are now several groups in other countries following the lead of the Anglo-American Graphic Medicine Group, including in Spain, India, and Japan. Our aim is to study and give value and visibility to graphic stories of illness and for some of us also to create useful material.

As part of the Spanish Graphic Medicine Movement (2), I tend to reflect and study comics and vignettes, always looking for new material that may be useful to patients and health care professionals. When the pandemic was finally declared here in Spain on March 13, 2020, I immediately turned to Graphic Medicine as a tool to help share valuable information to lay people and doctors and nurses. Like many others and almost immediately, I started drawing infographics to share information on issues like symptoms, what to do if you had them, how to do confinement at home, and how to protect yourself. But I also had the impression that we may come across a lot of stories told by a wide range of people. I suspected that many illustrators, either professional or amateur, would draw their experience of being ill, watching their ill or dying loved ones, or giving the perspective of symptoms and management of the disease. With this in mind and hoping to get a large collection of graphic medicine examples, I started to collect examples of cartoons related to COVID-19.

The criteria I used to select the pieces was as follows:

- Comics, vignettes, single panel illustrations, and cartoons.
- Mostly with text in Spanish or without any text.
- Filtered by social media. Meaning that these were images that I came across via Twitter, Instagram, and WhatsApp. I didn't actively search them out but considered that they arrived to my Timeline because someone else found them interesting. I didn't look for any of the images that I collected. I didn't perform any media search, although many of them came from there.
- I felt personally attracted immediately to them, either because of the artistic look, the information, the emotion, or the strength. This wasn't done in any careful way but in the middle of the madness of attending patients online and creating my own material. The criteria for selection if at all was subjective and performed unorthodoxly.
- The initial examination of the piece was never longer than a few seconds. No deep impressions on content were acquired.

Classification and First Impressions

Following weeks of informal collection of materials, I had the chance to give it a more analytic review. It was not until the healthcare crisis started to settle down around the end of May 2020 that I started to look at the hundreds of pieces I collected.

These were the initial personal conclusions drawn from my review of these materials:

- It was possible to use the cartoons to construct a timeline of events that happened during the pandemic. So many things have occurred, and the situation had been so chaotic and dynamic in various ways that the graphic material made a great chronographic reminder of the evolution of those first weeks. The fast-changing protocols and rules and the social response (or lack of it) were nicely reflected.
- There were many more drawn pieces of material than there were photographic ones. It became apparent that we hadn't had many witnesses taking photos. With full confinement or quarantine and patients in isolation (both in hospitals and in their own rooms within the homes), the cartoons and comics may have become the main graphic material in this unprecedented crisis.
- There was a pattern of issues that was repeated by the authors, themes that were catching the artists' imagination.
- There was very little Graphic Medicine in graphic pathographies as such. The material collected gave us the broken narrative of a whole world becoming ill with anxiety and loneliness but not the kind of material I was expecting. Illustrators and comic artists were not representing the experience of suffering from COVD-19, including symptoms, treatment, or vulnerability but rather the emotions and personal or social events related to everything around it. The comics and cartoons described everything around the illness except for the illness itself. This was an extraordinary finding. Our world turned upside down by a contagious infection, but most people haven't witnessed it and therefore it hasn't been drawn, there was hardly any graphic medicine but more of a story about a whole society going through a unique and traumatic common experience. A kind of social graphic medicine, covering the side effects of the pandemic, rather than the pandemic itself. The suffering of a whole society rather than the illness of the individuals.

The next step was to organize all these small pieces of covid art. I used the repetition and patterns that I had found to identify theme that would allow me to perform some deeper examination.

These were the main areas I identified:

- The virus
- Death
- Confinement

- "The curve"
- Healthcare professionals
- Patient suffering
- Face masks
- The vaccine
- Seasonal themes.

The Virus

Published in *Diarios Regionales de Vocento* (*Author* Rafael Vega "Sanson")

The virus causing COVID-19 belongs to the group of coronavirus, the same family of microorganisms that cause the common cold. The reason why it is called coronavirus is because it has peculiar spikes that give it the appearance of a crown (*corona* is the Greek origin for crown).

It can only be seen by electronic microscope and to get a proportional idea of its size, take the section of one hair, divide that surface into 100 pieces, and take one. The coronavirus is between 0,1 and 0,5 microns.

The images of this or any other virus shared in scientific literature are usually 2D, blurry with a slight appearance of been made of discolored chalk. There are some model 3D images of the virus but the very first real 3D picture of it hasn't become true until January 2021. In popular culture, it has become an irregular dark grey sphere with white and pink, cotton like spikes (3).

Knowing all this, it's fascinating to examine the way the coronavirus has been drawn by artists since the beginning of the pandemic. Surprisingly, it has been the artistic view of the drawn virus rather than the plain and less lively look of a 2D photo that has become popular. Owing to the explosion of illustrations all over the media, most people were able to visualize this enemy. Since the real images are scarce, the imaginary ones have given the virus a consistent physical appearance.

It's interesting to see how children are able to draw the virus and how most of them accurately provide them a round shape and spikes (4). Having an invisible creature as the protagonist of the history of the world has excited people's imagination.

It's also fascinating to notice the array of colors artists have been using. Deciding on a color to portray the virus is part of the creative process. In the work I've reviewed, the coronavirus was drawn in a spectrum of colors, although the predominant one was green followed by pink. Single color with a variation of tones appears to be the more popular look. The experiment of entering the term "virus" in an image search engine like Google Images throws a variety of chromatic possibilities, green not being a common one. Maybe we're used to the use of green to draw cartoons of extra-terrestrial creatures, as the most unlikely color for a human, so artists who have never drawn a virus before may have chosen it a sign of an unnatural or alien presence.

The size is another interesting feature to make note of. Many artists choose to use the image of one single oversized virus. In fact, a recurrent resource has been to relate the microorganism to the earth: the world as a huge coronavirus, the world inside the virus, planet earth with the spikes of the virus, a monster virus squeezing the world. This visual approach seems like a resourceful way to show how it is affecting everyone and everywhere. The fact that the shape is a sphere has been used to show it as a ball, a massive one rolling down menacing people or as a metaphor of the mythological Sisyphus. A ball that rolls uncontrollably creating devastation in its way or a ball used to play sports by a very active Death figure or even by health care professionals.

The anthropomorphism of the coronavirus is also a commonly used resource. The attribution of human traits by adding facial features and elongating spikes to look like arms and legs has produced some interesting cartoons. The most remarkable thing to notice is that when looking human, it doesn't look too scary or terrible but more of a cheeky, mischievous, naughty being. Surprisingly benign, looking tired, bored, or just absent-minded. A virus that sits with groups of unmasked people or in the back seat of their cars looking hopeful as humans seem to have unpreoccupied fun.

The level of creativity achieved using the virus as a source of inspiration is rather surprising. It has been turned into all sorts, from Donald Trump's head, emojis and stickers, musical instruments, cow udders, pac-man game, or part of the scenery in the Little Prince or in one shot film images like The Shining or Star Wars. It's impossible not to wonder, has the microorganism not been round and with spikes, would it have triggered this sort of artistic response?

Death

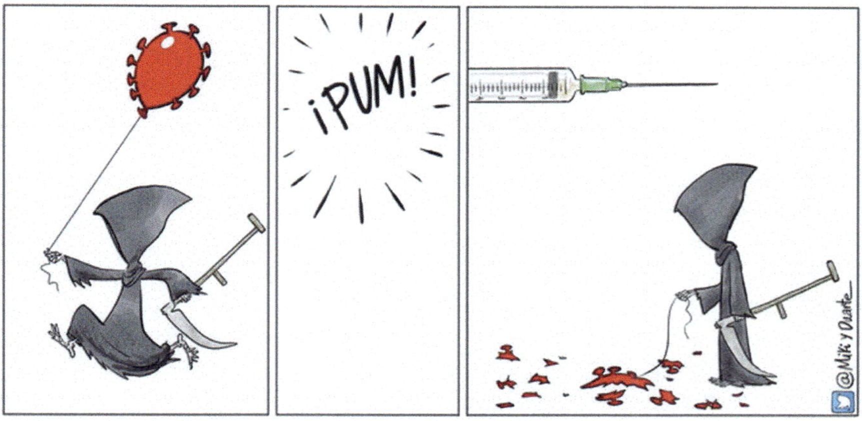

Published in the newspapers of *Grupo Joly* (*Authors* Miki y Duarte)

In spite of the millions of fatal victims during the pandemic, the images of dying or deceased people have been completely missing from the public eye. Families haven't witnessed the death of their loved ones as patients were made to stay alone in their own rooms at home or as inpatients in hospital wards and intensive care units. As people passed away, the bodies were rather quickly disposed of, and the relatives were presented with their ashes with limited or no witnesses. We have been spared the images of dead bodies or even coffins. The society hasn't been exposed to the appearance of the dead but still, death as such has been floating in the air, center of our fear and conversations.

This strange situation may have had an effect on the cartoons created. There are very few of them that represent dead bodies or even people dying but rather, the anthropomorphic figure of death has been a popular choice for artists to represent. The skeleton covered with a black cape with its oversized hood holding a scythe has been the main character in many cartoons. Yet again, like it's happened with the drawings of the virus, death appears more like a playful creature than that of a scary presence. We've seen death wearing face masks, playing sports or games against health care providers, sharing events in a watchful attitude, sneezing, or even using anti-septic gel. Death is illustrated as a character, sometimes gloomy but often not much of a reminder of our mortality but more of an artistic image, a presence.

Confinement

Living in confinement seems to be one of the dominant themes in the whole collection. Staying at home for weeks, being only allowed to come out to buy food or going to the pharmacy was a common event for all Spaniards. Spanish people make a particularly outgoing, street-lovers and social kind of society. Staying at home for so long was particularly complicated. Besides, we tend to live in flats rather than houses, so space is limited and fully indoors.

Cartoons have reflected in various issues related to confinement. Many vignettes portrayed full buildings where you could see many different people doing similar things through their own windows or balconies. In fact, coming out to the balcony for the only social gathering of the day to give a daily ovation with cheering and clapping to the healthcare workers has instigated quite a bit of drawing.

Reflections on different quality of life during the confinement depending on the kind of accommodation you have has raised some interesting reflections on different ways to survive the pandemic depending on social determinants. Peeping at people's lives through their windows is a perspective that seems new. And it wasn't only real windows but also virtual ones. In fact, there are a fair number of vignettes where you can see how social media has given us a certain amount of human connection. The little windows of meetings through Zoom, WhatsApp, or any other digital app.

Comics, vignettes, and cartoons have provided the means to show a lot of emotion. We've seen expressions of fear, anxiety, loneliness, and boredom. Working and socializing from home has given way to illustrations demonstrating the careful preparation of the area that would show in the screen, a tidy space and cared-for upper clothes as the rest of the house looks like a mess and the bottom clothes are pyjamas. Keeping up appearances during confinement in social media or work meetings has triggered some humorous cartoons. Exercising at home, shopping from home, too much eating, too much drinking, too many video games and binge-watching series and a list of other activities have been reflected as part of the confinement routine. But there have also been authors making a point about the nice things of the confinement: kids spending much more time with their parents, taking up hobbies, getting to do more reading, writing, drawing, more family meals or table games. We've seen the role played by pets, too. For weeks only dog owners were allowed to walk in the streets, so this fact gave way to funny cartoons with exhausted dogs taken for far too many walks and how the relationship with the owners changed.

As well as the house balconies, toilet rolls became the center characters of the first few days of the confinement as people feared a complete sell out in supermarkets and bought too many rolls.

"The Curve"

Flattening the curve became both an icon and a calling as a final aim for the whole society. A graphic showing the relationship between the pandemic progressing too fast and the healthcare system collapsing became viral all over the world. It was such a simple display of healthcare management and the importance of whole communities working together to save the hospitals from succumbing that it was understood and widely accepted. Flattening the curve became an expression that was used extensively. In the same manner, the curve became the muse for various interesting cartoons where people would be pulling the line in the graphics from their balconies and healthcare professionals would be pushing it from within. These cartoons gave us all a sense of common purpose, a clear idea that this was something that we had to do together, a feeling of community fighting against an impending doom. A line turned into a rope or sometimes into a sea wave has been used in cartoons and in small cartoony animations.

But the idea of the curve soon became inspirational to add new areas of interest to the cartoons. The curve of the extra weight caused by overeating and under exercising during the confinement, the curve of the diabetics daily sugar levels, the curve of the increasing unemployment as the economy started to suffer, the curve of toilet paper or baking powder sells or even the smile of people seen as a changing curve.

It seems extraordinary that a highly technical statistical graphic could trigger so many imaginative ideas. For weeks, flattening the curve became a motto, and it is not inappropriate to think that the explosion of illustrations about it may have helped.

Healthcare Professionals

Published in social media (*Author* Rafael Marrón)

Not surprisingly, healthcare workers have often appeared in comics and cartoons during the pandemic. There has been representation of three significant issues:

- The lack of protection they suffered for the first weeks
- Their own personal emotional suffering and their illness
- Their heroic attitude.

For reasons still to be clarified, the COVID-19 epidemic caught most countries unprepared, and Spain was one of them. On March 13th, we went into lockdown, thousands of people were becoming very ill. But the national provisions of personal protection equipment (PPE) was scarce. Furthermore, availability of face masks was also very limited. The common masks in the health care settings are the hygienic ones, not enough protection for a respiratory virus.

People's imagination and solidarity focused not only on helping workers to protect themselves but also creating vignettes to tell this tale. Garbage bags, swimming goggles, shower caps, or plastic aprons are common costumes in some of the pieces drawn during March and April 2020. But doctors and nurses were not only missing proper physical protection but there was a general use and abuse of what they could provide to the society. They were clapped for a few minutes daily but were uncared and exploited by the system 24 h.

There was a lot of suffering amongst the staff, working longer hours than usual, with a great amount of stress, often outside their area of comfort and even away from their own specialities. The physical toll was only part of it, the emotional suffering was huge. Not only the fear of getting infected but seeing so many patients dying and having to comfort patients who were alone in hospitals was emotionally devastating and that was well reflected in various comics. Many workers decided to leave home to avoid spreading the infection from the hospital to their loved ones. Not seeing their kids or partners was an extra source of suffering. This generosity at so many different levels has surely caught some artists' imagination.

Some cartoons created by health professionals have reflected one of the sad tragedies of this pandemic and the death of many colleagues. In fact, Spain is one of the countries where more health staff have become infected, and many have died. According to an estimation by Amnesty International, every half an hour a health care professional dies in the world due to COVID-19. Some strangely beautiful drawings have been made as a tribute to colleagues fallen in this unprecedented crisis.

For years we've being trying to stop the already common use of war metaphors to talk about illness. For example, the idea of cancer being a battle where you fight and you either win or lose is increasingly becoming unacceptable as it portrays illness as something you can control, and thus, it is quite unfair for the patient. Therefore, it has come as a big surprise that war metaphors are once again used to talk about covid. However, the soldiers in this scenario are not patients but healthcare professionals. The number of cartoons of doctors and nurses portrayed as brave soldiers fighting against either the virus or the figure of death is large. Presented as warriors with a mixture of health and war garments, the most common image is that of determination, heroism, and incredible courage. In fact, doctors are nurses have received superheroes' treatment either presented as supermen and women or like the owners of out of this world powers. We've seen quite a few vignettes with them as soldiers, superheroes, or angels, fighting face to face, fearless, and putting the patients' safety first.

In an era where doctors and nurses have long been treated as a social commodity and the doctor–patient has been turned into provider–client, seeing this surge of admiration and gratitude towards them, seems like one of the biggest surprises of the graphic narration of the pandemic. Social appreciation is something that long ago became obsolete and is now a refreshing idea. Tiredness, fear, and determination have probably been the more common facial expressions in these characters, not an easy task considering that face masks seem widely present in most of these drawings.

Patients Suffering

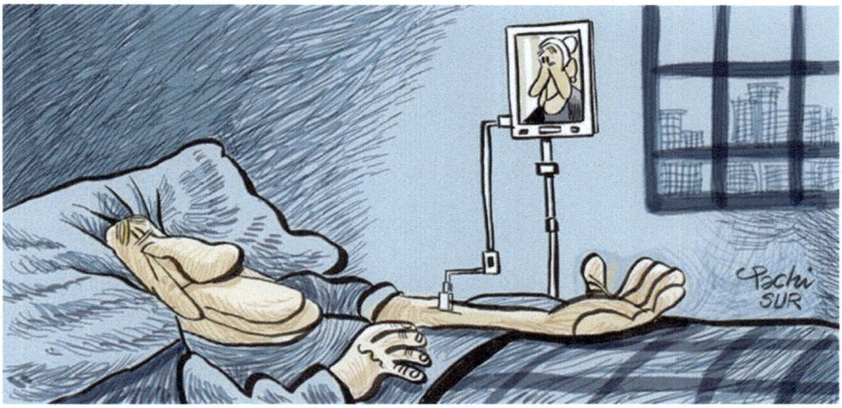

Diario Sur (*Author* Pachi)

This was the initial reason I started collecting cartoons at the beginning of the pandemic, however, I have come across very few examples that show symptoms, treatment, hospital stays, physical deterioration, and the way to recovery. The only symptom reflected in some of these vignettes is cough, but just as a trigger of fear around the "culprit."

The drawings of people dying have been missing too but we've seen several cartoons that use metaphors to tell the story of the void left behind by someone passing away. The most surprising one has been the image of missing elderly people from the parks and how the pigeons notice their departure and those empty benches. Quite a stereotype of the life of old people. Even if these cartoons are aesthetically pleasing, they deserve a deep reflexion about the role of our elders in this modern society. Pigeons as main witnesses of a tragedy feels initially sweet and sensible, but it may say a lot about us. Maybe we decided, as a society, to turn our backs to the most vulnerable members of our communities, who died terrible deaths in confinement and alone.

Loneliness is probably the side effect of covid effectively drawn and communicated during these months. Loneliness as a proof of love appears in various pieces, where grandparents appear pleased to not receive their kids and grandchildren as a true show of love. The more loved you are, the less visits you get. But technology turns into a rescue tool, and this is also depicted in these comics. Elderly citizens appear as avid social media users, keen to remain in touch with their families. Some sweet but still sad cartoons prove it.

Face Masks

Published in ABC (*Author* JM Nieto)

Together with the image of the virus, the drawings of face masks are the second single item that has generated more pieces. Face masks have been drawn in many ways and with different purposes. To criticize the clinical management of the pandemic, as a way to differentiate "good citizens" from bad ones, as a luxury item (when they were scarce), in political cartoons but also, they have gained anthropomorphic status having a life of their own.

Face masks have been turned into many other objects aiming for a metaphoric meaning, like flags, parachutes, slings, clothes, ornaments, or as sails of boats trying to reach the safety of the sore in a sea of viruses. Face masks to show the beginning and the end of life. Such a simple item as the source of so much art is quite surprising.

It is remarkable how the fabric of real masks has turned into a canvas for unusual art even for some sort of activism showing original encouragement to use them.

There have been plenty of official and nonofficial infographics showing the good use of a mask. Surely drawing facial expressions with only the eyes in view has been challenging for artists. Washing your hands, opening the window, keeping social distance, and wearing a face mask have been the most widely drawn scenes in the area of educational graphic medicine.

Vaccines

The graphic narration of the arrival of the anti-covid vaccines has been a truly fascinating story to follow. It is likely to represent the general feeling and the fast evolution of society towards them.

The cartoons first represented skepticism, doubts about two things, safety of the vaccine and "money-grabbing" as final aim of the vaccine development.

I collected various cartoons where people were not particularly happy to get "jabbed", suspicious of the vaccination and its goal. People hiding from human size needles and a lot of criticism of the pharmaceutical industry, even naming brands.

As weeks went by, the prejudice turned to vignettes that showed hope. This got mixed with a pre-Christmas spirit, where Father Christmas was portrayed fabricating vaccines and people hoping to get a vaccine as Christmas presents. That was followed by critical pieces on certain people jumping the queue, getting vaccinated before they were due and lately they transpire a sense of celebration.

Some late cartoons have depicted again heroic, determined, and tired looking healthcare providers fighting against large virus, vaccine in hand.

A huge syringe making do as a bridge to cross from the darkness to the sun, a vaccine looking like a missile about to hit a large flying virus or a vaccine, half syringe half champagne glass are some of the more original ones.

Some artists have become playful with the year numbers. The 0 of the 2020 looking like a coronavirus and the 1 in the 2021 like a vaccine.

Seasonal Themes

One more peculiarity of this graphic narration of the covid pandemic is the adaptation of the cartoons to different seasons. The event has lasted, so far, almost 20 months and this is showing in the cartoons too.

There have been pieces based on the look of the virus themed for Christmas or Halloween. Summertime has offered its own issues, like people suntanned all over except for the face mask area or people in the beach wearing full personal equipment protection.

One area that caught a lot of illustrators' imaginations was back to school time. Children attending school wearing masks and with strict distance rules were the underlying inspiration for several cartoons. Central and Northern Spain is a very cold country in winter, and the fact that kids attended schools where the windows had to remain open at freezing temperatures was portrayed in different ways.

Valentine's Day was another date that provided new ideas of representing the epidemics. A reminder of the new situation faced by people going out on new dates and the risk of contagion.

Conclusion

We have lived, and still do, in unprecedented times, and this has been transmitted in formal and informal communication both in words and in images. This single common issue, a global viral infection, has been drawn from many perspectives throughout a long period of time. There hasn't been a similar situation for us to compare with. The previous world pandemic was the swine flu back in 2009 caused by **H1N1 influenza virus**. The crisis was far less serious and much shorter in time and there were no critical measures such as confinement taking place. Also, communication was far more limited at the time since social media was just about starting to find its place. So, the social and health crisis were very limited, and although cartoons were surely created at the time, their reach was also limited. So, this time not only the critical and maintained situation has triggered the productions of thousands of cartoons all over the world but the active participation of most people in social media means that we have been widely exposed to them.

This informal collection and reflections started as a way to review prospective graphic medicine in view of an emerging pandemic. The expectation of collecting a large number of graphic testimonies triggered by infectious illness affecting the whole world was daunting. However, collecting and studying all this material turned out to be quite interesting. The collection of cartoons and comics and a critical look at them allow a very peculiar analysis of this world health crisis.

So many things happened, so many changes, so many unusual situations and extreme limitations to our normal lives that having some form of common graphic memoire of the events seem useful. This is possible by keeping a timely order of all these publications. By looking at them now you can get a fairly graphic and rapid feeling of the evolution of the pandemic. Most important events seem to be reproduced in the way of cartoons. Considering the lack of photographs all along, illustrations have filled a gap.

My expectations of finding pieces of graphic medicine were not fulfilled. However, I was met with cartoons expressing the experience of the side effects and collateral damage caused by the fight against the virus. Some have suffered the direct consequences of the infection, either in a minor way or as a serious illness or even death, but almost everyone have suffered the anxiety, the fear, the discomfort of the quarantines or lockdowns, the loneliness or complete separation from loved ones, the worry about work and economic consequences, the worsening of previous alignments, and so on. Saying that all these cartoons don't tell the story of a viral infection but the description of a society falling ill and even falling apart. Maybe there hasn't been graphic medicine but a kind of *social graphic medicine*, where comics witness not only a disease that not affects the health of individual citizens but the well-being of a whole society.

The special language of comics and cartoons has played its usual special roll in transmitting both information and emotion, as well as a critical view and even information. The use of metaphors and metonymies, the speech and thought bubbles, the possibility of moving freely in time and space, the use of color, the playful management of fonts, the capacity to make anything much bigger or smaller than real, the ability to capture any imaginable scenery, zooming in or out and choosing the preferred angle are some of the unique resources that make comics a fantastic medium. During this pandemic, the lack of photos, the multiple and complicated situations, and the microscopic nature of our world enemy have proved yet again that comics and cartoons can reach a level of communication where other mediums can't.

Part VII
Health Inequity in the Time of COVID-19

Chapter 15
"Their Lives Just Do not Matter?": "Racing" COVID-19 and Graphic Medicine

Sathyaraj Venkatesan

> *No republic is safe that tolerates a privileged class, or denies to any of its citizens equal rights and equal means to maintain them.*
>
> *Frederick Douglass*

Abstract Though COVID-19 is touted as a great leveler, it has disproportionately impacted the Black, Indigenous, and People of Color (BIPOC) communities putting them more at risk of getting sick and dying from COVID-19. Taking this cue, the chapter through close reading selected graphic medical narratives (intersection of comics and healthcare) seeks to foreground the experiential and affective responses of the BIPOC sufferers. In order to do so, the article utilizes the concepts of precarity (Judith Butler), syndemics (Merrill Singer), and intersectionality. The chapter concludes in showing how graphic medical narratives mark the health of BIPOCs as a decisive consequence of our relations within the complex architectures of various actors, networks, and (racial and colonial) histories.

BIPOC: An Introduction

While COVID-19 has impacted everyone in terms of economic, social, and psychological well-being, it has disproportionately impacted the BIPOC communities putting them more at risk of getting sick and dying from COVID-19. BIPOC refers to the Black, Indigenous, and People of Color communities. BIPOC is also expressed as BAME which includes Blacks, Asians, and Minority Ethnic (BAME) populations. Although there is not universal agreement about what BIPOC/BAME include, but to most, it is a millennial shorthand for Blacks, Asians, native American, native Hawaiian, Pacific Islander, and Latinx communities in the USA. Irrespective of

S. Venkatesan (✉)
Department of Humanities and Social Sciences, National Institute of Technology, Tiruchirappalli, Tamil Nadu 620015, India
e-mail: sathyaiitk@gmail.com

S. Venkatesan et al. (eds.), *Pandemics and Epidemics in Cultural Representation*,
https://doi.org/10.1007/978-981-19-1296-2_15

their racial origins, these ethnic groups experienced severe forms of exclusion, variable treatment (both legal and health), and inequality during the pandemics. As a character in Kay Sohini's "Pandemic Precarities: An Account from the Intersection of Two Worlds" (2021) in *COVID Chronicles: A Comics Anthology* insightfully muses, "this pandemic has exposed and amplified everything that is wrong with our world—obscene wealth disparity, ableism, racism, sexism, bigotry. The need for an intersectional future is direr than ever" (217). Elsewhere, Yascha Mounk in her essay "The Virus Will Win" (2020) published in *The Atlantic* succinctly puts it thus, "[p]andemics reveal the true state of a society. Ours has come up badly wanting," suggesting how COVID-19 not only has magnified systemic racism but also has exposed social inequalities and health disparities in the USA.

While there are several studies/reports[1] that focus on health disparities in relation to BIPOCs, only a few them have paid attention to how comics in general, and graphic medicine, in particular, develops and expands such issues. Taking these cues, while this chapter draws examples from various comics, it develops its analysis around "Between Two Worlds," a mini-comics by Julio Anta, illustrated by Jacoby Salcedo and lettered by Hassan Otsmane-Elhaou; "How to Have a Powwow in a Pandemic" by S. L. Rosenbaum and Arigon Starr and Kay Sohini's "Pandemic Precarities: An Account from the Intersection of Two Worlds." All these comics appear in the recently published *COVID Chronicles* (2021), a compilation of comics which covers the first wave of COVID-19. While "Between Two Worlds" represents the African American and Latinx experience of the pandemics, "How to Have a Powwow in a Pandemic" by S. L. Rosenbaum and Arigon Starr conveys native American perspective. On the other hand, Kay Sohini's "Pandemic Precarities: An Account from the Intersection of Two Worlds" functions as a metatext in that the mini-comics not only offers commentary on the current situation but also theorizes the crisis in visual and verbal terms. Irrespective of different art styles and rhetorical situation (audience, purpose, and context), all these comics reiterate the racist/colonial histories, its ongoing impact on BIPOC's lives, and, more importantly, the variable impact of COVID-19 on marginal communities. In so doing, this chapter through using comics as a source illustrates how political-economic forces and social disparities impact health, illness, and well-being of the marginalized communities. However, before analyzing how comics engages with such issues, it would be instructive to briefly trace the history of the term "graphic medicine."

[1] Several studies point to how the structural and racial pasts influence COVID-19 health outcome. See, for instance, Newkirk, V. R. I. I. (2020). The coronavirus's unique threat to the South. *The Atlantic*. Retrieved from https://www.theatlantic.com/politics/archive/2020/04/coronavirus-unique-threat-southyoung-people/609241/.

Graphic Medicine: An Introduction

Whether it is superhero comics or recent works such as Art Spiegelman's *Maus* (about holocaust), Joe Sacco's *Palestine* (plight of the Palestinian people in the West Bank and the Gaza strip) or Alison Bechdel's *Fun Home: A Family Tragicomic* (sexual orientation) comics not only have consistently illustrated injustice, trauma, and disasters but also have spoken back to (authoritarian) power structures. Comics's deft intermixing of words and images in a sequential order makes it an opposite medium to present trauma, outbreaks, and taboo themes (such as one's own illness/sexual orientation). Historically speaking, underground and/or alternative comics (aka. commix) as opposed to the mainstream comics have facilitated the growth of such anti-authoritarian, taboo, and impermissible themes. Chute (2017), a scholar in comics, picks up on this in her *Why Comics? From Underground to Everywhere* when she points out thus, "[comics make] readers aware of limits and also possibilities for expression in which disaster, or trauma, breaks the boundaries of communication, finding shape in a hybrid medium" (34).

Drawing strengths from the underground and alternative comics and capitalizing on health humanities and narrative medicine, graphic medicine, a recent development in the comics genre, concentrates on the issues related to health, illness, and well-being. Coined by a British physician and comics artist, Ian Williams in 2007, graphic medicine refers to the intersection of comics and concerns of healthcare. In recent times, graphic medicine has expanded to include issues related to climate, health disparities, ethnicity/race, displacement, capitalism, carcerality, and their impact on health among others. In so doing, graphic medicine advances an ecological dimension of (public) health in that it focuses on the role of medical and non-medical factors shaping one's/community health. For instance, in her keynote address in 2018 Graphic Medicine Conference (at White River Junction, Vermont), Susan Squier suggests the need to internationalize and expand the cultural work of graphic medicine, thus:

> We choose this name with very deep awareness of its limitations ... We see it as our mission to reframe, expand, and to reclaim both the term graphic and the term medicine to cast the widest net ... Our priorities will expand to facilitate more cross-disciplinary work, mentorship, and research, to focus on patient projects, environmental applications, and social activism. We strive to be intentional about diversity, inclusion, and international outreach, all without losing sight of our core mission which is to work at the intersection of the medium of comics and the discourses of health, illness, caregiving, and disability. (Comic Nurse, 2019)

Again, in a recent interview with Kasthuri and Peter (2021), M. K. Czerwiec (aka. Comic Nurse), the co-founder of graphic medicine, points out the preparedness of graphic medicine to embrace newer concerns thus, "[w]e would also like to feature more work on [the] environmental impacts on health. For instance, how climate change increases asthma rates. I think, in the future, graphic medicine can include more work about the environmental contexts in which we all live. In addition to environmental concerns, I would love to see comics about health and social activism, about how our political and financial environments impact our health, for example, issues around access to care, and the availability of resources that promote health and combat illness" (105). Such an expansive role for graphic medicine as anticipated by M. K. Czerwiec and Squier is not surprising; after all, graphic medicine is "a movement for change that challenges the dominant methods of scholarship in healthcare, offering a more inclusive perspective of medicine, illness, disability, caregiving, and being cared for" (Czerwiec et al., 2015, 2).

The recent inauguration of Graphic Mundi (spearheaded by the editor and publisher, Kendra Boileau), an imprint of The Pennsylvania State University Press, is a provocation to interpret health beyond the immediate realms of the individual and medicine. The very tag line of Graphic Mundi ("drawing our worlds together") and their vision statement[2] among others are a decisive departure from their previous publishing agenda of concentrating on individual illness conditions and caregiving practices. In fact, some of their recent publications such as *COVID Chronicles* and *Crude: A Memoir* among others emphasize the need to interpret health beyond the immediate realms of medicine/biology. Put differently, it lays emphasis on the social, cultural, political, environmental, and structural factors that influence health.

Overview of COVID-19 Comics

COVID-19 pandemics spurred various visual responses. For instance, there were memes, cartoons, animations, gifs, comics posters, webcomics, and editorial cartoons that proliferated in both traditional and non-traditional platforms. Comics, in particular, played several functions during these challenging times; as Venkatesan and Joshi (2021) argue, comics at once functioned "[a]s a coping mechanism, as a therapeutic medium, as a pedagogical tool, as an impactful science communication and finally, as a medium which can affect behavioral change in individuals and communities" (7). Some of the major online and print anthologies on COVID-19 are: Ethan Sacks and Dalibor Talajic's *COVID Chronicles*, *The Lockdown Lowdown*, *Graphic Narratives for Viral Times* published by Distant Connections, and, more importantly, The Penn State University Press's *COVID Chronicle: A Comics Anthology* (curation of more than sixty short comics from diverse creators). In addition to these visual narratives, graphic medicine website (<graphicmedicine.org>) collected comics, editorial cartoons, autobiographical cartoons, and social media posts under the heading "COVID-19 Comics." All these visual narratives exemplify what Priscilla Wald calls "the outbreak narrative." Outbreak narratives are "epidemiological stories" that not only "promote or mitigate the stigmatizing of individuals, groups, populations, locales (regional and global), behaviors, and lifestyles" (Wald, 2008, 2–3) but also "delineate the social, bodily, and geographic boundaries that have been impacted by the virus, the reconfiguring of social interactions, and emotional responses to such

[2] "[O]ur creators represent a broad range of voices and experiences, and the graphic novels we publish address important topics such as health and human rights, politics, race, the environment, science and technology".

things as physical distancing, isolation, and the risk of becoming infected" (Callender et al., 2020, 1061).

"Racing" COVID-19

COVID-19 confirms our collective vulnerability. As Slovic reminds, we are living in "the COVID-19 era" with a "magnified... sense of universal vulnerability" (Slovic, 2020, n.p.). Though all lives are inherently vulnerable and thus precarious, to paraphrase Judith Butler,[3] what bothers is the uneven distribution of precarity and risk. Precarity designates "politically induced condition in which certain populations suffer from failing social and economic networks of support and become differentially exposed to injury, violence, and death" (Butler, 2009, 25). Though everyone likes to believe that the COVID-19 "does not discriminate," the fact is "some communities are more vulnerable to, and more affected by, the virus than others" (Sohini, 2021, 214). Such differentially induced risk is mediated by pre-existing inequalities and longer histories of exploitation marked by class, race, gender, caste, ableism, age among other categories.

COVID-19 has disproportionately impacted the BIPOC communities putting them more at risk of getting sick and dying from COVID-19. Consolidating several reports, Hooper et al. (2020) in "COVID-19 and Racial/Ethnic Disparities" presents the following data:

> In Chicago, Illinois, rates of COVID-19 cases per 100 000 (as of May 6, 2020) are greatest among Latino (1000), African American/black (925), "other" racial groups (865), and white (389) residents. Mortality rates are substantially higher among African American/black individuals (73per100 000) compared with Latino (36 per 100 000) and white (22 per 100 000) residents. 3 New York City (as of May 7, 2020) reported greater age-adjusted COVID-19 mortality among Latino persons (187 per 100 000) and African American individuals (184 per 100 000), compared with white (93 per 100 000) residents.

Similarly, Holtgrave et al. (2020) argue that by the end of March 2020, Black adults were 5.38 times more likely to die of COVID-19 than White (non-Hispanic) adults, whereas Latinx adults were 3.48 times more likely to be impacted by COVID-19. Though select pre-existing conditions such as "older age, diabetes, obesity, and male sex" (Hooper et al., 2020, 2466) have made the virus deadlier for BIPOCs, the role of non-medical factors cannot be underestimated. In fact, in his essay titled "Micro-Immensity," Hsuan L. Hsu (2021) persuades us to consider the pre-existing conditions such as "environmental racism, occupational health hazards, nuclear colonialism, and American consumerism's addiction to "chemical regimes of living""(274) and their continuing legacies as cause for the health inequities among BIPOCs. Elsewhere, CDC in a brief report titled "Health Equity Considerations and Racial and Ethnic Minority Groups" (2021) elaborates on the various non-medical and structural factors

[3] As it "can be expunged at will or by accident; [and] their persistence is in no sense guaranteed" (25).

which determine the health and well-being of the racial and ethnic minority groups. Particularly, the report lists five major reasons that contribute to the increased risk to COVID-19, viz., discrimination, health care access and use, occupation, housing, educational, income, and wealth gaps.

Though developed to conceptualize the impact of HIV/AIDS racial communities (particularly Blacks), Merrill Singer's concept of syndemics is a useful lens to describe the current impact of the COVID-19 on BIPOCs. The term syndemics simultaneously conveys the "biological interactions between epidemics" and "between epidemics and the social conditions" (Gravlee, 2020, 2). As such, the term captures how racism/discrimination and such structural obstacles really underpins the socio-economic drivers of health outcomes. Though the default vulnerable groups such as the unemployed, children, uninsured and underinsured, gender non-conformists are more susceptible to the COVID-19, there are adequate research which reports how racial and ethnic minorities (as opposed to the Whites) are at the highest risk of infection and death. Put differently, BIPOCs are at the highest risk of infections and death not just because of their biological vulnerability but owing to several socio-political, historical, and structural issues. As Willen et al. (2017) argue, "upstream social, political, and structural determinants contribute more to health inequities than do biological factors or personal choices" (964).

The confluence of structural, historical, biological, and contemporary racial events (such as Black lives matter) creates an adverse biopsychosocial landscape for BIPOCs. Put together, the case of BIPOCs not only demonstrate how health is determined by medical factors but also by the socio-cultural, historical, and other non-medical factors. Such a synergetic understanding of public/individual health augurs well with planetary health, medical post-humanism and critical health, and race studies. Fostering transformative action, the comics series called *Confinada* by Triscila Oliveira and Leandro Assis throws light on such intersectional disparities (class, race, gender, and health) and its toll particularly on Black (women) community in Brazil. Though *Confinada* comics series is not the immediate concern of this chapter, what is relevant here is the near universal nature of the way vulnerable/poor, (racially) marginalized, and the subalterns' (farmers, immigrants among others) lives are made precarious during the pandemics. In the USA, such inequities have further deepened with the brutal killings of Blacks. Several comics analyzed here not only comment on biological vulnerability but also draw attention to socio-political factors that reduce BIPOCs to "discounted bodies" (Mbembe, 2021, 23). Elaborating the megaprocesses that structure the contemporary times, Mbembe (2021) explains how there has been "a bifurcation between life and bodies" (23) in recent times and how certain bodies are unimportant. Labeling them as "*Discounted bodies*" (italics in original), Mbembe describes them as follows: "Not each life and every breath has utmost value. *Discounted bodies* are believed to contain no life as such. They are, strictly speaking, bodies at the limits of life, vulnerable and porous. Trapped in uninhabitable worlds and inhospitable places, they generate multiple kinds of tumors" (23).

Intersectionality of Precarity

Concise and graceful in its exposition, Kay Sohini's mini-comics titled "Pandemic Precarities" via autobiography concentrates on the production of precarity and class inequity in American (New York) and in Indian (Kolkata) contexts. Sohini (2021), originally from Kolkata, is a South Asian immigrant settled in New York, and her work mainly focuses on the use of comics "to amplify marginalized voices" (277). Among other things, "Pandemic Precarities" addresses politicization of pandemics, the impact of pandemics on the livelihood of the working class/disabled, the issues related to race/religion, devaluation of disabled lives and, more importantly, the politics of livability and grieavability in the USA and in Indian contexts. Sohini's insightful mini-comics treats COVID-19 as a problem of precarity in terms of embodied experience of vulnerability and how certain class/racial groups are subjected to (in)visible health injustices and increased levels of risk. As a South Asian immigrant who "exist[s] on the threshold of two worlds" (212) and as someone who considers both countries as "home" (217), Sohini leverages her immigrant identity to offer insights into the parallel production of precarity and disenfranchisement. In so doing, the author justifies the subheading of her comics, "An account from the intersection of two worlds" (New York and India) (211).Though such a politically induced precarity is differently produced (in terms of geography, history, and cultural patterns) in the USA and in India, it also has certain transnational and comparative qualities to them. Granted that socio-economic and cultural realities of these two places are vastly different, however, what Sohini does is to expose how these vastly different worlds "make" precarious and marginal subjects, thereby deepening their already existing socio-economic inequities and helplessness. If in India, migrants and minority religious communities constitute the precariat class, then it is BIPOCs who are differentially exposed to risk in the USA.

The comics begins with the narrator attending a conference at Boston when she receives a hard-to-believe message from the Governor of New York about the closure of the state. In the next set of panels, Sohini through Ariel shots provides an "already emptying streets of New York" which has become "the epicenter of the [COVID-19] crisis" (211) (as opposed to the previous weekend which witnessed less than a hundred cases). As the COVID-19 crisis unfolds at the New York, the readers are informed of the death of Sohini's grandmother at Kolkata in India.

Although the mini-comics is a personal account of COVID-19, it quickly segues to political issues related to displacement, medical racism, internal migration, etc. In India, for instance, the migrant workers whose services are deemed essential are made to suffer the unplanned imposition of an overnight lockdown by Narendra Modi on March 24 producing hyper-precarity among workers. To concretize these dynamics, Sohini divides the page into four neat panels, and each one of them visually enunciates different aspects of precarity—for instance, in the first panel, we witness an exodus of migrant workers subjected to police violence as they were desperately trying to reach their hometown. In the next panel, the migrant workers are arbitrarily accused as superspreaders and subjected to chemical disinfectants at

Uttar Pradesh (a state in India). In the third panel, we see gendered violence and racial abuse unleashed on the Northeast Indians (who share Chinese features) demanding them "to go back to China" (215). Worse still, an international event of the Tablighi Jammat (a Muslim religious organization) in Delhi is "vilified for spreading the virus" (215). Although similar religious events were held during this period, it is only the Muslim community, as Kolachalam (2020) argues, which was held responsible for the spread of COVID-19. Though the panels quickly transition from one event to another, it synoptically covers the very process of producing precarious subjects through religiousing/racializing and politicizing COVID-19. Not only does such incidents reveal how specific religious minorities are made scapegoats but also expose virulent forms of Hindu nationalism, thus putting secular credentials of the ruling government at stake. Put differently, as a multisited issue, COVID-19 exacerbates the already existing marginalization as it creates new precariat subjects (such as Muslims, Northeast Indians). These newly produced subjects', as Suliman and Weber (2019) contend, emerge "from the spatial and temporal restructuring of the relationship between production, politics, and life" (528).

Sohini quickly moves to the USA and draws readers' attention to the neoliberal inflected market logic where human lives are "calculated based on its functionality" and "utility" (216). The broader point is how an essential worker is reduced to a mere "market actor" and as a "human capital" (3), to borrow Wendy Brown's (2016) expressions. To prove her point, the narrator references to the Blacks and Latinx Americans employed in the meat industry and how they are exploited on the existing socio-cultural inequalities to make sure the continuity of meatpacking supply chain in the USA. Later, in a different context, we are shown the shocked faces of essential workers whose hazard pay is cut, while the wealth of select people increased exponentially during the pandemics, suggesting the growing inequality in the USA. The scene quickly shifts to the USA where Sohini draws on several other representative instances such as the attack of a Target store by "Karen" Melissa Lively, the politicization of the use of mask, and the untimely death of Michel Hickson (a disabled Black person). Hickson's story is given special importance in a panel where we are shown Hickson is surrounded by his relatives to underscore the intersectional complexity of disability and race producing invisible injustices with severe consequences (here, the death of Hickson). In putting all of these discrete facts together in quick succession of panels, Sohini not only delineates affective dimensions of COVID-19 (such as loss, grieving) but also brings to light on how nation-state, industrial capitalism/neoliberal/necrocapitalism, and biological vulnerability synergetically produce, sustain, and maximize the precariousness of its subjects. Put together, the mini-comics presents interconnected articulations on the intersectional effects of the COVID-19 on BIPOCs and marginalized communities in the USA and in India.

Social Distancing En"force"ment

Inspired by true events, "Between Two Worlds" (2021) is a mini-comics drawn by Julio Anta, illustrated by Jacoby Salcedo and lettered by Hassan Otsmane-Elhaou. While Anto is an emerging New York-based Latinx writer interested in Latinx issues in America, Hassan Otsmane-Elhaou is an Eisner award-winning magazine editor of *Panel X Panel*. The mini-comic introduces two parallel narratives thereby justifying the title "Between Two Worlds." The two worlds alluded here are the West Village (a posh neighborhood in the western section of the larger Greenwich Village) and the Spanish Harlem at the New York City. In between these sections, news bytes generously borrowed from NYT, CNN are presented to provide the much needed context and setting for the action in the comics. Put differently, "Between Two Worlds" narrates two different events and two realities as they unfold in a typical pandemic day. Broadly, the mini-comics throws light on the racial bias in social distancing policies and the "arbitrary state violence" (Butler, 2009, 26) which maximizes the precariousness of select communities (here, Blacks and Latinos). While most of the comics in *COVID-19 Chronicles* iterate the health disparities and how it particularly impacts racial minorities in the USA, central to "Between Two Worlds" is the intersections between race and law, and the ideologies infusing social distancing policy enforcement. More specifically, "Between Two Worlds" demonstrates how every aspect of the COVID-19 operates via the exclusion of minority populations and is predicated on White privilege.

The two worlds are also structurally conveyed by dividing the page into two sections: While the upper section of the page traces the events in the West Village, the lower section of the page covers the happenings at the Spanish Harlem at the New York City. In the first set of panels, readers are introduced to Kinsey who is responding to a picnic brunch through her telephone, while the lower section of the page introduces Santiago who plans to meet his friend, Malik. Seeking permission from his mother and wearing mask and gloves as per the US government advisory, Santiago readies himself to meet his friend, Malik, an African American. On the other hand, Kinsey is seen walking in a street without wearing a mask and later meeting her friend, Cara. All these meetings happen at a time when COVID-19 caseload soars past ten thousand in New York and social distancing/wearing masks are legally imposed on the population.

The scene quickly shifts to an apartment, where Kinsey and her friends (Ashleigh and Cara) visit an ailing boyfriend and exchange pleasantries. What is striking here is that in spite of the strict orders on social distancing and masks, Kinsey and her friends flout all guidelines. For instance, they are seen in close proximity with their male friend without wearing a mask. And later, they huddle to discuss different aspects of pandemics and office life. The lower section of the page, on the other hand, shows the encounter of Santiago and Malik. Both of them are wearing mask, gloves, and even maintain social distance in spite of not meeting each other ("not gonna lie, I missed you, bro") for a longer period of time (Anta et al.,2021, 192). In between these two sections, the author inserts an actual news item published in *NY1:* "New

York City Faces First Major Social Distancing Test as Nice Weather Lures Many Outside" suggesting the challenge of keeping New Yorkers safe in nice weather. In fact, the entire page is presented in such a way that a contrast is created between the masked faces and the unmasked ones.

Suddenly, a White policeman, citing violations of emergency procedures, intercepts Santiago and Malik and insists that they should be inside their homes. When Santiago retorts that he is wearing "masks and gloves," the White policeman becomes irritated and ends up arresting Malik and Santiago. In the final panel, Malik and Santiago are inside a police vehicle, while a policeman warns them, thus: "Remember this moment next time you want to mouth off, Boy." This scene not only vindicates CNN's headline "80% of summons are sent to the blacks and Latinos" but also reminds of the brutal killings of George Floyd. Of particular interest is one such panel where a policeman is seen pressing hard on the neck of Santiago. Statistically speaking, "374 summonses were handed out from March 16 to May 5, averaging less than 10 summonses a day over the 42-day period. And of that total, 193 of those issued summonses were Black and 111 were Hispanic" (Morales & Joseph, 2020). The White women, on the other hand, have a festive time, while the policemen are shown distributing masks to them. Ironically, one of the women is even seen flirting with a policeman commenting thus, "Is it just me, or are these cops pretty cute?" (Anta, et al.,2021, 195).

The use of the slang, Amerikay (as opposed to America) in the last page of the comics uncovers the double speak of democracy: that is, how the laws/rules/social procedures are differently implemented for the African Americans/Latinos (racial minorities at large), on the one hand, and for the White Americans, on the other. In so doing, the comics scrutinizes the arbitrary and biased enforcement of law and how it works through race and class among other axes of oppression. Within such a framework, Whites are privileged and unsuspected class. While it is a definitive case of racial dynamics, the comics specifically exposes the intersections among race, law, and public health system.

Native Indian Realities

Native Americans, like other minorities, were disproportionately affected by COVID-19. Elaborating the impact of COVID-19 on the native Americans in an article titled "COVID-19 Ravaging the Navajo Nation," (2020) Mary Hasbah Rossel, a physician, not only highlights the immediate impact of COVID-19 on the native Americans but also underscores how historical and cultural factors influence the largest tribe in the USA making them more vulnerable to the virus. The article concludes with a note of hope, thus: "The Navajo are having to rely on the strength and resilience of traditional Navajo culture and philosophy to confront this monster, Dikos Nitsaa'igii' 19. We have relied on Western medicine and its limited resources but now need to empower the strength from our traditional ways of knowing." Many comics produced by the native Americans during the first wave of COVID-19 reflect many of these concerns and, more importantly, a range of health and social challenges unleased by pandemics.

Scripted by S. I. Rosenbaum and drawn by Arigon Starr, "How to Have a Powwow in a Pandemic" (hereafter, "How to Have a Powwow") is one such comics which narrates the challenges of organizing an indigenous dance festival (Powwow) during pandemics. Author of influential *Super Indian* and *Tales of the Mighty Code Talkers,* Starr is an enrolled member of the Kickapoo Tribe of Oklahoma. While the mini-comics is about the difficulties in organizing Powwow (a North American/Tribal American ceremonial event and the oldest recorded native gathering involving dancing, singing, and socializing), the comics interlaces the colonial and racial histories of native American displacement and exclusionist policies to signal the inherent structural inequalities created by the mainstream US society. For instance, in one of the panels, the author reasons the disproportionate impact on native Americans, thus: "COVID has been particularly deadly in native communities, for the same reasons as previous pandemics: poverty and oppression" (110). Later, the author also points out the structural problems, thus: "on many reservations people live with little access to health care or, sometimes, running water." Finally, the narrator ends the panel by reminding the continuing legacies of exclusion cultural mistrust, "broken agreements, and underfunded services from the US government" (110), as the prime reasons of suffering during COVID-19. Elsewhere, the author draws striking parallels to earlier pandemics when the native Americans had to endure different "wave upon wave" pandemics such as childbirth fever, smallpox, whooping cough (106). In fact, the comics begins with two verbal statements drawn in bold letters, thus: "Native communities in North America have been particularly hard-hit by COVID-19," and "This is not the first time" (105) suggesting not only the impact of COVID-19 but also how they have repeatedly endured such pandemics in earlier times. Though ethnographic in its scope, "How to Have a Powwow" underscores the continuing legacies of racial exclusion and the very vulnerability of the native American community. However, without losing hope, the native Americans organize an online Powwow movement and console themselves stating that "but the world has ended so many times before we are always being forced to transform" (110).

In a different way, compared to the Whites, essential services are disproportionately represented by racial and ethnic minority groups, making them more prone to COVID-19. Essential services, as the CDC defines, refer to "a range of operations and services in industries that are essential to ensure the continuity of critical functions in the USA." They include but are not limited to "keeping us safe, to ensuring food is available at markets, [and] to taking care of the sick" among others. According to the report by the Urban Institute, Black, native American, and Hispanic/Latinx workers are "more likely than White workers to have jobs that place them at greater risk of exposure to and transmission of the coronavirus" (Dubay et al., 2021, vi). Produced by the Indian Country Today, "The Navajo Nation Has Been Ravaged by COVID-19" (art by Arigon Starr, writing by Kalle Benallie and adapted by Eleri Harris) is a text book instance of how the employment of native Indians in essential services took a toll on ethnic communities. Besides memorializing the life of Lilly Tsosie, the main character in "The Navajo Nation Has Been Ravaged by COVID-19," the mini-comics narrates the family and professional life of Lilly Tsosie, a sixty-five-year-old phlebotomist at the San Juan Regional Medical Center in New Mexico. As "the backbone" of the family, Tsosie, a Navajo, loses her life to the COVID-19 and is survived by her daughter (Lacey Williams) and grandchildren. In spite of her family's expressed concern and hesitancy regarding Tsosie working during the pandemics, she dedicatedly continues to do her job in the hospital. To quote her, "This is my job, I have to go. I have to help others, and if I don't do it then who's going to do it?" Although Tsosie is given a break from her work by the hospital, she is later shifted to COVID-19 floor because of "short staffing." Fittingly, Tsosie is shown wearing hazmat suit and preparing for her duties. Later, Tsosie is shown coughing, while the narrative explains that she has been "experiencing symptoms" of the COVID-19. Put differently, she is exposed to the dangers of infection. Soon, we witness Tsosie bidding farewell in a hospital bed flanked by her daughter and granddaughters. Although the hospital arranges a well-attended procession and placard holding friends (#Tsosie Strong) in memory of Tsosie, the last panel closes with a plea requesting everyone to wear mask not only to protect oneself but also "somebody's loved one" and "health care workers (15)." Put together, both these comics capture how native American communities across the USA faced disproportionate rates of COVID-19 infection and causality.

Implications

While there are enough statistical data about the disproportionate impact of the COVID-19 on BIPOC communities, graphic medicine using comics (through perspectives, sequentiality, non-linear presentation of panels, and multimodal affordances) and through contextualizing BIPOC issues within rich historical and cultural contexts renders the lived experiences of health equities and structural racism. As such, the phenomenologically and affectively rich comics analyzed in this chapter are representative of the larger malaise plaguing the US society. While "Pandemics Precarities" charts the role of the state and other neoliberal forces in creating and maintaining precarity both in the USA and in India, then "Between Two Worlds" shows how Blacks and Latinos are differently treated in terms of social distancing policies. On the other hand, "How to Have a Powwow in a Pandemic" and "The Navajo Nation Has Been Ravaged by COVID-19" convey the perspective of the native Americans. Second, these graphic medical narratives imagine health as a negotiation and as a consequence of our relations within the complex architectures of various actors, networks, and histories. Such a network approach to health augurs well with the principles of global public health, relational ethics, and post-humanism which differently emphasize the role of structural racism, necrocapitalism, and continuing legacies of colonialism as a fundamental driver of health disparities.

References

Anta, J., Saledo, J., & Otsmane-Elhaou, H. (2021). Between two worlds. In: K. Boileau, & R. Johnson (Eds.), *COVID Chronicles: A Comic Anthology*. Graphic Mundi.

Brown, W. (2016). Sacrificial citizenship: Neoliberalism, human capital, and austerity politics. *Constellations, 23*, 3–14. https://doi.org/10.1111/1467-8675.12166

Butler, J. (2009). *Frames of war: When is life grievable?* Verso.

Callender, B., Shirlene Obuobi, M. K., & Czerwiec, and Ian Williams. (2020). COVID-19, comics, and the visual culture of contagion. *The Lancet, 10257*, 1061–1063. https://doi.org/10.1016/s0140-6736(20)32084-5

Centers for Disease Control and Prevention. (2021). Health equity considerations and racial and ethnic minority groups. https://www.cdc.gov/coronavirus/2019-ncov/community/health-equity/race-ethnicity.html. Accessed September 11, 2021.

Chute, H. L. (2017). *Why comics? From underground to everywhere*. Harper.

Comic Nurse. (2019). 'Scaling up graphic medicine'—Susan Squier's 2018 conference keynote address | graphic medicine. *Graphic Medicine*. https://www.graphicmedicine.org/scaling-up-graphic-medicine-susan-squiers-2018-conference-keynote-address/

Czerwiec, M. K., Williams, I., Squier, S.M., Myers, K. R., Green, M. J., & Smith, S. T. (2015). *Graphic medicine manifesto*. Penn State University Press.

Dubay, L., Aarons, J., Steven Brown, K., & Kenney, G. M. (2020). How risk of exposure to the coronavirus at work varies by race and ethnicity and how to protect the health and well-being of workers and their families. The Urban Report. https://www.urban.org/research/publication/how-risk-exposure-coronavirus-work-varies-race-and-ethnicity-and-how-protect-health-and-well-being-workers-and-their-families. Accessed September 11, 2021.

Gravlee, C. C. (2020). Systemic racism, chronic health inequities, and COVID-19: A syndemic in the making? *American Journal of Human Biology, 5*. https://doi.org/10.1002/ajhb.23482

Holtgrave, D. R., Barranco, M. A., Tesoriero, J. M., Blog, D. S., & Rosenberg, E. S. (2020). Assessing racial and ethnic disparities using a COVID-19 outcomes continuum for New York State. *Annals of Epidemiology, 48*, 9–14. https://doi.org/10.1016/j.annepidem.2020.06.010

Hooper, M. W., Nápoles, A. M., & Pérez-Stable, E. J. (2020). COVID-19 and racial/ethnic disparities. *JAMA, 323*(24), 2466–2467. https://doi.org/10.1001/jama.2020.8598

Hsu, H. L. (2021). Micro-immensity. *Manifest: A Journal of American Architecture and Urbanism, 3*, 273–275.

Kasthuri, R. R., & Peter, A. M. (2021). Twelve years of graphic medicine: A conversation with comic nurse on the future of comics and medicine. *Configurations, 1*, 97–106. https://doi.org/10.1353/con.2021.0004

Kolachalam, N. (2020). Indian muslims are being scapegoated for the coronavirus. Slate. https://slate.com/news-and-politics/2020/04/tablighi-jamaat-india-muslims-coronavirus.html. Accessed September 11, 2021.

Mbembe, A. (2021). Futures of life and futures of reason. *Public Culture, 33*(1, 93), 11–33. https://doi.org/10.1215/08992363-8742136

Morales, M., & Joseph, E. (2020). Over 80% of social distancing summons given to blacks and Latinos—CNN. *CNN*. https://edition.cnn.com/2020/05/08/us/social-distancing-stats-nyc/index.html

Mounk, Y. (2020). The virus will win. *The Atlantic*. https://www.theatlantic.com/ideas/archive/2020/06/virus-will-win/612946/. Accessed September 11, 2021.

Roessel, M. H. (2020). COVID-19 Ravaging the Navajo Nation. *Medscape*. https://www.medscape.com/viewarticle/932019. Accessed September 11, 2021.

Slovic, S. (2020). COVID world, COVID mind: Towards a new consciousness in the ongoing COVID reality. Bifrost Online. https://bifrostonline.org/scott-slovic/. Accessed September 11, 2021.

Sohini, K. (2021). Pandemic precarities: An account from the intersection of two worlds. In: K. Boileau, & R. Johnson (Eds.), *COVID Chronicles: A Comic Anthology* (pp. 211–217). Graphic Mundi.

Suliman, S., & Weber, H. (2019). Global development and precarity: A critical political analysis. *Globalisation, 16*(4), 525–540. https://doi.org/10.1080/14747731.2018.1463739

Venkatesan, S., & Joshi, I. (2021). Comics play the role of another COVID-19 helpline. *The Pioneer*. May 8, 2021. https://www.dailypioneer.com/2021/columnists/comics-play-the-role-of-another-covid-19-helpline.html

Wald, P. (2008). *Contagious*. Duke University Press.

Willen, S. S., Knipper, M., Abadía-Barrero, C. E., & Davidovitch, N. (2017). Syndemic vulnerability and the right to health. *The Lancet, 389*(10072), 964–977. https://doi.org/10.1016/S0140-6736(17)30261-1

Chapter 16
Caste in Epidemics in India: A Historical and Literary Reading

Antara Chatterjee

Abstract This essay examines how caste and its practices and mechanisms, like untouchability, spatial and social segregation, a birth-based, socio-economic hierarchical and graded inequality, and an ontological link between caste and work, form a crucial and textured subtext in perceptions and understandings of infectious disease outbreaks in India, both in societal and state responses during historical epidemics, as well as in as cultural representation. Reading multiple epidemic outbreaks, like the 1896 bubonic plague outbreak, the nineteenth-century cholera outbreaks and the 2020–21 COVID pandemic intersectionally with literary and cultural texts featuring epidemics, the essay uncovers how caste has configured meanings and trajectories of infectious disease in India, and in turn, been perceived and negotiated within epidemic outbreaks and their representations. It reflects on how notions of the body and its boundaries, touch, defilement, contamination, (im)purity, figure centrally in both caste and infectious disease discourses, and complexly and multifariously intersect within epidemic disease constructions in India. In foregrounding caste as a crucial index in experiencing and understanding epidemic disease, leading to caste-based vulnerability and marginalization, the essay highlights the differential experience of epidemics, based on socio-economic markers like caste, in turn intimately connected with parameters like class, economic status, and labor.

Introduction

Priscilla Wald has influentially argued that "Contagion is more than an epidemiological fact. It is […] a foundational concept in the study of […] society."[1] The ongoing COVID-19 global crisis has offered new insights into social and cultural formations and processes. It has also provided an opportune moment once more to reflect on

[1] Wald (2008), 2.

A. Chatterjee (✉)
Department of Humanities and Social Sciences, Indian Institute of Science Education and Research, Bhopal, India
e-mail: achatterjee@iiserb.ac.in

S. Venkatesan et al. (eds.), *Pandemics and Epidemics in Cultural Representation*,
https://doi.org/10.1007/978-981-19-1296-2_16

health disparities across the world and within nations. It has powerfully brought to light again the inequitable distribution of disease burden and vulnerability, mitigation and recovery processes, and resources for prevention, including in this case, as in the case of many other infectious diseases, opportunities for isolation, distancing, and vaccination. The current pandemic and its multiple, complex trajectories across the world offer an apt opportunity to refocus on health inequities based on the social and political markers of race, class, gender, ethnicity, religion, caste, and location, in their manifold, intertwined connotations.

Sociological disaster research has, over the last few decades, increasingly highlighted the complexities within the "risk" paradigm for understanding disasters. It has directed attention to the socio-structural and systemic risks which inhere within social formations, leading to a differential experience of vulnerability, with certain groups more vulnerable to disaster and its damages than others. Anthony Oliver-Smith, Ben Wisner, Kenneth Hewitt, and others foregrounded the need to address not just the "agents" of disasters, "but also the social, political, and economic *processes* that put particular groups at risk and underpin the scale of disasters."[2] As Anthony Carrigan asserts, "developing countries suffer the most from disasters," and "large-scale catastrophes [...] take a disproportionate toll on the world's poorest communities,"[3] pointing toward the intimate correlation between disaster and its effects, and social, political and economic inequality. Oliver-Smith emphasized that disasters are effectively manifestations of historical vulnerabilities, which cannot be disentangled from systemic power structures.[4] Rob Nixon influentially denoted these long-term modes of structural violence and power inequities, leading to disempowerment and unequal vulnerability in the face of catastrophe, "slow violence."[5] Hewitt too foregrounded the need to include "missing voices," "shadow risks and hidden damages,"[6] and "silent or quiet violence" in understanding disaster and its vulnerabilities, highlighting the need to acknowledge within disaster, the plight of "distinctly more vulnerable members of society."[7]

In the case of epidemic disease outbreaks, vulnerability to disease as a result of systemic inequities of race, class, gender, ethnicity, nationality, etc. is further complicated and exacerbated through the complex ways in which disease and its threats are socially and culturally constructed. This leads not just to increased vulnerability of some groups to disease due to their socio-economic situation but also to certain groups and communities being *perceived* to be *associated* with the disease in public imagination, and consequent marginalization, ostracism, and stigmatization.

---

[2] Carrigan (2015), 120.

[3] Carrigan, "Towards," 117.

[4] Anthony Oliver-Smith, "Global changes and the Definition of Disaster," in *What is a Disaster: Perspectives on the Question*, ed. E. L Quarantelli (London and New York: Routledge, 1998), 189.

[5] Nixon (2011).

[6] Kenneth Hewitt, "Excluded Perspectives in the Social Construction of Disaster," in *What is a Disaster: Perspectives on the Question*, ed. E. L Quarantelli (London and New York: Routledge, 1998), 80, 81.

[7] Hewitt (1995), 120.

In India, the COVID pandemic and some of the ways in which the state and the people have negotiated its multiple medical, social, cultural, political, and economic meanings and contexts have once again directed attention to the deeply entrenched and historically perpetuated practices of caste difference, discrimination, and caste-based division of labor, leading to caste-based vulnerability and stigmatization.[8] These ground realities of caste inequity are often obfuscated in public focus by discourses of *legal* and *constitutional* equality, through tenets of equal rights enshrined in the Indian constitution since 1950.

Taking the COVID-19 pandemic as a point of departure, this essay aims to delve into the intersections of caste and epidemic disease outbreak in modern India, starting from the end of the nineteenth century till the current pandemic of 2020–21, both in history and in cultural and literary texts. The aim is to uncover how caste and its constructions and practices have figured in perceptions and understandings of infectious disease outbreaks, both historically and culturally. Examining multiple epidemic outbreaks, like the 1896 bubonic plague outbreak and the 2020–21 COVID pandemic intersectionally with literary texts featuring epidemics, the essay will attempt to understand how caste has configured meanings and trajectories of infectious disease in India, and been perceived and understood within epidemic outbreaks and their representations.

Examining government documents, media reports, and imaginative depictions, I will investigate how multiple kinds of narratives, official and cultural, have configured caste within epidemic disease in India. I will also examine the connections between caste and other social indices like class, religion and gender, and their conjoined social and economic marginalization in the experiences of infectious diseases in India. The essay will focus, in particular, on how literary and cultural narratives of epidemics represent and reevaluate experiences of infectious disease in terms of social, political, and economic inequities like those of caste and class.

Locating "medical" phenomena like epidemics equally as social sites where social and cultural meanings are generated, I assert that epidemic disease in India is experienced differentially, based on socio-economic markers like caste, which in turn is intimately connected with parameters like class, economic status, and labor. Caste, in its practices of segregation, its connotations of purity/impurity, its focus on bodily boundaries and their im/permeability, touch, defilement, and pollution, becomes

[8] Caste is the Indian social system whereby people are categorised through birth into a hierarchical social structure, based on *varna*. Caste-based exclusion, particularly the historical practice of untouchability, whereby the lowest castes in the *varna* system, the *avarna* (without varna) as opposed to the *savarna* (with varna), were subjected to severe social and economic restrictions, was historically perpetuated in India for centuries, before it was legally abolished after Indian independence, ensuring equal status and rights to all. Caste-based marginalization was perpetuated through caste-based allocation of work, through endogamous marriage practices, and through untouchability, whereby the untouchable castes (later claiming the term Dalits, "the downtrodden" for themselves) were socially and spatially segregated from the upper castes. For more on this, see Gail Omvedt, *Understanding Caste: from Buddha to Ambedkar and Beyond*, (Hyderabad: Orient Blackswan, 2016), Anupama Rao, T*he Caste Question: Dalits and the Politics of Modern India*, (Ranikhet: Permanent Black, 2010), Eleanor Zelliot, *Ambedkar's World: the Making of Babasaheb and the Dalit Movement*, (Navayana Publishing, 2013).

a particularly meaningful and layered terrain, in understandings of and constructions of infectious disease. As in caste discourses, notions of the body, touch, the im/permeability of bodily (and by extension, communal and national) boundaries, impure or contaminated bodies figure centrally in constructions of infectious disease.

This essay situates epidemic disease outbreaks and their representations within the overlapping contexts of caste, community, nation, the body, and their contested and intersecting boundaries. Acknowledging the differential experiences of health and disease, based on social and economic disparities, is of crucial significance in understanding disease in its social connotations. The corporeal body is to be read in relation to the social body, firmly located within the body politic. Infectious disease, with its connotations of bodily integrity, immunity, external threat, borders and boundaries, becomes a potent site to configure the dialogic relationship between the corporeal body and the body politic.

The essay will be divided into three sections; the first section will examine ramifications of caste in nineteenth-century epidemics in India like the cholera epidemics and the 1896 bubonic plague outbreak; the second section will read the COVID-19 pandemic in relation to caste. Situating caste as a crucial dynamic within the socio-economic contours of these pan/epidemics from the nineteenth to the twenty-first century, these two sections will offer a close reading of different narratives of these diseases, to examine if caste and its ramifications have persistently shaped experiences of infectious disease in India, or been rethought significantly in public consciousness. They will ask whether state and societal responses to the COVID pandemic reestablish the links between caste and epidemics found in the nineteenth century, or lay the ground for "a newly configured social imagination" vis-à-vis caste and disease.[9] The third section will attempt a reading of some literary texts narrating epidemic outbreaks in Indian literatures, deliberating on how epidemic disease is represented in literary and cultural texts as complex, multifaceted medico-social and biocultural phenomena, situated within intersecting discourses of caste, religion, community, nation, and disease.

Caste in Nineteenth-Century Epidemics in India: Bubonic Plague and Cholera

During the bubonic plague outbreak in India between 1896 and 1914, Surgeon-General Harvey, Director-General of the Indian Medical Department, had memorably said that bubonic plague was "a disease of filth, a disease of dirt, and a disease of poverty."[10] As Rajnarayan Chandavarkar points out, the unprecedented panic surrounding the outbreak of the plague in the late 1890s, particularly in the Bombay province, gradually began to subside as the disease began "to acquire a specific social

[9] Swaroop and Lee (2021): 36.

[10] *Report of the Indian Plague Commission, 1898–99*, V, *PP.*, LXXH, 1902, 170.

character,"[11] as it became increasingly apparent, as plague spread across the subcontinent, that "the poorer classes suffered most severely."[12] As the disease increasingly became associated in European perception with the poor, it began to be seen by the colonizers as endemic in India, one among many deadly diseases thriving there, produced and spread by the unhygienic and insanitary living conditions and practices of the Indian population, especially of the poorer classes. In the British perception, "India was a repository of infectious diseases," and "the plague [...] became the focus of the most terrible anxieties which India evoked in the British imagination."[13]

Consequently, sometimes, investigations and research into the origin and spread of the disease "proceeded along lines determined less by scientific evidence than by social assumptions."[14] For instance, for a long time, British health officials and researchers subscribed to the view that the plague bacillus spread to humans from rat excreta through cuts and abrasions on the bare feet of Indians. This hypothesis was confirmed by observations like "patients mostly belonged to the lower classes who commonly go bare-footed and bare-legged."[15] While there was evidence that "halalcores who remove the night-soil from the houses, and who form probably the dirtiest portion of the population, were notably free from plague," questioning such a hypothesis, this hurdle was overcome with the suggestion that "they are a strong, well-nourished class, of whom only the fittest have survived."[16]

It is obvious that the plague became the center of the anxieties of the British about the threat of disease in India and converged many of their notions about the unhygienic lives and practices of Indians. However, what also comes to the fore here is that the perception of the disease, while being underpinned by racial and cultural anxieties, was also strongly inflected by class and caste. Due to the intimate historical nexus between caste and class in the Indian social order, as a result of historically perpetuated processes of social and economic marginalization and disempowerment of the lower castes, the poorest Indians would most often be of the lowest castes. Therefore, a disease of the poor would also necessarily have been associated with the lowest castes in the caste hierarchy.

Caste has also historically had connotations of im/purity and cleanliness/defilement, with the lower castes being identified as impure and unclean vis-à-vis the upper castes. Harvey's statement that the plague was a disease of poverty and "filth," and the observation by Captain J. K. Condon in 1900, quoted earlier, that "halalcores who *remove the night-soil from the houses, and who form probably the dirtiest portion of the population*" (my italics) were surprisingly free from plague because they were a "strong" community, reveal how the plague and its threat also came to be inflected by caste and its connotations of filth and impurity, not just literal,

[11] Chandavarkar (1992), 209.

[12] Condon (1900), 132.

[13] Chandavarkar, "Plague Panic," 211.

[14] See Footnote 13.

[15] Condon, *The Bombay Plague*, 104. L. F. Hirst, *The Conquest of Plague. A Study of the Evolution of Epidemiology* (Oxford, 1953), 111–19.

[16] Condon, *The Bombay Plague*, 70.

but metaphorical. Caste has traditionally been the basis of distribution of work in Indian society, with the lowest castes of the Dalits or untouchables being allocated the most "unclean" jobs like cremating the dead, sanitation work including cleaning garbage, toilets, and removing night-soil. Some of these jobs, while perceived to be unclean, and therefore allocated to the Dalits, were simultaneously often hazardous. Caste-based division of work persists insidiously in societal and state behavior even in the present, despite the Indian constitution abolishing this in 1950, a reality that has been glaringly brought to the fore during the ongoing COVID pandemic, a point that I will return to in the next section of this essay.

Therefore, the stigma attached to the work of the "halalcores" or sanitation workers, who at the turn of the twentieth century were considered to be the "dirtiest" section of the population, was born not only because of the literally unclean nature of their work but also by virtue of their polluted or unclean caste status, which assigned this work to them. The association of diseases with particular communities predicated on race, culture, religion, caste, etc., and their consequent exclusion and stigmatization, has been a regular feature of imagining and responding to epidemics globally. In the Indian context, the connotations of contamination and defilement, embedded within both discourses of disease and caste, explosively coincided in infectious disease outbreaks.

The stringent plague control measures, unleashed by the colonial government in Bombay in 1896–97, spread rapid alarm to the "municipal scavengers, whose flight, it was feared, would bring the city to a standstill."[17] The then municipal commissioner, P. C. H Snow, dramatically projected the effect, which a scavenger's strike would have on the city. Before the epidemic, Snow pointed out, Bombay had been praised by sanitary experts as "the best kept and most sanitary oriental town," "but its continued well-being relied upon the daily labor of six thousand sweepers [:] Halalkhors from Gujarat and Marathi speaking Bigarris, mainly Mahars."[18] As we observe, scavenging work was stipulated on the basis of caste. "On their presence or absence, respectively, depended the safety or ruin of this vast and important City," Snow pointed out; "on these men [...] hung the carrying out of every sanitary measure and [...] were they all to remove from the town [...] Bombay would be converted into a vast dunghill of putrescent ordure."[19] If these caste-ordained scavengers fled the city, Snow feared, "half the inhabitants would speedily follow them, and no single measure could be adopted against the plague [...], nor could even the Europeans, Parsis, and high-caste natives have remained in the City."[20]

Epidemic outbreaks saw the convergence of, on the one hand, caste-based allocation of sanitation work to the lowest castes and, on the other, heightened risk of disease among these communities due to the nature of this work. As we see from Snow's observations above, the lower-caste scavengers had to be in the city in order

[17] David Arnold, *Colonizing the Body: State Medicine and Epidemic Disease in Nineteenth Century India* (Berkeley, Los Angeles: University of California Press, 1993), 206.

[18] Snow (1897), 5, quoted in Arnold, *Colonizing the Body*, 206.

[19] Snow, *Report*, 5.

[20] See Footnote 19.

to enable the government to undertake sanitary measures, so that the Europeans and "high-caste natives" could continue to live in the city. Moreover, in addition to exposing the untouchable sanitation workers to a heightened risk of disease and death, these communities were then identified and ostracized as potential disease carriers. Marginalization of the lower castes during infectious disease outbreaks thus happened multidirectionally.

The serio-comic magazine *Hindi Punch*, published in Bombay during the nineteenth century, brought out a cartoon in March 1899 while the plague was raging in Bombay province, in which a municipal sweeper was shown with a pig's face, with customary equipment like a basket, broom, an axe slung on its back, waist-deep in dirty water. The figure had "Sewers" written on its body, with the accompanying caption: "Our Unwashed Soowars" (Our unwashed pigs), ironically playing on the juxtaposition of sewers/soowars (pigs).[21] Recognizing the unacknowledged work of the sanitation workers like sweepers, submerged in dirty water, trying to clean the plague-infested city, the cartoon draws attention, through its powerful visual metaphor, not just to the dirty work of cleaning a diseased city undertaken by the sweepers but also to the unclean and dirty form within which they are cast(e) in the public imagination. The pig is an animal usually considered to be unclean in Indian cultural sensibilities, particularly offensive to certain religious sentiments, often actually to be found loitering near sewers, and an origin of many infectious diseases.

Caste-based marginalization also became apparent during the plague epidemic in widely prevalent practices of identifying plague suspects and potential carriers, both by the government, and under its influence, by local communities, which often revealed underlying layers of caste prejudice and exclusion. During the outbreak, while identifying potential plague carriers, "the burden of [...] suspicion fell primarily upon those who could be identified as marginal and isolated as outsiders. With its vocabulary of "detection" and "search," "surveillance" and "informants," "suspects," and "fugitives," the plague administration served to criminalize the disease and its victims."[22] The Bombay Health Officer had famously proclaimed in 1897: "From the beginning the greatest attention was paid to the disinfection of houses and to the segregation of the poor."[23] The historical convergence between lower classes and castes has been pointed out. The identification of the poor as potential disease spreaders, therefore, also meant singling out and ostracizing the lower castes.

As Chandavarkar points out:

> [D]ominant groups and local majorities tried to deflect the impact of the plague measures onto weaker and more peripheral groups [...] those whose rights in the village or the neighbourhood appeared the most tenuous, or who could otherwise be defined as out of the "community", were the most likely to be reported for "suspicion" of sickness. It was among

[21] *Pickings from the Hindi Punch*, March 1899, Asiatic Library, Mumbai.

[22] Chandavarkar, "Plague Panic," 234.

[23] Report by Executive Health Officer, Bombay, in Snow, *Report*, 87.

> untouchables, "deviants" and outsiders that the disease was first identified in previously unaffected localities.[24]

Noteworthy here is how the untouchables were clubbed together with "deviants" and "outsiders" as potential disease carriers. When a search party visited Ghori village in Nasik district in 1897, the village schoolmaster directed them to suspicious cases in the Teli (oil-pressing caste) quarters of the village.[25] In Ankleshwar, the plague was said to have "originated among the Ghanchis, a very dirty class."[26] In Rajarpur in Ahmednagar district, plague was first purportedly detected among the Mahars and Chambars.[27] Teli, Ghanchi, Mahars, and Chambars are all low castes and, according to prevalent caste-based spatial demarcations in Indian villages, would have separate living areas.

In a similar vein of identifying certain sections of society as plague carriers, plague was purportedly brought to Broach not by rats but by butchers, and the "quarter mostly inhabited by Mohammedans" was "deeply infected with plague."[28] In Hubli too, in present day Karnataka, plague was associated with "the Mussalman community."[29] We thus see how the threat of infectious disease came to be associated within dominant narratives with marginalized or peripheral sections of society like minority communities and the lower castes.

Caste was also a crucial parameter in practices of segregation and hospitalization during the outbreak. The underlying connotations of contact, touch, segregation, threat of contamination, embedded in both disease and caste, coincided in these instances, and in the widespread public panic which the plague triggered. The measures adopted to control plague by the colonial administration in places like Bombay, Pune, Calcutta, and Karachi were harsh and almost draconian; in the words of W. C Rand, they were "perhaps the most drastic that had ever been taken in British India to stamp out an epidemic," measures which were eventually to cost Rand his life, and produced an unprecedented public panic gripping the country.[30]

Commenting on the public panic in Bombay, Brigadier-General W. F Gatacre, Chairman of the Plague Committee, pointed out the "dread entertained by all classes of the population of being removed to hospital, and the fear of having their houses invaded by a disinfecting staff."[31] He goes on: "Their prejudice against hospitals was so strong that it is no exaggeration to say that most patients actually preferred dying

---

[24] See Footnote 13.

[25] J. K. Kabraji, Assistant Collector, to Collector, Nasik, 26 November 1897, GOB in "Collector, Dharwar, to Commissioner, Southern Division", 13 March 1898, in GOB, General (Plague), Vol. 389, Compilation no. 298, 1898. MSA, quoted in Chandavarkar, "Plague Panic," 236.

[26] Condon, *The Bombay Plague*, 155.

[27] Condon, *The Bombay Plague*, 201–2.

[28] Report from Broach district, quoted in Chandavarkar, "Plague Panic," 235.

[29] Collector, Dharwad, to Commissioner, Southern Division (1898), quoted in Chandavarkar "Plague Panic," 235.

[30] *Draft of Report to Government of Bombay by the Late Mr W. C. Rand, I.C.S., Chairman, Poona Plague Committee* (n.p, n.d [1897?]), p. 3. MSA.

[31] W.F Gatacre, *Report on the Bubonic Plague in Bombay*, Bombay, 1896–97, 2.

in their houses to going to hospital [...] Part of this repugnance was [...] owing to the fact that there was only one hospital [...] *for the accommodation of all castes and classes*."[32] The reference was to the Arthur Road hospital, the first and for some time only plague hospital in Bombay, which was also the site of several riots protesting against the stringent plague control measures of the government.

Referring to the measures of the earlier Plague Commissioner, Gatacre observed that: "[N]o provision was made for the accommodation of different castes in different wards or hospitals. It was not explained [...] that the prejudices of the various castes in the matter of food, & c., would be respected," leading to "a storm of protest."[33] "In the absence of caste hospitals," he observed, "no one would bring their sick out of their houses."[34] To many Indians of the upper castes, hospital was "a place of pollution, contaminated by blood and feces, inimical to caste, religion, and purdah."[35] In April 1897, the *Kesari* (a Marathi paper run by Bal Gangadhar Tilak) "carried the story of a Brahmin who had been forced to live on milk while in hospital because the food he was offered had been polluted by being served by a Sudra"; the *Mahratta* also complained that caste observances were being violated in Pune's general hospital.[36] The population's fear of and repugnance toward having their houses disinfected was also caused, partly, by the difficulty of obtaining "high-caste coolies to carry out the measures."[37]

In a letter to the President of the Municipal Corporation, the Governor of Bombay, Sandhurst, had noted: "The first thing to be done was, to search for and discover the cases, and *provide hospitals for each caste*" (my italics); during the identification and segregation of the sick, "the caste and religious usages of the people should be treated with all consideration."[38] Accordingly, hospitals were built to accommodate separate communities and castes, supported by leading and affluent members of that community. The segregation of infectious disease fell in line with caste-based segregation in these "private caste hospitals."

Separate hospitals were thus built in Bombay in 1897 for the Bhattias, Dussa sa Banniyas, Marwaris, for the Bunniya, Halari Jogjari communities, the Bhangsali caste, other than separate hospitals for Jains, Parsis, Mahomedans, Telugus, and so on.[39] "The Staff, i.e., medical men and attendants for every private Hospital, were provided by the people of the caste themselves."[40] In his report, General Gatacre appreciated the efforts of the natives who supported the government's plague control endeavors: "[M]uch excellent service was done by these gentlemen; the mere moral effect of their associating themselves openly with the measures taken to stamp out the

[32] Gatacre, *Report*, 2.

[33] Gatacre, *Report*, 3.

[34] See Footnote 33.

[35] Arnold, *Colonizing the Body*, 213.

[36] See Footnote 35.

[37] See Footnote 32.

[38] Gatacre, *Report*, 5.

[39] Gatacre, *Report*, 13.

[40] Gatacre, *Report*, 12.

plague was of great value in convincing the poorer masses of their caste-fellows."[41] Photographs of the plague outbreak in Karachi from 1897 show quarantine or segregation camps differentiated on the basis of caste/community and/or work, which was in turn, caste-based. For instance, a photograph from 1897 of a segregation camp in Karachi, shows the names of different camps like "Pathan Camp" and "Sweepers' Camp."[42]

In the case of cholera too, whose multiple outbreaks across the nineteenth century proved to be the biggest and severest cause of mortality on the subcontinent, the epidemic came to be associated with the lower classes, necessarily mostly coinciding with the lowest castes. In 1820, James Jameson observed its high incidence among Calcutta's slum dwellers and the contrast with the "higher classes of Native, and Europeans generally," who, living in the "better raised and more airy parts of the town, suffered proportionately less than the lower ranks."[43] "In Bombay, it was noted that the disease was confined mainly to the poor and those most exposed to the elements."[44] In Madras, William Scot stated in 1824 that "only the lowest of the common people suffered from this disease."[45] To a much greater degree than in Europe and North America, where cholera made its home in industrial cities and urban slums, cholera flourished in India as a disease of the rural poor.[46] The parallels between caste and class hierarchies and their coinciding disempowerment, as explained before, meant that the rural poor were most often of the lower castes.

Cholera too, therefore, was inflected by both class and caste in the public perception. It was also situated within the contexts of race relations and the colonial encounter in the Indian mind. Villagers at Cunnatore in Nellore district of Madras, for instance, ascribed the outbreak of the disease to the coming of the "low-caste soldiers" and their encampment near a sacred tank.[47] The inhabitants of Ongole in the same district attributed it to the maltreatment by the British of the Raja of Gumsur.[48] According to Moreau de Jonnes, the Hindus attributed the 1817–21 epidemics to the resentment of the deity "Yagatha Ummah" against British rule in India.[49] Thus, Indian popular belief construed the disease as stemming from the British violation of Hindu taboos and through their military intervention on the Hindu cosmos.[50]

After delving into the ramifications of caste in epidemic outbreaks in India in the nineteenth and early twentieth centuries, in the next section of this essay, I will examine how caste-based vulnerability to infectious disease, and the stigma and

[41] Gatacre, *Report*, 16.

[42] "A group of plague staff preparing to do some white washing, Karachi, India," Photograph from the Wellcome collection, 1897. https://wellcomecollection.org/works/yhmncj4v

[43] Jameson (1820), 110–11.

[44] Arnold, *Colonizing the Body*, 166.

[45] Arnold, *Colonizing the Body*, 167.

[46] See Footnote 44.

[47] Scot (1824), 237.

[48] Samanta (2018), 66.

[49] Arnold, *Colonizing the Body*, 171.

[50] Samanta, *Living with Epidemics*, 66.

exclusion that follow, persist in the present times, through an analysis of caste and its multiple layers of exclusion and violence within the trajectories of the current COVID pandemic.

Caste and COVID-19

More than a century later, the COVID-19 pandemic is showing that caste-based stigma and discrimination, and unequal vulnerability to disease, based not just on class but also caste location, persist. As Pratik Chakrabarti asserts: "Covid-19 [...] [has] historical and colonial legacies," including "asymmetrical social and economic consequences and the violation of basic human rights, particularly those of the poor and marginalized communities."[51] The contemporary vocabulary of "social distancing," triggered by COVID-19, takes on a wholly different and more complex set of meanings when seen within the fraught context of centuries of caste discrimination in India, imposing caste-based spatial segregation and distance. The segregation imposed by both infectious disease and caste have once again coincided in the current pandemic, necessitating the need to critically rethink the paradigms of segregation and isolation, and their multifarious meanings within the intersecting discourses of caste and disease. As Kanthi Swaroop and Joel Lee affirm, "the pandemic has triggered a recrudescence of untouchability practices in the name of social distancing."[52] Subhajit Naskar observes "caste prejudice is [...] regenerated by social distancing regulations," reinvigorating the historical separation and isolation of lower caste Indians through centuries of practices of untouchability, in conjunction with endogamous norms of the higher castes.[53]

Linguist G. N. Devy expresses his discomfort with the term "social distancing," noting how the term appears to give "spurious legitimacy"[54] to practices of caste-based untouchability: "The concept of untouchability that operated under the caste system exercised social distance from particular communities."[55] He points out how Ambedkar and others had emphasized the need to fight the "pandemic of untouchability," the expression foregrounding the interfaces between the "untouchability" inherent in both caste and infectious disease discourses. Devy opines:

> The distance required for minimizing the spread of the virus is "physical distance" or "safe distance". On the other hand, "social distance" is a term that was used during the colonial times [...] prohibiting the English administrators from "mixing with the natives". It was also used in South Africa during the apartheid era [...] In India, since the times of *Manusmriti*, contact between the self-proclaimed "higher caste" and the unfairly branded "lower caste" was prevented [...] The principal instrument employed for the inhuman and discriminatory practice was "social distancing". It involved not sharing food, goods, space, rituals, gods,

[51] Chakrabarti (2020).

[52] Swaroop and Lee, "Caste and Covid-19," 36.

[53] Naskar (2021).

[54] Swaroop and Lee, "Caste and Covid-19," 37.

[55] Ghoshal (2021).

icons, habitats and blood between the so-called "higher caste" and the socially distanced "other castes".[56]

Other social scientists have also expressed unease about the term "social distancing" in the Indian context, fraught as it is with the underlying connotations of caste segregation and exclusion, preferring instead alternative formulations like "disease distancing," "hygienic distancing," or "corona distancing."[57]

Awanish Kumar reads the term "social distance" in relation to two concepts put forth and analyzed by Ambedkar: "social isolation"[58] and "social nausea."[59] Social isolation marked the interactions between castes in India, according to Ambedkar, whereby castes, especially the lower castes, were strictly restricted and isolated within the particular social spheres which individuals were born to, which was maintained through practices of endogamy and caste-based work; "social nausea," on the other hand, was the social boycott of one group (caste) by another in the form of untouchability and other modes of socio-economic marginalizations and disempowerment. "This social nausea takes the form of explicit social boycott against the Dalits, if they are seen to be getting too mobile from their social stations. Or [...] [it] results in the invisibility of Dalit interests or outright denial of their demands in policy."[60] The discursive problematics of the contemporary "social distance" of COVID-19 thus comes to the fore, dovetailing as it does with the social isolation and social nausea marking inter-caste transactions, as delineated by Ambedkar.

While the pandemic-induced isolation and social distance raises connotations of age-old practices of untouchability and caste segregation, leading to the resurfacing of traditional caste discourses on the one hand, on the other, the brunt of the effects on the ground, particularly the economic impacts of the pandemic, as well as increased risks of acquiring the disease and dying from it, are also borne, overwhelmingly, by the lower castes. A complex set of interconnected socio-economic marginalizations and vulnerabilities cause this, stemming from the conjoined disempowerment of caste-class location. Pointing out that the nationwide lockdown in India starting March 2020, which was among the strictest in the world, "affected the marginalized, stigmatized group of lower-ranked castes much more severely than the higher-ranked castes," Ashwini Deshpande avers that COVID-19 "has worsened labor market inequality among caste groups," and that "the caste-differentiated economic impacts of the pandemic [...] has deepened the preexisting fault lines of the caste system."[61] She elaborates: "the labor market or employment effects of the pandemic have been much more severe for the more disadvantaged groups—the Scheduled Castes and Scheduled Tribes, and the OBCs. Their early job losses

---

[56] Ghoshal, "G. N. Devy".

[57] See Footnote 56.

[58] *Dr. Babasaheb Ambedkar Writings and Speeches* (BAWS), Multiple Volumes, (Mumbai: Government of Maharashtra) Vol. 1, 249–50.

[59] BAWS, Vol. 7, 370.

[60] Awanish Kumar, "Reading Ambedkar in the time of Covid-19," *Economic & Political Weekly* IV, no. 16 (2020): 35.

[61] Deshpande (2021): 127–8.

[…] were three and two times higher, respectively, than Hindu upper castes experienced."[62] As Subhajit Naskar too points out: "The global outbreak of COVID-19 epidemic combined with social distancing norms has […] laid bare the fragile livelihood situation of lower caste communities in the midst of massive job loss in the urban locations […] [P]eople from higher income groups are less likely to be impacted while India's disadvantaged communities with lack of access to resources […] [are] intensely affected in the unprecedented global health emergency situation."[63]

The heavier losses in employment among the disadvantaged castes are produced by the combination of the following factors: "their higher representation in precarious, vulnerable daily wage jobs, and their lower levels of education."[64] According to the 2011–12 NSSO data, "the share of wage laborers among SCs was 63%"; corresponding values for others were "44% for Other Backward Classes (OBCs), 42% for upper castes."[65] Among wage laborers, SCs comprised a much greater share of casual wage workers, implying "higher job insecurity and bare minimum earnings."[66] During COVID, within the lower-caste informal sector workforce, an estimated 119 million workers lost their jobs within the first two weeks of lockdown in 2020.[67] In addition to unequal job losses and unemployment along caste-class parameters, the gaps in education, especially higher education, between the upper and lower castes, and unequal socio-economic resources like access to information technology, have also differentially impacted how different communities and castes have been able to cope with the lockdown and the unprecedented public health crisis triggered by the coronavirus outbreak.

Caste-based allocation of high-risk work during the pandemic, including sanitation and cremation, has also been observed since the outbreak of the pandemic in India in March 2020. Caste-determined sanitation work, evidenced during the nineteenth-century epidemics, has had a contested trajectory in the decades leading up to independence and subsequently in postcolonial India, through the problematic valorization of sanitation work by Gandhi, among others, without factoring in workers' actual demands on the ground for better work environments and resources. "In his […] writing on the issues of both sanitation and caste, Gandhi consistently conveyed sympathy for 'the Bhangi,' a [caste-marked] term he insisted on calling sweepers and scavengers, despite their stated dislike of the term. He emphasised the importance and dignity of sanitation labour, while at the same time firmly opposing sweepers' strikes, expressing reluctance to adopt new sanitation technologies, and asserting that 'the ideal Bhangi' must consider his work 'a sacred duty'."[68,69] "Gandhi's approach

---

62 Deshpande, "How India's," 128.

63 Naskar, "Social Distancing," 45.

64 Deshpande, "How India's," 129.

65 Naskar, "Social Distancing," 46.

66 See Footnote 65.

67 See Footnote 65.

68 M. K. Gandhi, "The Ideal Bhangi," *Harijan* 4, no 42 (1936): 336.

69 Swaroop and Lee, "Caste and Covid-19," 38.

to the sanitation question, institutionalized in the Harijan Sevak Sangh, set the agenda for the first several decades of sanitary policy in independent India."[70]

The COVID pandemic has revealed the persistence of caste-based allocation of sanitation and cremation work to Dalits in state policies (explicit or implicit) throughout the country, revealing the grim "necropolitical" logic formulated by Achille Mbembe, whereby the state controls not only the lives but also the death of individuals considered redundant in the state's project of disciplining bodies to make them useful,[71] deriving from the Foucauldian notion of "biopolitics."[72] Such practices are based on "[t]he presumption that waste and death work must and will be performed by Dalits, a viewpoint that assumes an ontological link between caste and occupation and naturalises the conscription of [...] [Dalits] as frontline workers in a pandemic."[73] Instances of such "conscription of Dalit labor for pandemic-related sanitation work" range from municipal authorities looking exclusively to hire Dalits for sanitation and disinfection work in different parts of the country like Rajasthan, Uttar Pradesh, Telangana, without even providing them basic safety gear for the job; this situates this kind of work primarily as "a form of caste labor" and points to the "continuity of the caste-based mentality that is pervasive and persistent in state practices."[74]

An order from the Department of Health and Family Welfare, Government of West Bengal, in May 2021 sanctions the appointment of "Domes," people of the erstwhile untouchable Dom caste, traditionally engaged, through caste-based work allocation, in scavenging, death and cremation work, in government hospitals in the state "for the disposal of COVID-19 positive dead bodies from the mortuary." This is part of the declared efforts of the government to contain "the outbreak of the [...] virus on war footing."[75] Some of the orders by the state government also sanction the building of separate "Dom" rooms beside the mortuary, which replicate the spatial segregation practices of untouchability. By officially appointing people of the Dalit Dom caste for this hazardous work of disposal and cremation of COVID positive bodies at a very meager pay, the state endorsed caste-based labor, while reasserting the ontological association between caste and certain kinds of work, traditionally considered "impure." The policies of the postcolonial state in the 2020s in negotiating disease and caste during public health crises, therefore, replicate and reify some of the exclusionary policies of the colonial state from more than a century ago in managing disease in native populations.

---

[70] Swaroop and Lee, "Caste and Covid-19," 38. For more on this, see Assa Doron and Robin Jeffrey *Waste of a Nation: Garbage and Growth in India* (Cambridge: Harvard University Press, 2018), 22, and Joel Lee *Deceptive Majority: Dalits, Hinduism, and Underground Religion* (Cambridge: Cambridge University Press, 2021), Chap. 4.

[71] Achille Mbembe, *Necropolitics* (Durham, NC: Duke University Press, 2019 [2011]).

[72] Michel Foucault, *Discipline and Punish* (Penguin Classics, 2020 [1975]).

[73] Swaroop and Lee, "Caste and Covid-19," 37.

[74] See Footnote 73.

[75] Bhaskar et al. (2021)

While the state has officially endorsed caste labor in the pandemic, caste-based boycott and intimidation of Dalits engaged in this kind of work as potential disease carriers by other dominant castes has also been rampant during this time throughout the country. Episodes have been reported of Dalits being suspected of spreading the virus, termed "corona," and blocked from stepping out, even to get essentials.[76] Dalits have allegedly been attacked as purported virus spreaders in different parts of the country, even for purportedly not responding to the Prime Minister Narendra Modi's national call to switch off lights for nine minutes in April 2020.[77] Mirroring caste prejudices at the turn of the twentieth century during the bubonic plague outbreak, there have been episodes of higher caste people refusing to consume food cooked by Dalits at quarantine facilities or hospitals, revealing the continuity in caste exclusion and prejudice over centuries. In Uttar Pradesh's Bhujauli Khurd in Kushinagar district, two men at a quarantine facility set up in a primary school in April 2020, went home twice every day, breaking the expected segregation stipulations, refusing to eat the food cooked by the Dalit woman village chief, claiming that they had never eaten food cooked by a Scheduled Caste person before.[78] At about the same time, at Siswa-Baruar village in Basti district, also in Uttar Pradesh, the police registered a case against ten people (of the Other Backward Classes, a heterogeneous group of historically disadvantaged communities who had, however, not been untouchables) under the Epidemic Diseases Act for refusing to eat food cooked by Dalit cooks.[79]

After examining the continuity of caste and its multiple ramifications in configurations of infectious disease on the Indian subcontinent, from the nineteenth century to the twenty first, through historical readings of the bubonic plague outbreak in the 1890s, cholera in the nineteenth century and COVID-19 in the 2020s, in the next section of this essay, I will offer a reading of some literary representations of epidemic outbreaks in modern Indian literature. The section will examine a selection of representative literary texts to uncover the connotations and configurations of caste within cultural and literary representations of epidemic disease. Such an intersectional historical and literary reading of the imbrications of caste and epidemic disease allows me to critically engage with the complexities and nuances of their intersections in multiple forms of narratives, through multiple modes of reading.

---

[76] Sur (2021).

[77] Modak (2021), "Covid-19: Dalit family in Haryana attacked for allegedly not following PM's call to turn off lights," *The Scroll*, April 9, 2020. https://scroll.in/latest/958665/covid-19-dalit-family-in-haryana-allegedly-attacked-for-not-following-pms-call-to-turn-off-lights (accessed November 10, 2021).

[78] Srivastava (2021).

[79] Srivastava, Piyush. "Untouchability."

Cast(e)ing Epidemics in Literary Representations

Rajinder Singh Bedi's short story "Quarantine" (1939) gives a glimpse into a terror-stricken town in some part of North India grappling with a devastating outbreak of plague. The story delves into the public terror of plague through the image of the townspeople in mortal fear of the "quarantine," more than the plague itself. It brings to the readers the intersecting predicaments and interactions between two people during the plague: Dr. Bakshi, a doctor treating patients at a quarantine center, and William Bhagu, a sanitation worker, employed in the work of cleaning and disinfecting the quarantine center and city at large, and burning corpses during the plague. Dr. Bakshi does his duty of trying to treat the dying patients, though himself overwhelmed by fear and panic. The sanitation worker Bhagu, on the other hand, goes much beyond the call of duty, to take care of and support the miserable patients with utmost dedication and selflessness, without any fear of the plague, jeopardizing himself and his family in the process.

Though there is no explicit mention of caste in the story, William Bhagu is a recent convert to Christianity, a "sweeper who cleaned the streets," who "gets up at three every morning [...] [and] starts sprinkling lime powder on the [...] lanes and drains, as instructed, to stop the spread of the disease."[80] Bhagu got up "early so that he could collect dead bodies scattered all over the marketplace and also to help out the residents of his colony, who were too scared of the epidemic to come out, by doing some odd jobs for them. He was hardly afraid of the plague."[81] Given the traditionally caste-marked nature of sanitation and cremation work in Indian society, as discussed in detail in the last two sections of this essay, it is clear that William Bhagu was a low-caste man who had recently converted to Christianity, a "sweeper" who had been engaged possibly throughout his life in sanitation and cremation work. Conversion to Christianity has had an inextricable link to practices of caste marginalization and untouchability in India, with the lowest caste people historically often having taken recourse to conversion to escape the violence of untouchability and caste exclusion. Literary texts tracing the trajectories of Dalit Christianity in different parts of the country, for instance in Kerala, have also shown the inexorable persistence of caste-based work allocation, oppression, and exclusion meted out to Dalits, even after conversion, by upper-caste converts.[82]

Bedi's short story reveals the contrasting predicaments and responses of the low-class (and most likely low caste) sweeper Bhagu and the upper-class, educated Dr. Bakshi to the plague outbreak. The story reveals how poor and caste-determined sweepers and cleaners are pushed to be frontline workers during life-threatening disease outbreaks, to protect the lives of other citizens, as has also been revealed in the narratives of sanitation work during the COVID pandemic, discussed in the

---

[80] Rajinder Singh Bedi, "Quarantine" 1939. https://coronabluescollective.wordpress.com/2020/07/11/quarantine/ (accessed November 21, 2021).

[81] Bedi, "Quarantine".

[82] See for instance, Thakazhi Sivasankara Pillai, *Randidanghazhi* (Two Measures of Rice) (1967), Paul Chirakkarode, *Pulayathara* (1963), Arundhati Roy, *The God of Small Things* (1997).

previous section. Bhagu's selfless devotion to the suffering plague patients without any fear for his own life inspires the panic-stricken Dr. Bakshi to overcome his own fears to serve the people. However, at the end of the story, we see how the privileged Dr. Bakshi is venerated, his efforts at combating the disease publicly celebrated and acknowledged, while the poor sweeper Bhagu, who has meanwhile lost his wife to the plague, remains in the shadows of obscurity, his contribution unacknowledged and forgotten in the annals of public history. Scores of other such Bhagus remain vulnerable and unacknowledged in the times of public health crises, the story revealing the differential predicaments, both in terms of vulnerability of the two central characters, separated in their caste-class situation, as well as in the respect accorded to their respective frontline work. The story ends with Dr. Bakshi's wrenching realization of this very difference: "The image of Bhagu's dying wife and their child flashed before my eyes. It seemed my neck would break under the weight of garlands and my pocket would burst with the weight of my wallet. Despite receiving so much honor, I suddenly felt worthless."[83]

Sarat Chandra Chattopadhyay's novel *Pandit Mashay* (1914) centers around the outbreak of cholera (*ola-utha* or *bishuchika* in Bengali) in Bengal's countryside. Cholera appears in the novel "with such a force that it becomes a character itself, pulling events in and out, and finally forcing the novel's inevitable end. Almost all the characters in the novel come into focus when cholera breaks out in the village in an epidemic form, killing people young and old within a very short span of time."[84] The novel shows how relationships in the village community are strongly determined and underscored by caste, which the outbreak of cholera brings to the fore. Cholera becomes the locus of inter-caste tensions and animosities in the community. The cholera epidemic breaks out in the village possibly due to the consumption of contaminated drinking water in the only pond in the area, owned and maintained by Brindaban, who is the eponymous *pandit mashay* (school master) of the novel, a non-Brahmin, lower-caste man. One of his neighbors, a Brahmin, dies of cholera. When Brindaban notices that the infected clothes of the deceased are being washed in the tank, the water of which is shared by many, he objects on the grounds of hygiene and public health. However, villagers believed that, according to Hindu scriptures, the water of a tank dug ceremoniously could not be polluted. Representing the voice of science and reason, against such superstitious beliefs, Brindaban contends, "this is the only pond in the village; I shall not let it get spoiled in this trying time."[85] Tarini Mukherjee, the concerned Brahmin neighbor, requests another elder of the village, Ghoshal, also a Brahmin, to adjudicate on the matter. Ghoshal, the self-proclaimed custodian of Hindu religious teachings and practices in the village (also widely acknowledged in the community), lashes out at Brindaban in anger, "this is your wrong insistence, Brindaban; according to *shastras* pond water can never be

---

[83] See Footnote 81.

[84] Samanta, *Living with Epidemics*, 70.

[85] Chattopadhyay (2008) p. 54, quoted in English translation in Samanta, *Living with Epidemics*, 71.

polluted or made unholy! You may have read a few pages in English, but how can you disbelieve our *shastras*?"[86]

The public health and safety debate in the wake of the cholera outbreak soon morphs into a caste issue, as Brindaban is repeatedly reminded of his lower-caste status by the Brahmins. His appeal to scientific and hygienic logic is therefore construed as a challenge of Brahmin dictums by a lower-caste man, the issue taking on the underlying connotations of caste relationships and hierarchies. The Brahmin Tarini Mukherjee takes Brindaban's appeal as a caste slur and an insult of Hindu religious codes, makes common cause with other village notables (necessarily all Brahmin), and metes out a Brahmin curse to Brindaban as punishment: His whole generation will be wiped out through cholera. Eventually, Brindaban's mother, an old pious lady, dies of cholera, and a few days later his only son Charan also succumbs to the disease. Sarat Chandra thus, in both *Pandit Mashay* as well as in his larger oeuvre, "provides us with [...] [a] different archive of knowledge, not just on cholera, but on the process how a society functions when a pestilence strikes terror among its inhabitants."[87]

In the story "Gadkhalir Haath" (The Hand of Gadkhali),[88] the vocabulary of epidemic disease in rural Bengal is enmeshed within contemporary socio-cultural codes, which reflect contemporaneous social norms of caste, gender and community, and the spaces of familial and domestic intimacy. "Gadkhalir Haath" is a story that has become a part of Bengali cultural legend since the late nineteenth century, whose narrative has evolved and transformed over the centuries in response to the different social milieus in which it was narrated.[89] Around the 1880s-90s, the oral legend narrated the story of a Brahmin who arrives at the village of Gadkhali in Khulna district of Bengal (now in Bangladesh) after a long journey and seeks shelter in a deserted village mansion. In the mansion, he is greeted by a man lying ill in bed of a fever, who extends his gracious hospitality to the Brahmin, as turning away a Brahmin from one's house and not feeding him would be a great sin. The man is eventually revealed to be a ghost, long dead, his and his family members' spirits wandering restlessly as they were not cremated after an epidemic of cholera or malaria which wiped out the village. This is revealed when the man extends his arm many feet away from the room to a lime tree in the garden to pick limes, which gives the story its title. The man pleads with the Brahmin to perform his and his family's last rites, as per Hindu scriptures, to release their souls from eternal torment, revealing how epidemic disease and death is cast(e) within the social and religious codes governing caste relations in a Brahminical society. The story morphs in later centuries to ones in which the ghost is the mother-in-law and later the dead wife of the Brahmin, replacing earlier caste-based social codes with those of family and domesticity.

---

[86] Chattopadhyay, *Pandit Mashay, in* Samanta, *Living with Epidemics*, 71.

[87] Samanta, *Living with Epidemics*, 70.

[88] "Gadkhalir Haath" a Bengali oral legend was narrated in written form by Surendranath Bhattacharya, and mentioned by folk researcher Biren Bandyopadhyay and linguist Sukumar Sen.

[89] Projit Bihari Mukherjee, "Mahamarir Bhut" (Ghosts of Pandemic), https://www.anandabazar.com/rabibashoriyo/ghosts-of-pandemic/cid/1286684 (accessed November 28, 2021).

In Mahasweta Devi's novel *The Glory of Sri Sri Ganesh* (2003), in the remote Barha village in the tribal areas of Jharkhand, the tyrannical village landlord "malik" Tritirthanarayan (Ganesh), himself a Rajput, shows "himself to be a well-wisher of the bhangis and dusads" (low-caste people, earlier untouchables), upon assuming charge of the village on the death of his father Medininarayan.[90] The following extract from the novel reveals how caste underpins the public construction of epidemic disease, with the higher caste Rajputs ascribing the threat of epidemics to the lower castes, adding a further layer of caste-based ostracism and marginalization to the already existing multilayered caste oppression and violence.

> In the cholera season, he [Ganesh] said, "When the government van comes, you people [bhangis and dusads] had better take the injections. You're the ones most likely to catch it. And you invariably do."
>
> "What about all of you Malik?"
>
> "Us? How can cholera enter a house where puja is done regularly? Isn't god's blessing upon us?"
>
> "We also do puja to the *deota*."
>
> "As you think best."
>
> [...] Ganesh [...] said to the BDO, "Who can help these people? Told them, take the injection for cholera, but no, they said, we'll do puja to god, and we won't get cholera."
>
> "They don't know what's good for them".

Reminiscent of nineteenth-century identification of plague with the outcastes and marginal people of society, caste hierarchy determines the construction of cholera here. In Rabindranath Tagore's novel *Chaturanga* (Quartet),[91] the outbreak of plague in the village is associated with the marginal Muslim community and the protagonist Sachis's uncle (jethamoshai) is blamed for mingling with them, thereby jeopardizing himself and the village.

In Fakir Mohan Senapati's Odia short story "Rebati" (1898), the outbreak of epidemic cholera in the village of Patapur in Cuttack district of Orissa unfolds within contexts of both caste, and more significantly, gender. When Shyambandhu Mohanty dies of cholera, his wife and mother and the young village schoolteacher Basudev, of the same Karan caste as Mohanty, have to singlehandedly perform the last rites; "Shyambandhu's was the only Karan family in the village, and help was neither expected nor forthcoming; the two women and Basudev had to carry the body to the cremation grounds and perform the last rites."[92] The response of the entire village, on the other hand, was "to bolt the doors and windows and keep out of the path of the demonic deity, as though the evil old hag was out with her basket and broom sweeping up heads,"[93] revealing the gendered construction of cholera in the image of an "old hag". When Shyambandhu's wife and after some time Basudev too succumbs

---

[90] Devi (2003), 78–9.

[91] Rabindranath Tagore, *Chaturanga* (1916).

[92] Senapati (1898).

[93] Senapati, "Rebati".

to cholera, Shyambandhu's grief-crazed mother ascribes the misfortunes to the over-reaching ambition of her granddaughter, Shyambandhu's daughter Rebati, to educate herself by learning to read and write. As cholera becomes a crux of caste animosities in the village in Sarat Chandra's *Pandit Mashay*, in "Rebati" cholera is interpreted through the gendered societal norms of nineteenth-century rural India. Rebati, in her grandmother's view, transgresses her gendered social boundaries by attempting to educate herself through the help of her parents and Basu, and thereby invites the divine wrath that ultimately destroys her family. In the end, Rebati herself and her grandmother also perish, thereby completely annihilating the Mohanty family as punishment for its purported transgressions.

The connotations of defilement, contamination, and (im)purity inhering in both discourses of disease and caste, centering on the notions and imbrications of the body, touch, intimacy, and distance, powerfully inscribe U. R Ananthamurthy's Kannada novel Samskara (1965).[94] *Samskara*'s narrative hinges on the death by plague of the heretic Naranappa who had always flouted Brahmin norms and scriptures while he was alive, and led the most "un-Brahmin" life by having a low-caste mistress, among his multiple other transgressions, like eating meat and consuming liquor. Naranappa's death by plague (not known till almost the end of the novel) in the Brahmin *agrahara* (Brahmin settlement surrounding temples) Durvasapura, and the anxieties of the other living Brahmins in the community about both performing the cremation of the un-Brahmin, as well as leaving his dead body uncremated, ushers in an acute crisis in the *agrahara*. This moral, religious crisis, of faith and human values, centers on Praneshacharya, the most learned and respected Brahmin of the community, and his own human, corporeal awakening and transformation, in the wake of the crisis. The notions of bodily, religious, and spiritual (im)purity and defilement are situated, developed, and tested in the novel within the intersecting contexts of caste, disease, death, and the body. Naranappa's plague-ridden "un-Brahmin" dead body becomes the locus of the multiple, layered anxieties of the Brahmins surrounding purity, defilement, and consequent loss of caste, and ultimately, death, as other Brahmins too gradually start succumbing to the disease.

Plague is located in the novel not just as a medical, public health crisis, but as a social, religious, spiritual crisis, unfolding at both the individual and community level. The Brahmin lives, strictly regulated and controlled through the observance of scriptures and rituals in the *agrahara*, are exposed, tellingly, by the death of the un-Brahmin Naranappa, to be spiritually depraved and hollow, while exhibiting the external manifestations of Brahminism. This realization dawns upon Praneshacharya, triggering his own spiritual and identity crisis. These overlapping crises are mapped against the backdrop of the outbreak of plague in the *agrahara*, showing how caste, death and disease are interwoven in the novel, reflecting on the complex intersections of the body, touch, purity and defilement vis-à-vis both disease and caste.

[94] U. R. Ananthamurthy, Samskara, (1965) translated by Ramanujan (1975).

Conclusion

Over the last three sections, this essay has reflected on caste as a central and crucial marker in the constructions of epidemic disease in India, from the nineteenth century to the twenty-first. Through examining the configurations of caste, its multiple connotations of (im)purity, defilement, cleanliness, and its ramifications of touch, distance, segregation, the body (physical and social) and its boundaries within societal, political and cultural constructions of epidemic disease in India, the essay has attempted to make an intervention in understanding the intersections between epidemic disease and caste. Through an intersectional reading of literary, cultural, media, and historical narratives, extending from literary accounts to the nineteenth-century epidemics of bubonic plague and cholera and the current COVID-19, I have demonstrated how caste and its multilayered meanings have crucially informed and determined constructions of epidemic disease in India. Through my deliberations on caste and its significance in understanding epidemic disease in India, I have offered insights into the differential notion of vulnerability to disease and to disaster at large, based on categories like caste, class, gender, race, religion, ethnicity etc., on identity categories within the social construction of disease, and on disparities in public health within nations and communities.

References

Ananthamurthy, U. R. (1965). *Samskara*, translated by A. K. Ramanujan. (1975).

Bedi, R. S. (1939). Quarantine. https://coronabluescollective.wordpress.com/2020/07/11/quarantine/. Accessed November 21, 2021.

Bhaskar, A., et al. (2021). West Bengal: Selectively hiring Doms To cremate COVID victims reflects casteist mentality. *The Wire*, 24 May 2021. https://thewire.in/rights/west-bengal-dom-cremate-covid-19-victims-casteist-mentality. Accessed November 15, 2021.

Carrigan, A. (2015). Towards a postcolonial disaster studies. In E. DeLoughrey, J. Didur, & A. Carrigan (Eds.), *Global ecologies and the environmental humanities: Postcolonial approaches* (pp. 117–139). Routledge.

Chakrabarti, P. (2020). Covid-19 and the spectres of colonialism. *The India Forum*, 10 July 2020.

Chandavarkar, R. (1992). Plague Panic and Epidemic Politics in India, 1896–1914. In T. Ranger & P. Slack (Eds.), *Epidemics and ideas: Essays on the historical perception of pestilence* (pp. 203–240). Cambridge University Press.

Chattopadhyay, S. C. (2008). *Pandit Mashay, Sarat Rachana Samagra*. Basak Book Store, Kolkata.

Collector, Dharwar, to Commissioner, Southern Division. (1898, March 13) In GOB, General (Plague), Vol. 389, Compilation no. 298. MSA.

Condon, J. K. (1900). *The Bombay Plague being a history of the progress of plague in the Bombay presidency from September 1896 to June 1899*. Bombay.

Deshpande, A. (2021). How India's caste inequality has persisted—And deepened in the pandemic. *Current History, 120*(825), 127–132.

Dr. Ambedkar, B. Writings and speeches (BAWS), Multiple volumes. Government of Maharashtra, Mumbai.

Devi, M. (2003). *The Glory of Sri Sri Ganesh, translated by Ipsita Chanda*. Seagull Books.

Foucault, M. (2020). *Discipline and punish*. Penguin Classics [1975].

Gandhi, M. K. (1936). The ideal Bhangi. *Harijan, 4*(42).
Ghoshal, S. (2020). G. N. Devy on the evils of 'social distancing'. *Live Mint*, 29 April 2020. https://www.livemint.com/mint-lounge/fea-tures/gn-devy-on-the-evils-of-social-distancing-11588131950357.html. Accessed November 16, 2021.
Hewitt, K. (1995). Sustainable disasters? Perspectives and powers in the discourse of calamity. In J. Crush (Ed.), *Power of development* (pp. 115–128). Routledge.
Hewitt, K. (1998). Excluded perspectives in the social construction of disaster. In E. L. Quarantelli (Ed.), *What is a disaster: Perspectives on the question* (pp. 71–88). Routledge.
Hirst, L. F. (1953). *The conquest of plague. A study of the evolution of epidemiology*, pp. 111–19. Oxford.
Jameson, J. (1820). *Report on the Epidemick Cholera Motbus, as it visited the territories subject to the presidency of Bengal in the years 1817, 1818, and 1819*. Government Gazette Press.
Kumar, A. (2020). Reading Ambedkar in the time of covid-19. *Economic and Political Weekly LV, 16*, 34–37.
Modak, S. (2020). One held for beating up Dalit man, spreading virus rumours. *The Indian Express*, April 11, 2020. https://indianexpress.com/article/cit-ies/mumbai/coronavirus-one-held-for-beating-up-dalit-man-spreading-virus-rumours-6357207/. Accessed November 10, 2021.
Mbembe, A. (2011). *Necropolitics*. Duke University Press, Durham, 2019 [2011].
Mukherjee, P. B. (2021). Mahamarir Bhut (ghosts of pandemic). https://www.anandabazar.com/rabibashoriyo/ghosts-of-pandemic/cid/1286684. Accessed November 28, 2021.
Naskar, S. (2021). Social distancing, caste discrimination and vulnerability amidst COVID-19 pandemic. https://mu.ac.in/wp-content/uploads/2020/06/44-48-Social-Distancing.pdf. Accessed November 20, 2021.
Nixon, R. (2011). *Slow violence and the environmentalism of the poor*. Harvard University Press.
Oliver-Smith, A. (1998). Global changes and the definition of disaster. In E. L. Quarantelli (Ed.), *What is a disaster: Perspectives on the question* (pp. 179–196). Routledge.
Omvedt, G. (2016). *Understanding caste: From Buddha to Ambedkar and Beyond*. Orient Blackswan, Hyderabad.
Rao, A. (2010). *The caste question: Dalits and the politics of modern India*. Permanent Black, Ranikhet.
Samanta, A. (2018). *Living with epidemics in colonial Bengal 1818–1945*. Routledge.
Scot, W. (1849). *Report of the Epidemic Cholera as it has appeared in the territories subject to the presidency of Fort St George* (2d ed.). Asylum Press, Madras. 1824. Edinburgh: Blackwood and Sons, 1849.
Senapati, F. M. (1898). "Rebati" https://www.alephbookcompany.com/rebati-by-fakir-mohan-senapati-from-the-greatest-odia-stories-ever-told/. Accessed November 21, 2021.
Snow, P. C. H. (1897). *Report on the outbreak of bubonic plague in Bombay, 1896–97*. "Times of India" Steam Press, Bombay.
Srivastava, P. (2020). Untouchability, even in quarantine. *The Telegraph* (Online), Lucknow 11 April 2020. https://www.telegraphindia.com/india/coronavirus-outbreak-untouchability-even-in-quarantine/cid/1764186. Accessed November 11, 2021.
Sur, P. (2021). Under India's caste system, Dalits are considered untouchable: The coronavirus is intensifying that Slur. *CNN World*, 16 April, 2020 https://edition.cnn.com/2020/04/15/asia/india-coronavirus-lower-castes-hnk-intl/index.html. Accessed November 12, 2021.
Swaroop, K., & Lee, J. (2021). Caste and Covid-19: Notes on sanitation in a pandemic. *Economic & Political Weekly, 56*(13), 35–41.
Tagore, R. (1916). *Chaturanga*.
Wald, P. (2008). *Contagious: Cultures, carriers, and the outbreak narrative*. Duke University Press.
Zelliot, E. (2013). *Ambedkar's world: The making of Babasaheb and the Dalit movement*. Navayana Publishing.

The manufacturer's authorised representative in the EU is Springer Nature Customer Service Centre GmbH, Europaplatz 3, 69115 Heidelberg, Germany. If you have any concerns regarding our products, please contact ProductSafety@springernature.com

Printed and bound by CPI Group (UK) Ltd, Croydon, CR0 4YY
30/09/2026
02225248-0003